International Conference on Advances in Energy Industry and Power Generation (AdvEnGen-2020)

IOP Conference Series: Materials Science and Engineering
Volume 860

Almetvevsk, Russia
12 - 14 February 2020

Editors:

Akhmetsalim S. Galeev

ISBN: 978-1-7138-3173-0
ISSN: 1757-8981

Printed from e-media with permission by:

Curran Associates, Inc.
57 Morehouse Lane
Red Hook, NY 12571

Some format issues inherent in the e-media version may also appear in this print version.

Institute of Physics
Dirac House, Temple Back
Bristol BS1 6BE UK

Phone: 44 1 17 929 7481
Fax: 44 1 17 920 0979

techtracking@iop.org

Additional copies of this publication are available from:

Curran Associates, Inc.
57 Morehouse Lane
Red Hook, NY 12571 USA
Phone: 845-758-0400
Fax: 845-758-2633
Email: curran@proceedings.com
Web: www.proceedings.com

TABLE OF CONTENTS

IOP Conf. Series: Materials Science and Engineering **860** (2020) 011001 doi:10.1088/1757-899X/860/1/011001

PREFACE

International Conference on Advances in Energy Industry and Power Generation

AdvEnGen-2020

The International Conference on Advances in Energy Industry and Power Generation was held in Almetyevsk, Republic of Tatarstan, Russian Federation on 12-14 February 2020. Almetyevsk State Oil Institute participated as an initiator and an organizer of the event.

The three days' conference was attended by several delegations from leading Russian petroleum engineering and power engineering institutions: Ufa State Petroleum Technological University (Ufa, Republic of Bashkortostan), Gubkin Russian State University of Oil and Gas (Moscow), Kazan State Power Engineering University (Kazan, Republic of Tatarstan). Attendees from several oil and electrical companies took part in the discussion as well: Belorusneft (Rechitsa, Republic of Belarus), Tatneft (Almetyevsk, Republic of Tatarstan), Tatenergo (Kazan, Republic of Tatarstan). Also seven individual guest researchers from Russia and Vietnam participated in the sessions.

More than fifty studies related to the following topics were reported at the conference

- Electrical machinery and drilling equipment,
- Automation for oil and gas engineering,
- Engineering mathematics application,
- Power generation and distribution.

All the received manuscripts were reviewed by an editor and invited referees to qualify the most significant ones for publication. The single-blind peer-review method was applied.

We express our appreciation and deep gratitude to all the participants and the organizing committee for their contribution to preparing the AdvEnGen 2020 proceedings.

With the best regards,
Editor, prof. Akhmetsalim S. Galeev, Almetyevsk State Oil Institute

IOP Conf. Series: Materials Science and Engineering **860** (2020) 011002 doi:10.1088/1757-899X/860/1/011002

Peer review statement

All papers published in this volume of *IOP Conference Series: Materials Science and Engineering* have been peer reviewed through processes administered by the proceedings Editors. Reviews were conducted by expert referees to the professional and scientific standards expected of a proceedings journal published by IOP Publishing.

IOP Conf. Series: Materials Science and Engineering **860** (2020) 012001 doi:10.1088/1757-899X/860/1/012001

Simulation of proppant flowback from hydraulic fractures

A E Verisokin[1], D Yu Serdyukov[2], V A Vasil'yev[1], T A Gun'kina[1] and R E Shesterikova[1]

[1] Institute of Oil and Gas, NCFU, 1, Pushkin Str., Stavropol, 355009, Russia
[2] NGDU Rechitsaneft, RUP PO Belorusneft, 43, Lenin Str., Rechitsa, 247500, Belarus

E-mail: verisokin.aleksandr@mail.ru

Abstract. Hydraulic fracturing has become one of the main and most effective methods of intensifying oil production. A method of bench-scale study with the use of modern measuring equipment, automation of experimental variable measurements according to specified algorithms ensuring high-level works was developed in order to obtain reliable data on proppant migration in a fracture. The laboratory facility makes it possible to determine the speed of proppant flowback from a fracture, proppant permeability depending on layer elevation. The tests show the nature of permeability change as the pressure increases. The conditions under which the height of a filling layer varies depending on pressure were identified. The test bench greatly facilitates and accelerates the experiments. The use of modern information and measurement systems makes it possible to significantly increase the volume of received information and improve its quality. The use of experimental results obtained through hydraulic fracturing will allow for additional increase in hydrocarbon production.

1. Introduction

Hydraulic fracturing (HF) is currently the most efficient well intervention (WI) ensuring multiple increase of oil production and development efficiency improvement of low-permeable reservoirs [1-3]. The expanded application of HF has led to an increase in consumables, including proppants, which are the main components in the technological complex of this method [4-5].

Today in Russia there are many types of proppants. They have different properties. Quite often proppant manufacturers when demonstrating the advantages of proppant try to hide the disadvantages of their products, and oil and service companies having no necessary equipment to test proppants cannot assess the purchased product in a timely manner [6]. Incorrect choice of proppant can lead to problems when fixing the HF fractures.

A method of bench-scale study with the use of modern measuring equipment, automation of experimental variable measurements according to specified algorithms ensuring high-level works was developed in order to obtain reliable data on proppant migration in a fracture. A test bench was developed to study the proppant migration from hydraulic fractures.

This laboratory facility makes it possible to determine the speed of proppant flowback from a fracture, proppant permeability depending on layer elevation. Proppants were tested using modern equipment:

- optical microscopy systems;
- electronic analytical balance;
- bench unit to measure hydrodynamic characteristics of proppants;
- filter or testing sieve.

The test bench greatly facilitates and accelerates the experiments. The use of modern information and measurement systems makes it possible to significantly increase the volume of received information and improve its quality.

During experimental studies according to the proposed method, a fracture filler is first selected. Depending on the study objectives, sand, various types of proppants and other fillers may be used.

2. Materials and methods

The proppant conductivity system was designed to obtain data that allow assessing the efficiency of proppant during HF. In this unit, the experiment was conducted at various pressures and temperatures that model formation conditions. Permeability of proppant is determined at different flow rates and pressure drops [7].

Figure 1 shows the layout of proppant conductivity test unit.

A conductivity cell (Figure 2, 3) – the cell is made of metal. The cell is designed for compression pressures of up to 80 MPa. Two parts of a rock are placed in the cell, the space between which models a fracture. The cell design allows pressure monitoring during fluid filtration through proppant fracture.

The core used in the test represents a 0.5 cm sandstone tile treated to the cell size. The elasticity modulus of a test piece was about 25,000 MPa and permeability of about 0.25 $\mu m^2 * 10^{-3}$. The rock is insulated with high temperature seals.

Figure 1. Layout of proppant conductivity test unit

Pressure is provided by a small press. The press ensures the compression pressure from 5 to 80 MPa. The load increases at 0.5 MPa per minute, which can be seen with a pressure gauge mounted in a hydraulic line. The use of the press also allows for proppant destruction experiments.

The supply is ensured by a hydraulic booster and a reservoir. Cell inlets and outlets are equipped with mesh filters.

The width of samples is determined by micrometers. The average value is used for calculations. The obtained width is included into the computer and used in the calculation of experimental findings.

Throughout the test, the liquid flow is weighed using an electronic balance. Using the density of the sample, the flow rate is calculated. All data from the balance are recorded.

The temperature of the cells was controlled by the liquid flow through the heated proppant column. To avoid heat losses, all equipment is isolated.

During the experiment, it is necessary to measure the pressure parameters – differential pressure along the length of the measurement unit and compression pressure. The pressure drop in the cells is measured using a water column of known height to set the required interval. The zero mark is checked before each data read, and the interval is checked periodically.

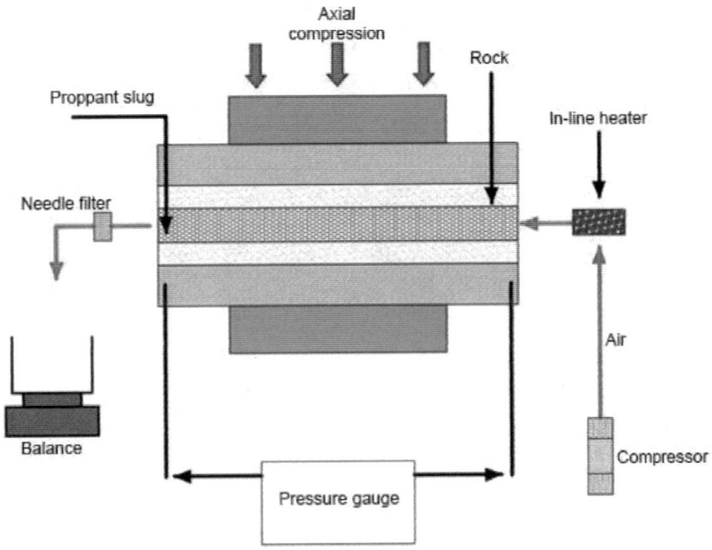

Figure 2. Filter cell configuration

Figure 3. Conductivity cell

The schematic diagram of the bench is shown in Figure 4.

The unit consists of a body 1 (a bench itself) made in the form of a cylinder 2, in an axial channel 3 of which there are two pistons 4, with a gap 5, in which wedging layer – proppant – is located. Gap 5 simulates a rock fracture. Cylinder body 2 has transverse slots isolated by branch pipes 6 with swivel nuts 7. Filter 8 is installed in an outlet branch pipe 6 for collection of released proppant particles. Axial channel 3 of cylinder 2 is covered from above and from below by caps 9 in which adjusting screws 10 are installed. A binding unit and a power drive of the bench are shown in Figure 5.

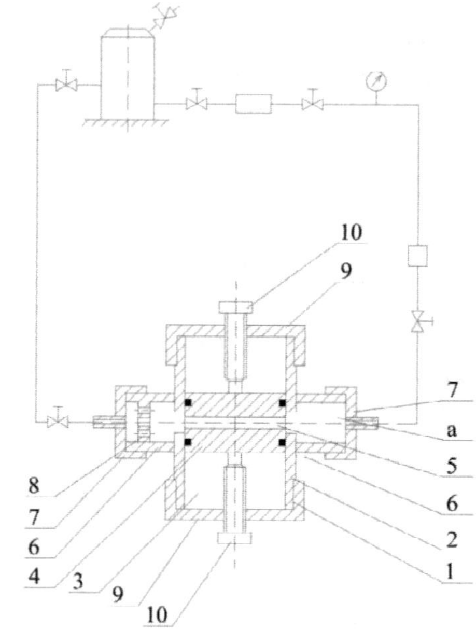

Figure 4. Test bench design

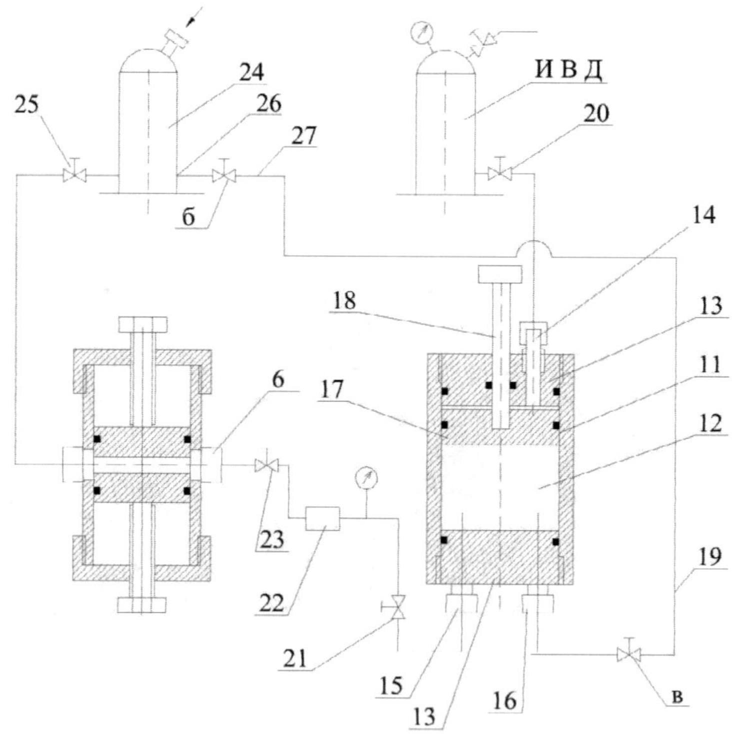

Figure 5. Binding unit and power drive

To ensure the supply of the working fluid under pressure (at preset flow rate) into a gap 5, the power hydraulic cylinder 11 is designed, which represents a chamber 12. The chamber 12 is closed hermetically from above and from below by caps 13 in which chokes 14, 15, 16 are installed. The axial channel of chamber 12 accommodates a power piston 17 with a rod 18. The choke 14 is connected through a hose 19 with high pressure source representing inert gas bottle.

A three-way valve 20 is installed on a gas high pressure line. The choke 15 is connected through a valve 21, a flow regulator 22 and a valve 23 with an inlet branch pipe 6 of a bench 1. The outlet 6 of a bench 1 is connected through high-pressure tube to a reservoir 24 through a valve 25. The outlet 26 of the reservoir 24 is connected via high-pressure tube 27 to a choke 16 on a lower cap 13.

Unit operaiton

When valves 25 and 23 are closed, pistons 4 are installed into reservoir 24 in cylinder 2, in gap 5, between which a proppant layer of certain fraction is placed. The power piston 17 in power hydraulic cylinder 11 is in extreme upper position. With closed gate 21 a cavity 12 of power cylinder 11 is filled with working liquid by opening of gates a and b. The extreme top position of a rod 18 is controlled. Valves b and c are closed.

Valves 21 and 23 are opened with the formation of hydraulic connection of gap 6 between pistons 4 in bench I. Valve 20 is opened from the source of high pressure of gas, with supply of the power cylinder 11 to the axial channel 12, as well as with the action on the area of power piston 17.

The movement of power piston 17 is controlled along rod 18. The flow of the working fluid through the tube with valve 21 gets through inlet branch pipe 6 of bench 1 and moves in gap 6 between pistons 4 affecting the proppant particles. Besides, proppant particles not attached in the gap migrate to the outlet branch pipe 6, where they are caught by a filter 8. After extrusion of the working fluid portion from power hydraulic cylinder 11, rod 18 is in extreme lower position. Closing a three-way valve 20 with inert gas, high pressure source is cut off and gas is released from the axial channel 12 of the power hydraulic cylinder 11 above the power piston 17. At the same time valves 23 and 25 on bench 1 are closed. The working fluid is supplied from reservoir 24 at open valves b and c, simultaneously filling axial channel 12 of power hydraulic cylinder 11 and moving power piston 17 upwards thus controlling the output of the rod 18. The cover 7 is disconnected on the outlet branch pipe 6 with removal of the filter 8 and estimation of the amount of removed proppant. The flow rate of the working fluid is ensured through the gap 6 between pistons 4, and estimation of flow rate is carried out after measurement of the gap 5 and counting the filtration area. The flow rate of the working fluid is also carried out by measuring the time of movement of the power piston 17, together with the rod 18, with recalculation of flow speed parameters in the gap 6. After installation of pistons (4) in the bench 1, the preparatory works are carried out for the next test cycle. The operation can be carried out with different fractional composition of proppant particles, simulating different flow rates in the gap 6, and estimating the results associated with proppant particles deposited in the filter 8.

By changing the flow rate of the inert gas it is possible to simulate the different flow rates of the working fluid through the gap 6 between the pistons 4. Test and measuring tools of the bench shall provide measurements of pressure, temperature, supply and head of the pump.

Since the studies are based on a closed cycle of circulation of the working fluid, it gradually heats up and its viscosity decreases. Fracture parameters are measured by the measuring system in continuous mode at a temperature varying over time.

3. Research results

Russian oil workers use proppants of four fractions to secure the HF fracture, of which two are the main fractions: 16/20, 12/18 and two auxiliary fractions: 16/30 and 20/40. The choice of the desired size of proppant grains is determined by a whole set of factors. The larger the granules, the more permeable the proppant package in the fracture. However, the use of a large fraction proppant has additional problems when it is transported along the fracture [8]. This study examined the proppant of fraction 16/20. For the first tests 5 samples (a), (b), (c), (d), (e) are presented on the unit. In terms of their bulk densit, proppants (b), (c), (d), (e) are considered light proppants because they have low bulk weight (less than 1.6 kg/m^3). Proppants (a) are medium-density.

The test results are shown in Table 1.

Table 1. Proppant test results

Indicator	Obtained averages				
	A	b	C	d	e
Crushing resistance (mass fraction of crushed granules), %, under pressure, MPa:					
7	1.5	11	5.5	4	5
14	2.5	15.5	7.5	5.5	8.5
28	7	25	17	11	15.5
42	12	31	30	18	19
49	16.5	38	37.5	27	23
Filter residue, %: No. 1 (4.5 mm)	6	4	2.5	3.5	0.5
No. 2 (5 mm)	9	6	4	5	2
No. 3 (5.5 mm)	14	11	7	9	6
No. 4 (6.0 mm)	20	24	15	19	10
No. 5 (6.5 mm)	26	26	19	23	15
No. 6 (7.0 mm)	30	32	25	28	18
No. 7 (7.5 mm)	44	39	31	31	21
Sphericity	0.9	0.9	0.9	1	0.9
Circularity	0.8	0.9	0.9	0.9	0.9
Packing density, g/cm^3	1.8	1.55	1.4	1.5	1.3

The measurements of hydrodynamic characteristics of proppants – permeability, layer width depending on pressure are shown in Figures 6, 7. The experiment showed that proppant (a) had the best permeability of the package, then in descending order (b), (c), (d) and (e) (Figure 6). The permeability of proppant (e) was very low at high pressure. The reduction in the height of the proppant layer (Figure 7) shows the fraction of broken particles, i.e. the ability to withstand loads. The permeability and conductivity of the package is reduced due to the height of the proppant layer and the reduction in the number of filter pores. According to Figure 7, it can be noted that a larger number of granules were broken down with pressure increase (decrease of the layer height), thereby significantly reducing permeability and conductivity of the proppant package [9].

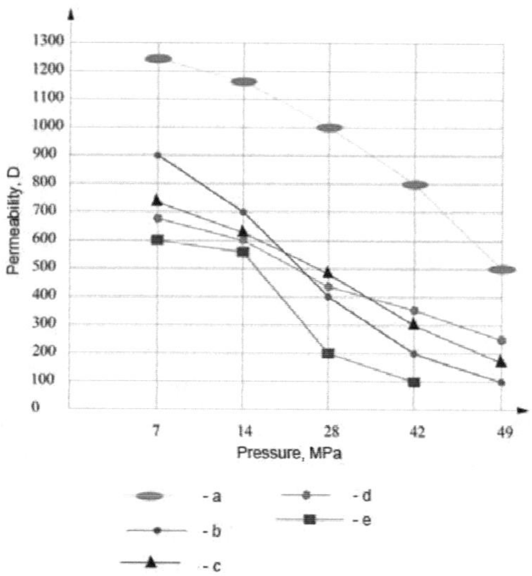

Figure 6. Permeability of proppant pack depending on pressure

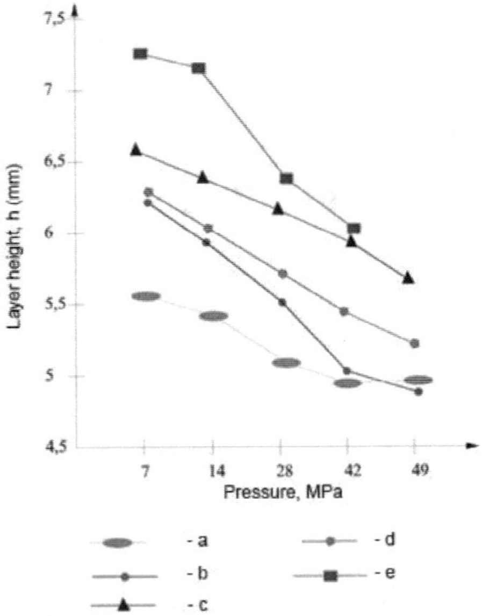

Figure 7. Height of proppant filling layer depending on pressure

4. Conclusion

1. Light proppants do not always meet consumers' expectations. The studies showed that a decrease in the weight of light proppants (b), (c), (d), (e) used for HF leads to a significant decrease in the permeability of their packaging compared to the medium-strength proppant (a).
2. As the fracture closing pressure increases, the permeability and conductivity of the proppant pack decreases. The main peculiarity if the general change in conductivity as the pressure increases.
3. The developed test bench was introduced into the training process at the Institute of Oil and Gas of the North Caucasus Federal University to test various types of proppants.

References

[1] Verisokin A 2018 Reduction of well development period after hydraulic fracturing *Science. Innovations. Technologies* **3** 99-114
[2] Kudryashov S 2006 Hydraulic fracturing as a method of development of low-permeable collectors *Oil industry* **7** 80-83
[3] Verisokin A and Serikov D 2019 Technology of hydraulic fracturing using titanium nickelide *Territory of oil and gas* **9** 20-24
[4] Kanevskaya R 1998 *Foreign and domestic experience of hydraulic fracturing* (Moscow – VNIIOENG)
[5] Mashkov V, Verisokin A, Dimitriadi Yu, Lukianov V and Geraskin V 2017 Analysis of used structures of anchor assemblies and various packer seals used in hydraulic fracturing *Science. Innovations. Technologies* **3** 167-180
[6] Skurichin V and Migal V 2008 Proppant selection – selection criteria *Oil Industry* **1** 90-93
[7] Magadova L, Mikhailov S and Magadov V 2011 *Study of long-term conductivity of proppants of various manufacturers Grant of British Petroleum, I.M.* (Moscow – Gubkin National University of Oil and Gas)
[8] Mozhzherin A and Korzhavin A 2017 Which proppant is more profitable: 16/30 or 16/20? *Drilling and oil* **07-08** 34-38
[9] Kupavikh K S and Nutskova M V 2016 Ecological features of oil well repair at low - permeability reservoir. *International Journal of Applied Engineering Research*

IOP Conf. Series: Materials Science and Engineering **860** (2020) 012002 doi:10.1088/1757-899X/860/1/012002

Energy-efficiency measures of oil and gas industry building using optimum conditions of faltering heating

K V Osintsev, D P Korabelnikova and T A Pshenitsyna

South Ural State University, 76, Lenin Ave., Chelyabinsk, 454080, Russia

E-mail: pte2017pte@mail.ru

Abstract. Nowadays the problem of use of energy resources is the most relevant. The actions directed to increase in energy efficiency of use of energy resources at the mining utilities should be developed and introduce to solve this problem. The refusal of the systems of the centralized heat supply and transition to the decentralized systems is offered as the solution of an objective. In the centralized systems, development of warmth is carried out in separate sources. The giving of the heating agent in the systems of heat consumption is taken place through thermal networks. Thermal networks have the considerable extent and diameters at the same time. They are also equipped with thermal points, pump stations, automatic equipment and a control system.

1. Introduction

The system of heat supply is called decentralized one if it does not have the developed thermal networks and the source of warmth is located directly near the objects consuming warmth. This system is widespread at the utilities of a mining complex, for example, in administrative and household complexes [1, 2, 3] of working personnel on mines and trade platforms.

Now there are technical solutions of implementation of autonomous [4] heat supply of an object: classical – using of autonomous boiler houses, modern – using of solar energy as a warmth source and using of low-potential energy of soil, water and other.

Today there is a set of measures for decrease in energy consumption by buildings. This measures concern both constructions and engineering networks. One of the most effective measures for economy of heat energy is implementation of the faltering heating mode [5]. It allows saving resources and simplifying the structure.

The purpose of research is the definition of actual heat economy using the mode of faltering heating during all heating season for administrative and household rooms of trade platforms.

2. Faltering heating concept

The faltering heating mode is provided for economy of thermal energy introduction into an object. The mode of faltering heating consists in the following: in time off work– at night or in holidays when the office building is not operated. Beside, the required temperature must be ensured by the time of using the room.

Such system allows not only saving thermal energy and demanding small financial expenses concerning other energy saving actions [6, 7, 8].

At the same time, the task is set: to operate so the mode of heating of the building that the expense of thermal energy was minimum and internal temperature at the beginning and at the end of the non-working period corresponded to rated value. This task was considered in [9, 10, 11].

The simplified heat exchange model based on the discrete one-dimensional equations of heat conductivity was used in this article. The model describes evolution of temperatures on surfaces of enclosing structures and internal partitions and temperatures of internal air indoors. The operating parameter of model is heating system power. Functional of I (W) from the power of W(t) of a heating system was applied during regulation. Functional represents the sum of two non-negative parts composed to assessment of efficiency of heating (1):

$$I(W) = E(W) + \gamma(\Delta T_{\kappa}(W))^2. \quad (1)$$

3. Methodology of the calculation method increasing efficiency of buildings due to faltering heating

This methodology was solved based on the mathematical models. The description of this model of thermal mode of the building for the optimum control algorithm for the non-working period was developed. The developed algorithm ensures the required temperatures of internal air at the minimum flow of thermal energy. Mathematical models were considered in [10, 12, 13].

The thermal mode of the building [11, 14, 15], strictly speaking, as object of management is an object with the distributed parameters therefore the problem about optimum control for this representation of an object was solved.

The analysis of literature [16, 17, 18] shows that the mathematical description of the thermal mode of the building can be well submitted as follows (2)-(7):

$$\frac{\partial t(x,\tau)}{\partial \tau} = a\frac{\partial^2 t(x,\tau)}{\partial x^2}, 0 < x < L, \tau > 0; \quad (2)$$

$$t(x,0) = t^0(x), 0 \le x \le L; \quad (3)$$

$$-\lambda\frac{\partial t(0,\tau)}{\partial x} = \alpha_в[t_в(\tau) - t(0,\tau)]\ \tau > 0; \quad (4)$$

$$-\lambda\frac{\partial t(L,\tau)}{\partial x} = \alpha_н[t(L,\tau) - t_н(\tau)]\ \tau > 0; \quad (5)$$

$$c_в m_в\frac{dt_в(\tau)}{d\tau} = u(\tau) - \alpha_в F_{cm}[t_в(\tau) - t(0,\tau)] - k_{OK}F_{OK}[t_в(\tau) - t_н(\tau)], \tau > 0; \quad (6)$$

$$t_в(0) = t_в^0(x), \quad (7)$$

$t(x,\tau), t(0,\tau), t(L,\tau)$ – temperature respectively in a point with coordinate x on thickness of an external wall, on an internal surface and on an external surface of an external wall of the building in time point τ; λ – coefficient of heat conductivity of material of an external wall; $\alpha_в, \alpha_н$ – thermolysis coefficient respectively for internal and external surfaces of a wall; $t_в(0), t_в(\tau)$ – temperature of internal air in initial timepoint and timepoint τ respectively; $t_н(\tau)$ – temperature of external air in timepoint τ; $u(\tau)$ – management, heating system power in timepoint τ; $c_в, m_в$ – the specific heat and mass of internal air in the building respectively; a – thermal diffusivity; F_{cm}, F_{OK} – respectively, the area of the exterior walls of the building and windows; k_{OK} – window heat transfer coefficient.

On a set of admissible decisions, the functionality the I was found. The first component I was defined defines proximity of reference and final temperature of internal air, and the second was defined an expense of thermal energy (8):

$$I = [t_в(\tau k) - t_в^0]^2 + \beta \int_0^{\tau k} G[u(\tau)]d\tau, \tag{8}$$

τ_k – the set period; G – the function estimating the current expense of warmth the building on heating; β – some weight coefficient.

The conditions of optimality are presented by the following ratios: for a formulation of conditions of optimality [19, 20, 21], function is entered (9):

$$H = \psi_0(\tau)u(\tau) - \beta G[u(\tau)], \tag{9}$$

$\psi_0(\tau)$ satisfies to the following system of the equations (10)-(11):

$$-c_в m_в \frac{d\psi_0(\tau)}{d\tau} + (\alpha_в F_{cm} + k_{ок}F_{ок})\psi_0(\tau) - \frac{a\alpha_в}{\lambda\psi(0,\tau)} = 0; \tag{10}$$

$$\frac{d\psi(x,\tau)}{d\tau} + a\frac{\partial^2\psi(x,\tau)}{dx^2} = 0; \tag{11}$$

with entry conditions (12)-(13):

$$c_в m_в \psi_0(\tau k) = -2[t_в(\tau k) - t_в^0]; \tag{12}$$

$$\psi_0(x, \tau k) = 0. \tag{13}$$

Boundary conditions for function $\psi(x,\tau)$ were set in a look (14)-(15):

$$a\frac{\alpha_в}{\lambda}\psi(0,\tau) - a\frac{\partial\psi(0,\tau)}{\partial x} - \alpha_в F_{cm}\psi_0(\tau) = 0; \tag{14}$$

$$-\lambda\frac{\partial\psi(L,\tau)}{dx} = a_н\psi(L,\tau). \tag{15}$$

4. Scientific novelty of research

Efficiency of the following control algorithms was estimated by the received results.

Figure 1 shows scheme of efficiency of control algorithms.

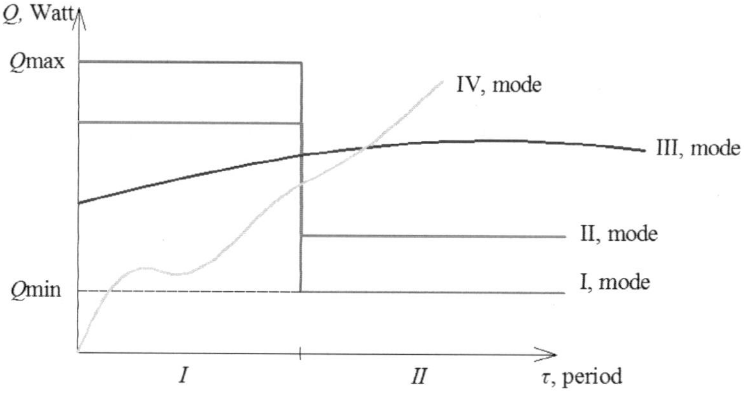

Figure 1. Scheme of efficiency of control algorithms: *Q*max – maximum power of a system; *Q*min – minimum power of a system; IV mode – connection to thermal networks; III mode – with the stabilizing regulator ; II – internal temperature comes to the border of 12 °C ; I – during the I period of time the power of a heating system is equal to the maximum (established) value, and during the II period – minimum [9]

Figure 1 presents an analysis and comparisons of the 4 modes. The intensive warming up of the building is carried out only on the final site of time off (mode II). This mode was established economic. The mode IV is most uneconomical widely using now. In addition, the building is connected to the central thermal networks and regulated warmth only on a source is carried out.

5. Calculation method of faltering heating

Thus, the application of the faltering heating mode in buildings with an autonomous source of warmth is expedient. On the other hand, the connection of them to the central thermal networks is irrational. This is due to the fact that building is connected to thermal networks and the central regulation of warmth on a source is made and the necessary power for a building warming up is function of temperature of external air. The energy from the central thermal networks is spent for compensation of the current heat losses of the building. A warming up of the building, or rising temperature of internal air in the building, does not happen because of lack of a stock of warmth.

6. Conclusion

In the research, the method of increase in energy efficiency of use of energy resources and energy saving actions is analyzed.

This task was solved on the basis of the mathematical models describing the thermal mode of the building. The optimum control algorithm of the thermal mode of the building during the non-working period was developed. It provides the required temperatures of internal air at the minimum flow of thermal energy [21, 22].

The application of the mode of faltering heating in buildings with an autonomous source of warmth is expedient. Their connection to the central thermal networks is irrational.

For object model with the distributed parameters the optimality conditions are defined. They are formulated in the form of the principle of a maximum. The general view of optimum control is established.

References

[1] Rudig W 1986 Combined heat and power for district heating *Physics in Technology* **17** 125
[2] Hammond G 1977 Combining district heating and power generation *Physics in Technology* **8** 163
[3] Panferov V I and Anisimova E Y 2008 The analysis of a possibility of economy of thermal energy at the faltering mode of heating *Bulletin of the SUSU* **112 (12)** 30-37
[4] Kiyanets A V 2017 Improving Energy Efficiency of Buildings in the *Urals IOP Conf. Ser.: Mater. Sci. Eng.* **262** 012068
[5] Panferov V I and Anisimova E Y 2009 About economy of energy at optimum control of the mode of faltering heating *ABOK NORTHWEST* **5 (44)** 32-35
[6] Hrvoje K and Mihaela T 2017 *IOP Conf. Ser.: Mater. Sci. Eng.* **245** 042049
[7] Sharovarova E, Alekhin V and Shcheklein S 2018 *IOP Conf. Ser.: Mater. Sci. Eng.* **451** 012045
[8] Danilov O L, Garyaev A B and Yakovlev I V 2011 *Energy Saving in Heat Power Engineering and Heat Technologies* (Moscow: MPEI Publishing house)
[9] Panferov V I and Anisimova E Y 2011 Effective operation of heating systems *Conf. Heat and gas supply: state, problem and prospects* (Orenburg: LLC Nikos) pp 129-133
[10] Gershkovich V F 2000 About a possibility of implementation of regulation of heat consumption of buildings by method of periodic interruption of a stream of the heat carrier *News of heat supplying* **02** 63-71
[11] Mozgalev K M 2011 *Energy Efficiency of Buildings* (Chelyabinsk: REKPOL)
[12] Golovnev S G 2010 *Modern Construction Technologies: a monograph* (Chelyabinsk: South Ural St. Univ.Publ.)
[13] Malyavina Y G 2011 *Heat loss of buildings: reference book* (Moscow: AVOK-PRESS)
[14] SP 23-101-2004. *Design of thermal protection of buildings*
[15] SP 50.13330.2012 *Thermal performance of the buildings.* Revised edition SNiP 23-02-2003
[16] SP 60.13330.2012 *Heating, ventilation and conditioning.* Revised edition SNiP 41-01-2003

[17] Z Y Li, S B Zhang, Y B Xiao, Q Q Shi, Y Q Zhao and J L Gao 2018 *IOP Conf. Ser.: Earth Environ. Sci.* **188** 012105
[18] Paiho S, Hedman Å, Abdurafikov R, Hoang H, Sepponen M, Kouhia I and Meinander M *2013 Energy saving potentials of Moscow apartment buildings in residential districts Elsevier Energy and buildings* **66** 706-13
[19] Dimevska L, Cvetkovska M and Gavriloska A T 2019 *IOP Conf. Ser.: Earth Environ. Sci.* **222** 012028
[20] Khayatian F, Sarto L and Dall'O' G 2016 Application of neural networks for evaluating energy performance certificates of residential buildings Energy and Buildings *2018 Annual Conf. IISE **125** 45-54*
[21] Osintsev K V, Prikhodko I S and Zavyalova M I 2018 Methods for improving energy efficiency of air handling unit using factor analysis of data *Earth and Environmental Science* **194** 052019 7-12
[22] Osintsev K V, Prikhodko I S and Psenitsyna T A 2018 Choice of Chelyabinsk and Kuznetsk coals as main fuel for steam generator PK-14 using elements of cluster analysis *Earth and Environmental Science* **194** 052018 6-10

Analysis of energy efficiency improvement of boiler-houses in oil and gas industry using method for deep chilling flue gases

K V Osintsev, Y S Bolkov and M I Zavyalova

South Ural State University, 76, Lenina Ave., Chelyabinsk, 454080, Russia

E-mail: pte2017pte@mail.ru

Abstract. The article is devoted to the methods of increasing boiler-house efficiency. This current problem is considered in the article by two reasons: big part of volume flue gases enter the atmosphere, utilization of this gases is difficult process. The scope includes a description of the sample of heat exchanger. The analysis method for deep chilling flue gases for increasing the energy efficiency of the boiler- house was carried out. This method can be used in enterprises with big energy consumption, also boiler- house with small capability. The calculation of a boiler-houses with different was given, as an example. Average increasing energy efficiency of boiler-houses was found.

1. Introduction

The improvement of the energy efficiency of energy equipment is of paramount importance for most branches. The reason for this is the reduction of reserves of fossil fuels. Designing new energy equipment with high-energy efficiency is an important task. However, the most important task is increasing the energy efficiency of existing equipment. To increase energy efficiency of boiler- house need to analyze using equipments. Main factor influencing on energy efficiency is efficiency of boiler.

The heat balance of the boiler was considered [1]. The studies of technical literature showed that in a modern boiler, it was possible to minimize heat loss due to dry flue gases, incomplete combustion and unburnt flue. For example, E A Biryuzova [2] considers the heat loss value from external cooling of the boiler unit surface. The authors obtained dependence for determining the value of this loses. The article [3] presents the analysis of the boiler performance thermal and technical indicators, and depending boiler energy efficiency of exhaust gases temperature, and furnace volume thermal stress. However, heat carried away by flue gases stay the main losses of heat in the boiler. Flue gas losses account for 5-10 percent. For comparison, other losses (Q_3, Q_4, Q_5, Q_6) are no more than 5 percent [4, 5, 6]. Flue gases have high temperature around 100-200°C. High temperature helps to avoid condensation moisture in fuel gases on the inner area of exhausted tube. It means that hidden heat and sensible heat are lost. A number of methods for reducing losses heat carried away by flue gases are listed in [7, 8, 9]. In article [7] presented concept of heat recovery from exhaust gases use it to prepare hot water. This solution is relevant for the industry with a large load and coal boilers. The authors of article [8] also consider the possibility of using heat recovery from flue gases. However, they also estimate limit of heat power of the heat exchanger. The method deep chilling of the flue gases was chosen of offered process engineering solution. It is because efficiency of this method can be reached and set up in the boiler-house. The literature [10] also was presented calculation methods for determining the effectiveness. The basis of the method is to install a recuperative heat exchanger at the flue gas outlet. This technical solution has shown well in other areas. For example, the authors of article [11] consider installing a recuperative heat exchanger in a ventilation system to increase efficiency. Also to achieve increasing efficiency can use other method, for example by water regime boiler [12], and other presented in [13, 14, 15, 16, 17].

The problem of this article is the improvement of efficiency of low power boiler by deep chilling the temperature of the flue gases with the return of part of the heat for secondary using. The idea of reusing the heat of flue gases was developed for a long time, but there is no basic research in this area. To achieve improving energy efficiency can by installing condensing recovery unit of surface heat.

2. Principle of operation of condensing surface heat recovery unit

There are several options for the design of heat exchangers to reduce the flue gas temperature. They can be divided into two groups. The first group is contact type heat exchangers were widely used in the past. The second group is condensing surface heat exchangers [18, 19].

Condensing surface heat exchanger is better than contact heat exchanger. They possess the improved technical characteristic. The using contact heat exchanger is complicated by the absorption by water of carbon dioxide and oxygen. In this case, the water acquired corrosive and aggressive properties. The use of the condensing surface of the heat exchangers allows avoiding this [20].

Figure 1 shows scheme of reusing the heat of flue gases in condensing surface heat exchanger. Flue gases enter from the economizer to the distributive valve. The valve divides flue gases flow into two flows. The main flow passes through the mesh filter into the heat exchanger. The second stream goes along the bypass line to the gas duct.

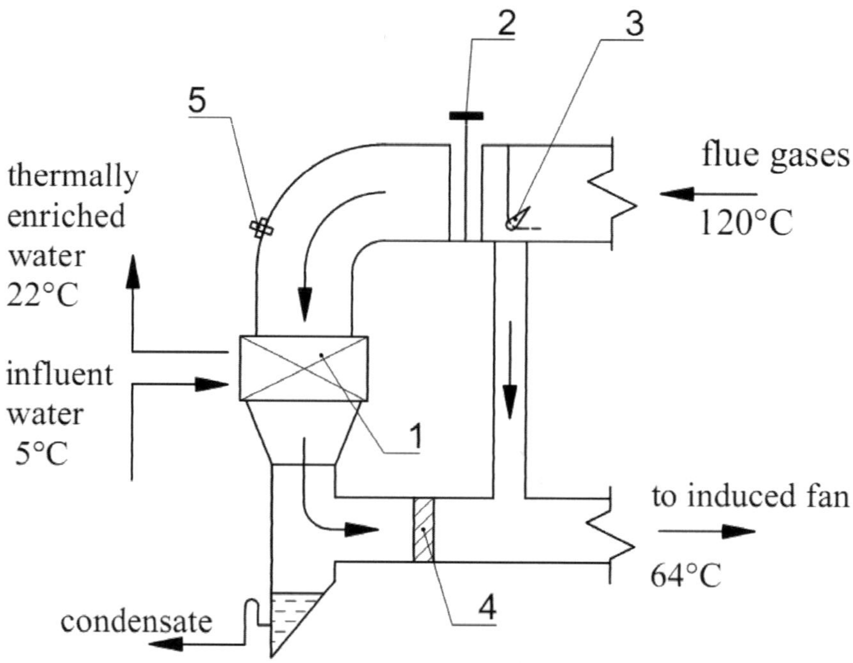

Figure 1. Scheme of reusing the heat of flue gases in condensing surface heat exchanger: 1- condensing surface of heat exchanger; 2 – mesh filter; 3 – distributive valve; 4 – drip collector; 5 – hydro-pneumatic blower [10].

The heat exchange has been reducing the moisture content of combustion products and condensate. After mixing, temperature of flue gases maintain at higher then condensing point. Accordingly, condensation in the gas duct is excluded.

Condensate is collected in a pan. Then it goes to the decarbonised water tank, bypassing the water treating plant. Pumps are fed water (condensate) from the tank into the deaerator when heat supply system is closed type [10].

3. Methodology of the calculation method increasing efficiency boilers

Several options for hot-water boilers of different capacities were considered. Each boiler house has two Logano sk755 boilers of different capacities. Calculation data was taken from the technical dossier of the boiler. The main characteristics of the boiler are presented in Table 1.

Table 1. Technical characteristics of boilers and boiler- houses

Characteristic	Boiler- house		
	1	2	3
Boiler-houses power, kW	1200	2080	3700
Boiler power, kW	600	1040	1850
Unit efficiency,%	90	90	92
Temperature of flue gases, °C	200	200	200
High heat value of heating agency(QHV), kJ/m^3	8867		

Natural gas is used as fuel of Urengoy field. The calculation of the condensation surface heat exchanger was made by the method given in [10]. The presented technique was helped to evaluate the efficiency of the installation of a condensing heat exchanger.

Calculate volume of flue gases of the inlet heat exchanger under standard conditions, V_N(m^3/s):

$$V_N = (1.135 + 1.32 \cdot \alpha_{FG}) \cdot Q_B \cdot 10^{-3} \cdot \eta_{LB}^{-1} \quad (1)$$

α_{FG}– the excess air factor of the flue gases; Q_B –heating capacity of boiler (MW); η_{LB} – efficiency of boiler under lower heating value.

Actual volume of flue gases of the outlet heat exchanger under field conditions,V_{1t}=V_{FG}(m^3/s) is:

$$V_{1t} = V_{FG} = V_N \cdot (t_{FG} + 273) \cdot 273^{-1} \quad (2)$$

t_{FG} – flue gases temperature of the outlet heat exchanger (°C).

Average volumetric flow of flue gases of the outlet heat exchanger V_{AVG} (m^3/s):

$$V_{AVG} = (V_{1t} + V_{FG}) \cdot 2^{-1} \cdot 3600^{-1} \quad (3)$$

Mass flow rate of flue gases, G_{FG}(kg/s) is:

$$G_{FG} = V_N \cdot \rho_N \quad (4)$$

ρ_N – density of the flue gases average composition under standard conditions, (1,295 kg/m^3).

Enthalpy gases of the inlet heat exchanger I_1 (kJ/kg) is:

$$I_1 = (C_{CP} + C_{WV} \cdot \chi_{CP}) \cdot t_{CP} + 2491 \cdot \chi_{CP} \quad (5)$$

C_{CP} – heat capacitance dry combustion products (kJ/(kg·°C);C_{WV} – heat capacitance of combustion water vapour (kJ/(kg·°C); t_{CP} – combustion products temperature of the outlet heat exchanger (°C); χ_{CP} – water content combustion products of the outlet heat exchanger(kg/kg$_{CP}$).

Find water content combustion products of the outlet heat exchanger χ_{CP} (kg/kg$_{CP}$):

$$\chi_{CP} = (0.13 + \chi_{AIR} \cdot \alpha_{FG}) \cdot (\alpha_{FG} - 0.058)^{-1} \quad (6)$$

χ_{AIR} – water content air (kg/kg).

Enthalpy gases of the inlet heat exchanger I_{FG} (kJ/kg) is:

$$I_{FG} = (C_{CP} + C_W \cdot \chi_{FG}) \cdot t_{FG} + 2491 \cdot \chi_{FG} \tag{7}$$

χ_{FG} – water content flue gases of the outlet heat exchanger(kg/kg).

Find water content flue gases of the outlet heat exchanger χ_{FG} (kg/kg$_{CP}$):

$$\chi_{FG} = (0.0006382 + 0.004 \cdot \alpha_{FG}) \cdot (0.199 + \alpha_{FG})^{-1} \cdot \exp(0.062 \cdot \alpha_{FG}) \tag{8}$$

Heating capacity of heat recovery unit Q_U (kJ/s) is:

$$Q_U = G_{FG} \cdot (I_1 - I_{FG}) \tag{9}$$

The amount heated water G_W (kg/s) is:

$$G_W = Q_U \cdot [C_W \cdot (t_{W2} - t_{W1})]^{-1} \tag{10}$$

C_W – heat capacitance water under average temperature (kJ/(kg·°C); t_{W2}, t_{W1} – temperature of heated water (°C).

Average temperature drop Δt is:

$$\Delta t = (\Delta t_{\max} - \Delta t_{\min}) \cdot \ln^{-1}\left(\frac{\Delta t_{\max}}{\Delta t_{\min}}\right) \tag{11}$$

Δt_{max} – difference between t_{FG1}, t_{W1}; Δt_{max} – difference between t_{FG2}, t_{W1}.

Heat transfer characteristic of heat exchanger K (W/(m^2·°C) is:

$$K = (\alpha'^{-1}_{FG} + \delta \cdot \lambda^{-1} + \alpha^{-1}_{IN})^{-1} \tag{12}$$

α'_{FG} – reduced heat transfer coefficient from the flue gases to the outer surface of the condensation heat exchanger (35, W/(m^2·°C));δ – wall thickness (m); λ – thermal conductivity of the material (W/(m·°C); α_{IN} – heat transfer coefficient from the inner surface of the pipe to the heated water(1000, W/(m^2·°C)).

Heat transfer surface of heat recovery unit F(m^2) is:

$$F = Q_U \cdot K^{-1} \cdot \Delta t^{-1} \tag{13}$$

Number of tube of heat recovery unit N is:

$$N = F \cdot f_i^{-1} \tag{14}$$

f_i – heat transfer surface of one tube(0.0025m^2).

Number of bank of tubes z is:

$$z = N \cdot n_i \tag{15}$$

n_i – number of tube in one bank(50).

High of heat recovery unit H(m) is:

$$H = d_{OD} \cdot z \tag{16}$$

d_{OD} – outside diameter of tubes (m).

The amount of useful heat exchanger's heat Q_{UU} (kJ/s) is

$$Q_{UU} = G_W \cdot C_W \cdot (t_{W2} - t_{W1}) \tag{17}$$

Find boiler efficiency without heat recovery unit under high heat value:

$$\eta_B = Q_B \cdot B^{-1} \cdot Q_{HV}^{-1} \cdot 100\% \tag{18}$$

B – fuel burn(m^3/s).

Increasing boiler efficiency with heat recovery unit $\Delta\eta$ is

$$\Delta\eta = Q_B \cdot B^{-1} \cdot Q_{HV}^{-1} \tag{19}$$

Total efficiency of one boiler η_T (%) is:

$$\eta_T = \eta_B + \Delta\eta \tag{20}$$

Calculations are presented in table 2.

Table 2. Result of calculation increasing efficiency boiler- houses.

Characteristic	Boiler- house		
	1	2	3
Boiler-houses power, kW	1200	2080	3700
Boiler actual efficiency without heat exchanger, η_B%	84.45	91.68	85.43
Increasing boiler efficiency, $\Delta\eta$	1.17	1.27	1.2
Total boiler efficiency, η_T%	85.62	92.95	86.63

4. Conclusion

Thus, the general range of increasing of boiler efficiency maximum values calculated by the proposed method is possible to be pointed out. The obtained calculations showed that the power of the boiler does not affect the increase of boiler efficiency. Average value of increasing boiler efficiency is 1.2 percent. Installing a heat exchanger is one of the most affordable methods for improving the efficiency of a boiler room that is especially important now with high demand for high efficiency equipment. This method of improving energy efficiency can be considered for low power boilers, for example, for boiler rooms located near residential areas. As a result, the installation of the heat exchanger was increased the boiler efficiency and reduced the amount of pollutants in the flue gases.

References

[1] Sadhu S 2011 *Handbook of Mechanical Engineering* (S.Chand & Company Ltd)

[2] Biryuzova E A 2018 Investigation of Methods for Determining Heat Losses from External Cooling of Surfaces of a Small-Capacity Hot-Water Boiler *Materials Science and Engineering* **463** 032056 1-4

[3] Batrakov P A, Mikhailov A G and Ignatov V Yu 2018 Fire-tube boiler optimization criteria and efficiency indicators rational values defining *Journal of Physics: Conf. Series* **944** 012009

[4] Lummi A P and Munz B A 2009 *Calculation of the hot-water boiler* (Yekaterinburg: SEI HVE USTU–UPI)

[5] Roddatis K F 1977 *Boiler plants* (Moscow: Energiya)

[6] Esterkin R I 1989 *Boiler plants* (Leningrad: Energoatomizdat)

[7] Bukowska M, Nowak K, Proszak-Miąsik D and Rabczak S 2017 Concept of Heat Recovery from Exhaust Gases *Materials Science and Engineering* **245** 052057 23-25

[8] Shatskikh Y V, Sharapov A I, Byankin I G 2017 Analysis of Deep Heat Recovery From Flue Gases *Journal of Physics: Conf. Series* **891** 012188 1-5

[9] Paschchenko D I, Semenov B A, Shchelokov A I, Naplekov I S and Mustafin R M 2018 Thermodynamic analysis of thermochemical recuperation of high-temperature flue gas heat *Journal of Physics: Conf. Series* **1111** 012009 6-10

[10] Kudinov A A and Ziganshina S K 2011 *Energy saving in power engineering and heat technologies* (Moscow: Mashinistroenie)

[11] Osintsev K V, Prikhodko I S and Zavyalova M I 2018 Methods for improving energy efficiency of air handling unit using factor analysis of data *Earth and Environmental Science* **194** 052019 7-12

[12] Khavanov P, Chulenyov A 2017 Boiler water regime *Earth and Environmental Science* **90** 012199 8-13
[13] Zhang X, Ye Y, Wang H, An X 2018 A new method for boiler combustion calculation *Earth and Environmental Science* **199** 032083 154-157
[14] Roslyakov P V and Proskurin Y V 2017 Increase of efficiency and reliability of liquid fuel combustion in small-sized boilers *Journal of Physics: Conf. Series* **891** 012243 1-8
[15] Wu Y-H, Hsiao C-C and Chen C-H 2018 Thermal stress analysis of a damaged waste heat boiler *Materials Science and Engineering* **372** 012025 9-16
[16] Koubogiannis D, Nouhou C 2016 How much Energy is Embodied in your Central Heating Boiler? *Materials Science and Engineering* **161** 012094 156-164
[17] Roslyakov PV, Proskurin Y V, Khokhlov D A and Zaichenko M N 2018 A study of burning processes of fossil fuelsin straitened conditions of furnaces in low capacity boilers by an example of natural gas *Journal of Physics: Conf. Series* **980** 012031 238-242
[18] Thulukkanam K 2013 *Heat Exchanger Desing Handbook 2nd ed.* (CRC Press)
[19] Danilov O L, Garyaev A B, Yakovlev I V et al 2011 *Energy Saving in Heat Power Engineering and Heat Technologies: studies. for high schools* (Moscow: MPEI Publishing house)
[20] 1998 *Thermal calculation of boilers (Standard method)* (Saint-Petersburg: NPO CKTI)

Angle impact on drill life during hole drilling land mine explosion at Quangninh area, Vietnam

Dinh Van Chien[1], Nguyen Khac Linh[1,2] and N S Golikov[2]

[1] Ha Noi University of Mining and Geology, 18, Pho Vien Street, Duc Thang Ward, Bac Tu Liem District, Ha Noi, 90000, Vietnam
[2] Saint-Petersburg Mining University, 2, 21st Line, St Petersburg 199106, Russia

E-mail: khaclinhhumg@gmail.com

Abstract. Rock drilling and blasting holes created during construction of underground mining in general and in Quangninh are the interaction of the drill into the rock. Drilling efficiency and productivity depend on many parameters such as power dam, the rock solidity and the rotational speed of drilling and geometric parameters of the drill head. These parameters determine and affect the durability of the drill head. In this report, the authors present the results of experimental studies of the influence of sharp angles on the durability of the cross drill head during the drilling process and blasting rock in the Quangninh coal region. According to the results, it is established that the wear intensity f varies positively with firmness i_h and inversely with the sharp angle of the drill head in the range of studied sharp angles. Nevertheless, the variation of wear intensity with regard to the change in sharp angles is not significant compared to the variation of wear intensity according to the rock strength. Research results can be used to calculate, design and choose appropriate drill rock quarry that increases the productivity and efficiency of drilling, contributing to reducing product costs.

1. Introduction

In this report, the author presents experimental results that determine the effect of sharp angle on the life of drill bit of the boring machine to drill blasting holes in underground coal mining in Quangninh.

The dam drill is used to break up rock and soil to form a drill hole. The drill bit is made of tool steel carbon with carbon content from 0.7 to 1% by forging and then sharpening. When the blade section is fitted with a hard alloy, the drill bit can be made by casting. When drilling, the drill is worn with a blade and the diameter due to friction with the rock and soil, the diameter becomes smaller, the sharp angle becomes obtuse, without grinding or replacing the new drill, it cannot continue to drill. The wear of the drill depends on many parameters such as: sharp angle, solidity of rock and soil, impulse, rotation angle after each dam, drill material, dam frequency of drill bits has scientific and practical significance to increase productivity, lower drilling costs and contribute proactively in the planning of production, maintenance and repair of drilling equipment [1-8].

Among the parameters mentioned above, sharp angle is the parameter that most affects the drill head wear and drill life. Therefore, in this report, the author delves into the effect of sharp angle α on tip life.

2. Research content

2.1. Basic structure of dam drill

The tool of the drill is the tip of the drill bit and the drill bit is made of special steel, one end is used to break the rock called the tip of the drill, and the other end is fitted with a machine called a tail ring. The casing has a hole to guide compressed air or water to the drill head to blow chips. The body of the cylinder is on duty: The axial transmission force includes thrust and thrust to the drill head; Orientation for drill holes; Exit Chips. There are two types of drill rigs [9]:

Instant drill bit: Made from special steel with a drill bit tip attached to the shaft as shown (Figure 1).

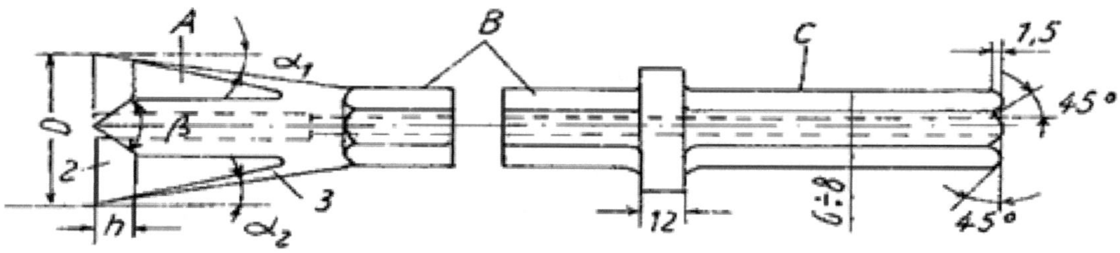

Figure 1. The drill bit has a cross-shaped drill bit
Including 3 parts: A- Drill head; B - Drill core body; C-Tail

2.2. Drills with removable drill bits:

It is made of special steel including a drill bit head (Figure 2), which is manufactured separately from the shaft. The drill bit is connected to the stem by means of a thread or taper (Tapered angle $3^0 30'$). Hard-alloy drill bits. Depending on the strength and texture of the rock, we choose the tip of the drill with the sharp angle as follows: To drill soft rock, the sharp angle of the blade is $\alpha = 90^0$, medium hard rock $\alpha = 100^0 \div 110^0$ and hard rock $\alpha = 120^0$.

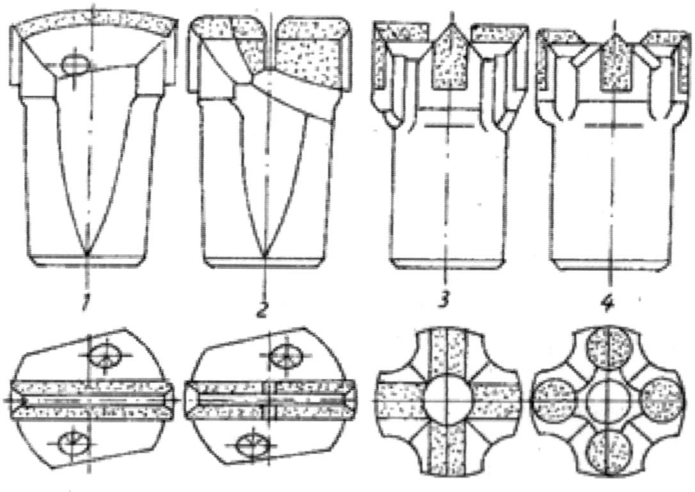

Figure 2. Geometric shape of the dam head
1) and 2) The tip of the drill has the most blade shape
3) and 4) The drill bit has a cross-shaped blade

Depending on the strength and structure of the rock, people use drill bits of different shapes as (Figure 2). The most common type of drill bit has a cross (+) blade (figure $2_{(3)}$) and most words (-) (figure $2_{(1)}$). The drill bit has the largest blade to ensure maximum drilling speed in dense, less cracked

IOP Conf. Series: Materials Science and Engineering **860** (2020) 012004 doi:10.1088/1757-899X/860/1/012004

rock. The drill has a cross-shaped blade used to drill cracked rock. Removable drill bits include the following diameters: 28, 32, 36, 40, 42, 44, 46, 52, 60, 65, 75, 85 mm, according to [9].

During the drilling process, the tip of the drill is easy to wear the blade and wear the diameter, the diameter becomes smaller, the sharp angle becomes obtuse, can be restored by grinding but must ensure to keep its geometric parameters and to create in front of the appropriate wear area of about 0.2 mm.

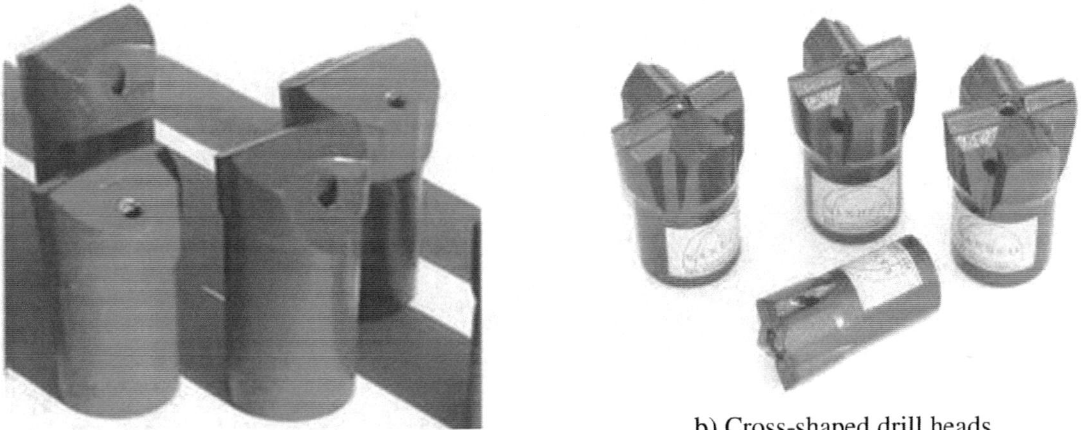

a) Drill bit has the most rectangular shape

b) Cross-shaped drill heads

Figure 3. Shape of dam tip

2.3. Research factors

- Drilling equipment and heads [8]:

+ Pneumatic drills, mounted on drilling racks with electrical control devices, hydraulic equipment and measurement equipment (see Figure 4);

+ Cross-shaped pneumatic drill head (Figure 3b);

- The original parameters

+ Rocks in Quangninh area, of sandstone type often have hardness f = (6 ÷ 8). Stone samples were analyzed according to Appendix 4;

+ Drilling machine: pneumatic pressure p = (0.4 ÷ 0.48) MPa, beating frequency (1880÷2000) times / minute, drilling speed (360 ÷ 600) rpm; thrust (80 ÷ 90) kN;

+ Drill diameter d = 42mm; sharp angle α = (100 ÷ 120) degree.

2.4. Experimental models and how to proceed

To test and evaluate the influence of technology parameters on the efficiency and service life of drilling tools, the authors designed and manufactured testing equipment using air compressors as a power source. The specific gas distribution element is the distribution valve, it is responsible for distributing compressed air flow to the working device, which is the power cylinder and drilling machine. One-way valve is used to guide the flow of air in one direction and block the flow in the opposite direction, the distribution valve for the pneumatic path to the cylinder in certain direction and other compressed air elements (air tank compression system; pipeline system; control valves; safety valves and gauges). Set of drilling parameters (converters, trip sensors, measurement signal processors, computer monitors) and load, all installed on a sturdy steel frame [10, 11, 12].

The purpose of designing and manufacturing a system of equipment to measure the impact force, compressed air pressure and rock drill journey, in order to study and determine some parameters of rock drilling machine.

The test equipment consists of the main components: dam drill mounting equipment, dam parameters measurement system and control software, data collection and processing.

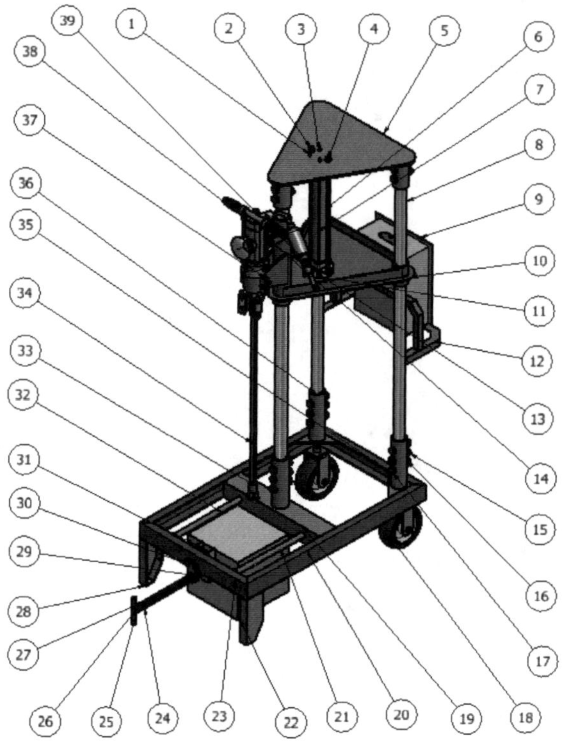

Figure 4. Drilling test device

1- Bolts; 2- bolts; 3- Bolts; 4- bolts; 5- The upper surface; 6-Cylinder Rod; 7 - Cylinder; 8- Moving axis; 9- Electric control box; 10- The moving block; 11- Cylinder cushion; 12- Support bar; 13 - Bolts; 14- Bolts; 15- bolts; 16- Left support post; 17- Bolts; 18- Wheel; 19- The middle bracing; 20- Vertical bracing; 21- Sliders; 22- Leg support; 23- Horizontal bracing; 24- Bolts; 25- Bolts; 26. Turning hand control; 27- Tamarind screw; 28- Leg support; 29- Tamar screw nut; 30- Cushion; 31- Moving bar; 32- Stone pattern; 33- The head of a rock drill; 34- Drilling rigs; 35- Horizontal bars; 36- Right support; 37 - Drill holder clamp; 38- Impact drill; 39. Cylinder bracket.

The test device has the basic structure as shown in Figure 4 with the following main parts and details:

- Pneumatic dam 38 has a synchronous structure: has a pneumatic cylinder mechanism 7 for pressing the drill into a hole when drilling and taking the drill out of the hole; drill head mounted on drilling rig 10 guide steel structure (drilling rack and pneumatic cylinder can adjust according to actual requirements). Slide drill head assembly with drilling racks.
- Cluster head drill racks linked to the bracket with bolts, can rotate around the center line.
- All of the above-mentioned assemblies are connected with frame frames made of bolts, couplings and silver;
- The frame of equipment model is made of shaped steel box and sprayed with paint, the whole device model is moved by wheel 18.

3. Experimental research results

In order to assess the impact of sharp angle and rock solidity on drill bit wear intensity, a test was conducted to determine the abrasive strength of the bit tip when changing the sharp angle of the bit tip at 5 values degrees, corresponding to 21 test rock samples with a f-strength of 6 to 8 [8, 13]. Each test is performed 05 times, after filtering out abnormal values, values The measurement is averaged over the measured values. Summarize the results of measuring the wear intensity of the drill corresponding to the sharp angle values and f as in table 1.

Table 1. Intensity of wear i_h (%) of drill bits with sharp angle α and strength f

Sharp angle (α) / *Durability (f)*	100^0	105^0	110^0	115^0	120^0
6.0	0.1312	0.1350	0.1412	0.1480	0.1531
6.1	0.1315	0.1355	0.1415	0.1485	0.1535
6.2	0.1320	0.1360	0.1420	0.1490	0.1540
6.3	0.1325	0.1365	0.1425	0.1495	0.1545
6.4	0.1330	0.1370	0.1430	0.1500	0.1550
6.5	0.1335	0.1375	0.1435	0.1505	0.1555
6.6	0.1340	0.1380	0.1440	0.1510	0.1560
6.7	0.1345	0.1385	0.1445	0.1515	0.1565
6.8	0.1350	0.1390	0.1450	0.1520	0.1570
6.9	0.1355	0.1395	0.1455	0.1525	0.1575
7.0	0.1360	0.1400	0.1460	0.1530	0.1580
7.1	0.1365	0.1405	0.1465	0.1535	0.1585
7.2	0.1370	0.1410	0.1470	0.1540	0.1590
7.3	0.1375	0.1415	0.1475	0.1545	0.1595
7.4	0.1380	0.1420	0.1480	0.1550	0.1600
7.5	0.1385	0.1425	0.1485	0.1555	0.1605
7.6	0.1390	0.1430	0.1490	0.1560	0.1610
7.7	0.1395	0.1435	0.1495	0.1565	0.1615
7.8	0.1400	0.1440	0.1500	0.1570	0.1620
7.9	0.1405	0.1445	0.1505	0.1575	0.1625
8.0	0.1412	0.1450	0.1523	0.1582	0.1631

Let us develop an empirical regression equation that reflects the influence of sharp angle and hardness of the rock on the strength of the drill bit tip wear.

Based on the measurement test data to determine the strength of the drill tip wear according to the rock solidity and sharp angle in Table 1. we select the experimental regression function of polynomial form of two variables. using regression method Experimental minimum squares least squares. determining the empirical regression formula showing the relation of the wear intensity function i_h according to the strength f and sharp angles α as follows:

$$i_h = 0{,}06279 - 7{,}248.10^{-5}\alpha + 0{,}001948.f + 2{,}857.10^{-6}\alpha^2 + 8{,}883.10^{-5}\alpha.f - 0.00051.f^2 \quad (1)$$

Comparing the experimental regression errors as shown in Table 2 and the coefficients in the empirical regression equation have been tested for compatibility according to Fisher standard.

Table 2. Compare errors and check compatibility

Sharp angle α (degree)	f	i_h TN (%)	i_h HQ (%)	Error Δi_h(%)*	Sharp angle α (degree)	f	i_h TN (%)	i_h HQ (%)	Error Δi_h(%)*
100	6.0	0.1312	0.1308	0.0030	110	7.0	0.1460	0.1465	0.0034
100	6.1	0.1315	0.1312	0.0023	110	7.1	0.1465	0.1469	0.0027
100	6.2	0.1320	0.1317	0.0023	110	7.2	0.1470	0.1474	0.0027
100	6.3	0.1325	0.1321	0.0030	110	7.3	0.1475	0.1478	0.0020
100	6.4	0.1330	0.1326	0.0030	110	7.4	0.1480	0.1482	0.0014
100	6.5	0.1335	0.1330	0.0037	110	7.5	0.1485	0.1486	0.0007
100	6.6	0.1340	0.1334	0.0045	110	7.6	0.1490	0.1490	0.0000
100	6.7	0.1345	0.1338	0.0052	110	7.7	0.1495	0.1494	0.0007
100	6.8	0.1350	0.1342	0.0059	110	7.8	0.1500	0.1498	0.0013
100	6.9	0.1355	0.1346	0.0066	110	7.9	0.1505	0.1502	0.0020
100	7.0	0.1360	0.1350	0.0074	110	8.0	0.1523	0.1506	0.0112
100	7.1	0.1365	0.1353	0.0088	115	6.0	0.1480	0.1469	0.0074
100	7.2	0.1370	0.1357	0.0095	115	6.1	0.1485	0.1475	0.0067
100	7.3	0.1375	0.1360	0.0109	115	6.2	0.1490	0.1481	0.0060
100	7.4	0.1380	0.1364	0.0116	115	6.3	0.1495	0.1487	0.0054
100	7.5	0.1385	0.1367	0.0130	115	6.4	0.1500	0.1492	0.0053
100	7.6	0.1390	0.1370	0.0144	115	6.5	0.1505	0.1498	0.0047
100	7.7	0.1395	0.1373	0.0158	115	6.6	0.1510	0.1503	0.0046
100	7.8	0.1300	0.1376	0.0585	115	6.7	0.1515	0.1509	0.0040
100	7.9	0.1305	0.1379	0.0567	115	6.8	0.1520	0.1514	0.0039
100	8.0	0.1412	0.1382	0.0212	115	6.9	0.1525	0.1519	0.0039
105	6.0	0.1350	0.1360	0.0074	115	7.0	0.1530	0.1524	0.0039
105	6.1	0.1355	0.1365	0.0074	115	7.1	0.1535	0.1529	0.0039
105	6.2	0.1360	0.1370	0.0074	115	7.2	0.1540	0.1534	0.0039
105	6.3	0.1365	0.1375	0.0073	115	7.3	0.1545	0.1539	0.0039
105	6.4	0.1370	0.1380	0.0073	115	7.4	0.1550	0.1544	0.0039
105	6.5	0.1375	0.1385	0.0073	115	7.5	0.1555	0.1548	0.0045
105	6.6	0.1380	0.1389	0.0065	115	7.6	0.1560	0.1553	0.0045
105	6.7	0.1385	0.1394	0.0065	115	7.7	0.1565	0.1557	0.0051
105	6.8	0.1390	0.1398	0.0058	115	7.8	0.1570	0.1561	0.0057

Table 2. Continuation

Sharp angle α (degree)	f	i_h TN (%)	i_h HQ (%)	Error Δi_h(%)*	Sharp angle α (degree)	f	i_h TN (%)	i_h HQ (%)	Error Δi_h(%)*
105	6.9	0.1395	0.1402	0.0050	115	7.9	0.1575	0.1566	0.0057
105	7.0	0.1400	0.1407	0.0050	115	8.0	0.1582	0.1570	0.0076
105	7.1	0.1405	0.1411	0.0043	120	6.0	0.1531	0.1526	0.0033
105	7.2	0.1410	0.1415	0.0035	120	6.1	0.1535	0.1532	0.0020
105	7.3	0.1415	0.1419	0.0028	120	6.2	0.1540	0.1538	0.0013
105	7.4	0.1420	0.1422	0.0014	120	6.3	0.1545	0.1545	0.0000
105	7.5	0.1425	0.1426	0.0007	120	6.4	0.1550	0.1551	0.0006
105	7.6	0.1430	0.1430	0.0000	120	6.5	0.1555	0.1557	0.0013
105	7.7	0.1435	0.1433	0.0014	120	6.6	0.1560	0.1563	0.0019
105	7.8	0.1440	0.1436	0.0028	120	6.7	0.1565	0.1569	0.0026
105	7.9	0.1445	0.1440	0.0035	120	6.8	0.1570	0.1574	0.0025
105	8.0	0.1450	0.1443	0.0048	120	6.9	0.1575	0.1580	0.0032
110	6.0	0.1412	0.1414	0.0014	120	7.0	0.1580	0.1585	0.0032
110	6.1	0.1415	0.1419	0.0028	120	7.1	0.1585	0.1591	0.0038
110	6.2	0.1420	0.1425	0.0035	120	7.2	0.1590	0.1596	0.0038
110	6.3	0.1425	0.1430	0.0035	120	7.3	0.1595	0.1601	0.0038
110	6.4	0.1430	0.1435	0.0035	120	7.4	0.1600	0.1607	0.0044
110	6.5	0.1435	0.1441	0.0042	120	7.5	0.1605	0.1612	0.0044
110	6.6	0.1440	0.1446	0.0042	120	7.6	0.1610	0.1616	0.0037
110	6.7	0.1445	0.1451	0.0042	120	7.7	0.1615	0.1621	0.0037
110	6.8	0.1450	0.1455	0.0034	120	7.8	0.1620	0.1626	0.0037
110	6.9	0.1455	0.1460	0.0034	120	7.9	0.1625	0.1631	0.0037
					120	8.0	0.1631	0.1635	0.0025

Error of abrasion intensity according to the regression function and experimental results:

$$\Delta i_h = \frac{i_h TN - i_h HQ}{i_h TN}.100,\% \cdot \tag{2}$$

From the empirical formula (1) draw a 3D graph showing the relationship of wear intensity i_h at the same time solidity f and sharp angles α as shown in Figure 5. and 2D graphs that reflect the dependence of wear strength i_h at sharp angles at some different solidities as shown in Figure 6 and depending on the solidity at some different sharp angles as shown in Figure 7.

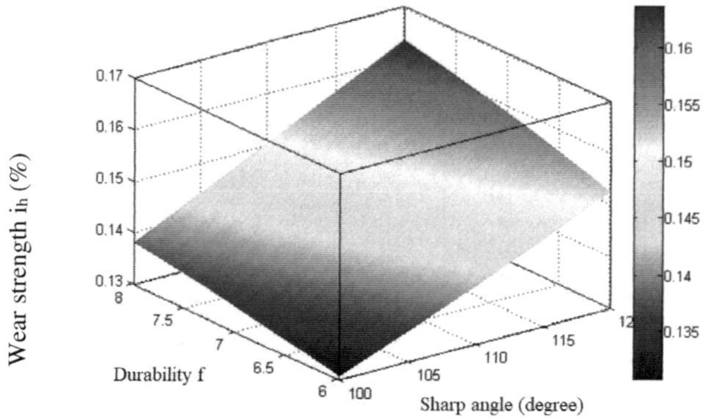

Figure 5. Graph of relationship to abrasion intensity i_h with sharp angles α and solidity f.

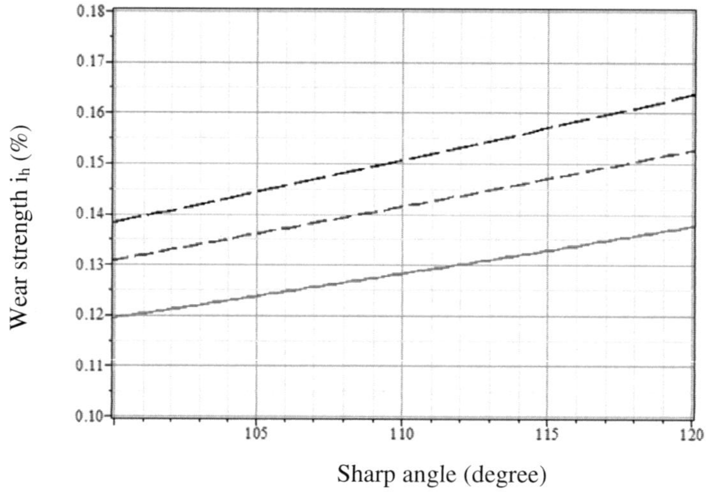

Figure 6. Relationship graph of abrasion intensity with sharp angle when different hardness

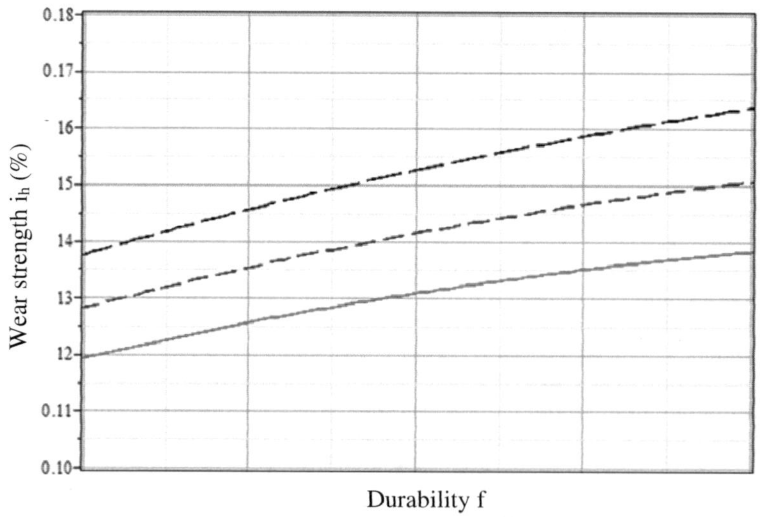

Figure 7. Relationship diagram of abrasion strength with strength when sharp angle is different

From the empirical formula (1) shows. the coefficient value of the solidity *f* is $0{,}001948>0$. means corresponding to the strength under consideration. the intensity of wear i_h Variable variable with solidity *f*. ie when the strength increases. the intensity of wear increases. The coefficient value of sharp angle α is $-7{,}248.10^{-5} < 0$. means that in the sharp angle range the intensity of wear is being considered i_h inversely variable with sharp angle. ie when sharp angle increases. the intensity of wear tends to decrease. but the absolute value of this coefficient is very small compared to the coefficient of solidity. so the decrease of intensity the sharpness of the sharp angle is negligible compared to the increase of the wear intensity with the strength of solidity. Also in the empirical formula (1). the quadratic coefficients of α and f have absolute values that are much smaller than the absolute values of the first order coefficients. that is. the degree The influence of quadratic quantities on wear intensity is negligible.

The influence of solidity and sharp angle on abrasion intensity is more clearly shown in 3D graphs in Figure 5 and 2D graphs Figures 6 and 7. Quantitatively. for each rock type soil has the most solidity When the sharp angle increases from 100-120 degrees. the intensity of wear increases by 0.024%. As for each type of drill with a certain sharp angle. when the solidity of rock and soil increases from 6 to 8. the intensity of wear increases by about 0.022%. specifically with a sharp angle of 100 degrees. the increase is 0.019%. sharp angle of 110 degrees the increase is 0.023% and the sharp angle of 120 degrees is 0.026%. Thus. for rocky soil in Quangninh coal mine (with common solidity in the range of 6 ÷ 8). it is possible to choose a drill with small sharp angle. preferably under 115 degrees.

Comparing with the fair value area of sharp angle according to the stable condition of cutting edge and the body of drill bit defined [14]. the adjacent sharp angle area of 110 degrees has been determined to be perfectly reasonable. not only ensure the durability but also ensure the longevity of the drill bit in solid rock condition (6÷8) in Quangninh coal mine.

4. Conclusion

The aforementioned results showed the influence of sharp angle and hardness of stone on the intensity of drill bit wear.

With the above-mentioned research results. it is possible to calculate. select and determine the wear tip of the drill bit in the direction of increasing the durability. ensuring that the drilling equipment works according to the requirements set out when drilling and blasting holes in service digging basic kilns in construction and mining in Quangninh region. In order to increase productivity. it is necessary lower costs and contribute proactively in production planning and repair of drilling equipment.

References

[1] Yungmeyster D A and Urazbakhtin R Yu 2017 Rescue complex for coal mines *IOP Conference Series: Earth and Environmental Science* **87** 092032

[2] Nguyen K L Gabov V V Doan V G Pham V T 2019 Study of the influence of structure and parameters of loading and transporting devices of a cleaning combine on the efficiency of coal loading *Journal of Physics: Conference Series (JPCS)* **1384** 012036

[3] Osman Hekimoglu 2017 A pick force calculation method suggested for tool lacing of mechanical excavators employing drag tools *International Journal of Mining. Reclamation and Environment* 10.1080/17480930.2017.1317946

[4] Bolobov V I and Le Thanh Binh 2018 Influence of deep cryogenic treatment on structure and wear resistance of materials of hydraulic breaker chisels. *IOP Conference Series Materials Science and Engineering* **327** 042016

[5] Nasonov M Yu. Lykov Yu V. Do Duc Trong 2020 The study of the resource and durability of metal structures of excavators after the expiration of the service life *Ugol'* **2** 13-17

[6] Aleksandrov A N. Rogachev M K. Thang Nguyen Van. Kishchenko M A. Kibirev E A 2019 Simulation of organic solids formation process in high-wax formation oil *Topical Issues of Rational Use of Natural Resources* **2** 779-790

[7] Zhang Qianqian. Han Zhennan. Zhang Mengqi and Zhang Jianguang 2015 Prediction of tool forces in Rock cutting using discrete tlement vethod *EJGE* **20** 1607-1625

[8] Gabov V V. Zadkov D A and Nguyen Khac Linh 2019 Features of elementary burst formation during cutting coals and isotropic materials with reference cutting tool of mining machines *Journal of Mining Institute* **236** 153

[9] John P M and Khamis Y H 1988 *Probe-hole drilling: high-stress detection in coal* (Pittsburgh Pa. : U.S. Dept. of the Interior. Bureau of Mines)

[10] Skryabin R M and Timofeev N G 2016 Development of an innovative shneko-heat-sink boring shell for drilling of shurfo-wells in the conditions of a kriolitozona *Eurasian mining* **1** 33-36

[11] Daolong Yang. Jianping Li. Liping Wang. Kuidong Gao. Youhong Tang and Yanxiang Wang 2014 Experimental and theoretical design for decreasing wear in conical picks in rotation-drilling cutting process. *The International Journal of Advanced Manufacturing Technology* **75** 3-11

[12] Bochkov V, Bolobov V and Chupin S 2017About increasing wear resistance of rock-breaking tool to abrasion by using mechanical and thermo-mechanical treatment *International Review of Mechanical Engineering* **11** 301-304

[13] Gabov V V and Zadkov D A 2016 Energy-saving modular units for selective coal cutting. *Eurasian mining* **1** 7-40

[14] Le Quy Chien 2015 *Researching and determining some reasonable parameters of drill bit heads to drill blasting holes in underground mining in Quang Ninh*

IOP Conf. Series: Materials Science and Engineering **860** (2020) 012005 doi:10.1088/1757-899X/860/1/012005

Methodology of ecological and geological assessment of technogenic oil product deposits (on the example of the territory of the Burnakovsky lowland, Nizhny Novgorod)

O S Izosimova[1], I I Podlipsky[1], P S Zelenkovsky[1], S V Lebedev[1], A M Belyaev[1], S V Dubrova[2], V E Kriyt[3] and Yu N Sladkova[3]

[1] St. Petersburg State University, 7-9, University Embankment, St. Petersburg, 199034, Russia
[2] Department of Geology and Geoecology, Herzen State Pedagogical University of Russia, 48, Building 5, Moika Rivers Embankment, St. Petersburg, 199034, Russia
[3] Federal State Budgetary Institution Scientific Center for Hygiene and Public Health, 4, 2nd Sovetskaya Str., 4, St. Petersburg, 191036, Russia

e-mail: dubrova.stanislav@gmail.com

Abstract. The present study is devoted to the assessment of hydrogeological conditions in the field of discharge of the water-bearing horizon. The study is carried out on the territory of the Burnakovo Lowland tank farm, Nizhny Novgorod, and the adjacent areas. These areas are potentially adverse areas with disturbed surface runoff conditions. Within these anthropogenically modified areas of megacities, there is pollution due to additional infiltration nutrition. In this case, it is possible to form a layer of liquid oil products on the surface of the water-bearing horizon and subsequent migration of pollutants within the catchment areas of rivers and water bodies.

1. Introduction

The research area is located on the Burnakovsky lowland in the city of Nizhny Novgorod. The width of the investigated area is 2.5 km. Lakes indent the surface of the lowland, and peat bogs, alternating with sand manes directed parallel to the river bank. Volga (Figure 1). Administratively, the worksite is located in the Moscow and Sormovsky districts of Nizhny Novgorod. Burnakovskaya lowland is the traditional name of the territory adjacent to the Volga River, covering an area of more than 700 hectares. Approximate borders – Kuibyshev street, Sormovskoe highway, Komintern street.

In the study area or the immediate vicinity of it, a limited amount of survey work was repeatedly performed. The data obtained are fragmented and do not contain some specific characteristics. However, in general, the available data is very informative and allowed the authors of this study minimizing the amount of work carried out.

Various specialists studied these territories and made many conclusions. The main characteristic feature of their conclusions is the convincingly proven multiplicity of sources of pollution of underground water with liquid petroleum products. Existing studies have provided data on the content of petroleum products in the study area. Existing data unequivocally indicate pollution both within and above the ground flow relative to the studied object.

IOP Conf. Series: Materials Science and Engineering **860** (2020) 012005 doi:10.1088/1757-899X/860/1/012005

According to the data of analogous objects, we have established [1, 3], oil products for a long time and systematically enter the Volga River. This conclusion is based on establishing the dispersion of pollution sources, a long period of their manifestation. The high filtration parameters of the host soils do not create conditions for significant local accumulations of liquid petroleum products on the surface of groundwater.

Figure 1. Scheme of the research area and the object of accumulated environmental damage in the territory of the former Sormovo oil depot

Thus, the objective of the research is to establish the volume of petroleum products and prepare recommendations on the creation of an in-system protection system from the pollution of the Volga River with liquid petroleum products [2, 4].

The territory of the tank farm and the adjacent area are confined to the right-bank floodplain of the Volga River. Absolute elevations of the floodplain surface are 69-73 m while maintaining a general slope towards the river.

The development of the territory in many places sprinkled with soils of various compositions. The combination of alternating lakes, swamps, and sand manes parallel to the river bank creates an incredibly complex surface structure. The particularities of the territory contribute to the transformation of surface runoff into groundwater and the entry into the groundwater of various kinds of substances, including oil products. The excess of the surface of the territory above the water edge in the river is at least 8.0 m.

The territory of the tank farm is built up with one- and two-story structures for various purposes, saturated with a large number of communications. A significant part of the territory is occupied by a tank farm for storing petroleum products. From 1985 to 1988, the company reduced its turnover by more than 18 times. In 1988, gasoline, diesel fuel, and fuel oil were wholly withdrawn from sales. Kerosene transshipment is completely stopped. Since 1996, the operation of the tank farm in its first direction has been completely stopped.

In the geological structure of the territory, to an exploratory depth of 25.7 m, sands and loams of the Quaternary age, covered by bulk soils, take part. Within the study area, in the thickness of alluvial sandy deposits, lake-bog loam and clay are developed. The thickness of the sandy deposits is 17.9-24.1 m, of loam and clay 1.5-3.1 m. The total thickness of the Quaternary deposits is 24.1 m.

Water-bearing horizons are confined to the sandy-clayey thickness of the Quaternary. Aquifers are formed on relatively water-resistant deposits, namely, loams and clays. Soil properties are confirmed during their sinking. Namely, the smell of petroleum products is felt. The smell indicates the infiltration of lost oil products through the thickness of these soils.

The horizon surface lies at a depth of 0.88-1.7 m, which corresponds to total marks 71.1-71.4 m. The thickness of the horizon is 2.2-3.4 m. The flow direction is towards the Volga River. Infiltration feed due to leaks from communications within areas with disordered surface runoff, as well as infiltration of leaks from the fire reservoir system. As a result, the regime of the water-bearing horizon is quite stable. You are unloading the horizon in the underlying stratum of sandy sediments. An oil products accumulation with a thickness of up to 0.01 m, it was noted on the horizon free surface.

The aquifer in alluvial deposits within the study area is ubiquitous, pressure-free. Pressure properties are due to the occurrence in the thickness of poorly permeable loams and clays. The filtration parameters of water-containing sands are characterized by values of the filtration coefficient from 12.7 to 22.3 m / day. The depth of the level during the period of work was 5.4–8.3 m, which corresponds to absolute elevations of 63.9–66.3 m. The thickness of the horizon is 13–18 m. The reddish-brown Perm clay of the Tatar layer serves as the lower confining layer. Horizon nutrition is infiltration, i.e., atmospheric nutrition, leaks from communications, overflow from the overlying horizon. Within the area with disordered surface runoff, increased infiltration nutrition is observed. Unloading is the Volga River. The horizon is heavily polluted within the entire territory of the tank farm, which is confirmed by previously completed work.

2. Materials and methods

The research methodology based on assessing the hydrogeological conditions in the area of discharge of the water-bearing horizon in the Quaternary sediments. To this end, 11 exploratory wells are drilled and equipped on the territory of the Sormovsk oil depot. Wells are located in the most unfavorable areas with disturbed surface runoff conditions. Within these areas, it was proposed to investigate existing pollution. It is assumed that pollution arose as a result of additional infiltration nutrition, the formation of a layer of liquid petroleum products on the surface of the water-bearing horizon in Quaternary sediments.

The need to minimize the volume of fieldwork led to the development of the next research strategy. The obtaining of areal characteristics and information about the regional structure, retrospective materials were used, both descriptive and mining.

3. Results and discussion

The distribution of the pollutants in the area is influenced by the sources' multiplicity and their action time extension. In this case, interference of the pollution fronts occurs. As a result, it is practically impossible to establish the role of each source reliably.

The present studies are focused directly on the industrial site of the tank farm, as an object directly adjacent to the Volga coast. It is within its limits that specific protective measures should be taken.

Currently, there is an accumulation of liquid petroleum products in all drilled and equipped wells, the thickness of which varies from the film on the surface of the water-bearing horizon to 0.4 m.

As can be seen from the obtained results, the accumulation of liquid petroleum products on the surface of the water-bearing horizon and their spreading and subsequent unloading in the area of the existing berth are observed within the discharge area of the aquifer. During the construction of the berth, the mud layer of the Volga river channel was destroyed. At the same time, favorable conditions have been created for unloading, including accumulated oil products.

According to laboratory studies, petroleum products accumulated on the surface of the aquifer are a mixture of kerosene and motor oil and can be classified as "stove fuel."

Corresponding calculations of the total possible volume and a recoverable fraction of liquid hydrocarbons can be performed only tentatively. Reasons are the lack of approved methods for assessing the oil saturation of soils in the aeration zone, the fractionation composition of oil products, and many other indicators necessary to justify the assessment of the "reserves" of oil products within the limits of pollution spots.

The initial data in this case are:

• the total volume of oil-saturated soils within the limits of the spots of pollution, determined by the map of iso-capacities;

• the porosity of sandy loam soils (average 0.35%);

• effective porosity, determined by the Vetsinsky formula, and corresponding to a value of 0.17.

The volume of contaminated soils, defined as the product of the average between the capacities and the areas limited by them, gives a value of 3068 m^3. The product of this value by the porosity value (0.35) gives 1074 m^3. The data are overstated, as not all pore space is filled with liquid petroleum products. The product of the same volume by the effective porosity indicates the most probable volume of liquid hydrocarbons. Liquid hydrocarbons can move directly with the flow of groundwater under the action of gravity and are 521 m^3.

When calculating, the various density of oil products was taken into account. Estimates of possible tonnage are approximate. At a conditional density of 0.6, the weight of the gravitational reserves of liquid hydrocarbons is not more than 412 tons. Under the conditions of the gravitational regime, the recoverable reserves do not exceed 0.5. If this fact is taken into account, the volume of oil products for which extraction can be calculated is not exceeded 200-210 tons. Naturally, an approach to solving environmental problems based on self-sufficiency, in this case, is not possible.

4. Conclusions

An analysis of the geological and hydrogeological conditions of the site indicates the presence of a particular zone of unloading of the soil flow in the tank farm area. Up along the Volga valley, this zone is limited by the discharge channel of the thermal power station, which is the boundary of the first kind. Down the valley, the discharge front is rather conditionally limited by streamlines determined from the map.

Primary protective measures should be carried out in the allocated lane. It should be noted that, in principle, the contaminated areas within the large industrial zone can extend much further down the valley [2, 4].

Based on the previous, we can recommend the creation of a system of active engineering protection against pollution of the Volga by liquid petroleum products. The protection method is a penetration path cutting off a series of wells located along the banks of the Volga River. This penetration path should ensure the formation of a very elongated depression funnel on the surface of the groundwater, which prevents liquid petroleum products from entering the river.

References

[1] Dubrova S V, Podlipskiy I I, Kurilenko V V and Siabato W 2015 *Environ Pollut* **197** 165-172
[2] Marzougui A and Mammou A *2006 C R Geosci* **338(16)** 1176-1183
[3] Podlipskiy I I 2015 *Ecological and geological assessment of the territory of landfills for household waste* (LAP Lambert Academic Publishing)
[4] Vorenhout M, VanStraalen N M and Eijsackers H J P 2000 *Environ Toxicol Chem* **19** 2161-2163

IOP Conf. Series: Materials Science and Engineering **860** (2020) 012006 doi:10.1088/1757-899X/860/1/012006

Impact of production wells operational changes on high-viscosity oil recovery

E M Almukhametova

Branch of Ufa State Petroleum Technological University, 54a, Devonskaya str., Republic of Bashkortostan, Oktyabrsky, 452607, Russia

elikaza@mail.ru

Abstract. As light oil reserves are developed, the share of heavy high-viscosity oil (HVO) reserves, which are considered difficult to recover, increases in the total balance of oil reserves. Therefore, the future of the oil industry is inevitably linked to the development of highly viscous heavy oil deposits. Today, according to various estimates, epy reserves of high-viscosity oil and bitumen range from 790 billion tons to 1 trillion tons. This is more than 5 times more than the residual recoverable reserves of low and average viscosity oil. The world experience of developing high-viscosity oil deposits shows that the cost-effective development of most of them is limited due to low efficiency of oil wells and low oil recovery achieved through natural operation mode or flooding. If the first of these problems is successfully solved in recent times by drilling horizontal and multilateral wells, the second one requires the introduction of various technologies to influence the formation (thermal methods, miscible displacement, etc.), which are not always highly efficient. In this work we will consider the impact of changing the operation modes of production wells on the production of high-viscosity oil reserves on the example of the Northern Buzachi field (Republic of Kazakhstan).

1. Introduction

One of the most common geological and technical measures aimed at improving development efficiency is optimization of production well operation modes. At the same time, this type of well interventions means the intensification of oil reserves withdrawal – increase of well fluid flow rate due to the installation of larger pump or change of pump operation parameters. In the conditions of the Northern Buzachi field, the optimization process includes forced fluid withdrawal (FFW). The FFW technology allows creating high pressure gradients with step-by-step increase of production well flow rates (reduction of bottomhole pressure). This leads to intensification of oil production in non-uniform (compartmentalized) watered layers, allows drainaging unrecovered oil [1, 3, 4]. FFW is particularly important is in the development of non-Newtonian oil deposits, where significant pressure gradients are required to create moving volumes of oil in the reservoir.

2. Materials and Methods

The questions raised were considered on the basis of scientific analysis and synthesis of experience in the development and operation of high-viscosity oil deposits. Processing of initial field information during the efficiency analysis of changing the operation modes of production wells was carried out using modern methods of statistical information processing [8]. The obtained results were put into practice and the implementation results confirmed their efficiency.

3. Results

The analysis of oil reserves recovery from the first production facility of the seventh block of the Northern Buzachi field shows that the current oil recovery index (ORI) amounted to 0.054 unit fraction. Besides, the current water cut in the field as a whole was 82.3%. The trend of water cut change shows that it is impossible to reach the approved ORI at saturated water cut (Figure 1).

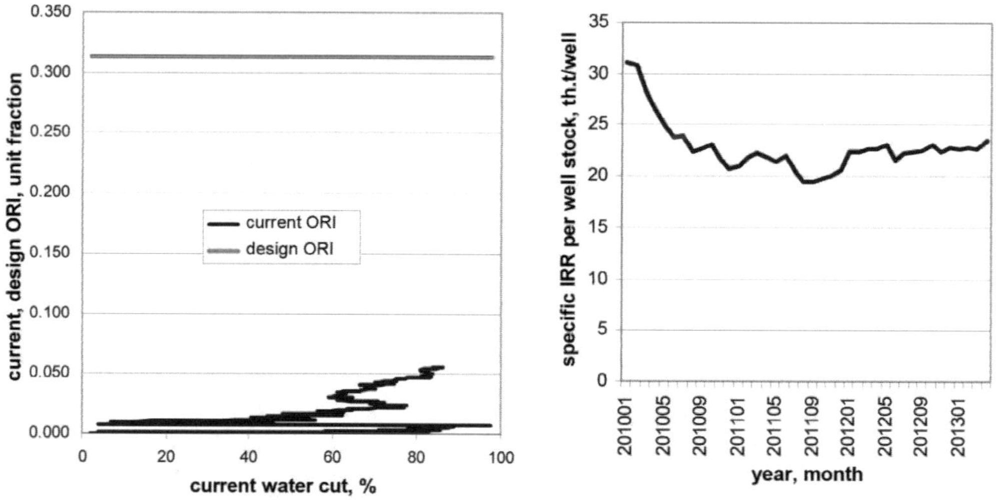

Figure 1. Dependence of current ORI on current water cut of extracted product (a) and dynamics of IRR brought into production per one production well (b) of the seventh block of the first operational facility of the Northern Buzachi field

The dynamics of specific initial recoverable reserves (IRR) brought into production, shown in Figure 1b, indicates a very slow increase in the indicator, despite the fact that during this period the production well stock was increasing and the formation-pressure maintenance system was formed. In other words, further "extensive" development path of the site development system makes it impossible to hope for significant improvement of technological indicators of development.

Thus, in the first production facility of the seventh block of the Northern Buzachi field, the operating development system does not allow achieving the approved values of the oil recovery index. At the same time the extracted product is characterized by high water content, water cut exceeds 82%, which indicates low efficiency of oil recovery.

Let us look at the analysis of changes in technological modes of production well operation.

The results of changes in the process modes of 42 production wells producing liquid from the formations of the first and second production facilities of the Northern Buzachi field were analyzed. The optimization process was limited to changing the reciprocating speed (rpm) of the pump [11, 12]. By way of example, Figure 2 illustrates the changes of well performance indicators as a result of changing the operating conditions of the pumping equipment.

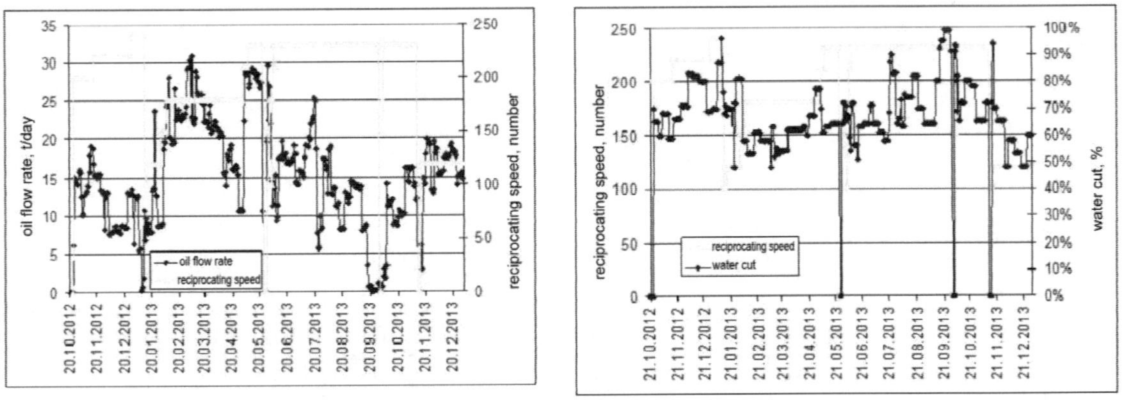

Figure 2. Change of oil flow rate (a) and water content of the extracted product (b) depending on the reciprocating speed. Well No. 1017-1 of the second production facility of the Northern Buzachi field

The analysis of dependencies of technological parameters of well operation (oil flow rate, water cut) on the change of pump operation mode showed that unique dependencies linking these values are absent or not determined by the operation data.

It was determined that a change in the reciprocating speed (rpm) of the pump may result in the following:

1. reduction of water cut and increase of oil flow rate with the increase of the reciprocating speed (for example, well 1017-1, May 2013, change of the reciprocating speed (rpm) from 180 to 230, well 1088, August 2013 change of the reciprocating speed (rpm) from 160 to 240).
2. increase of water cut and decrease of oil flow rate with the increase of the reciprocating speed (rpm) (for example, well 1017-1, October 2012 - January 2013, successive increase of the reciprocating speed (rpm)).
3. reduction of water cut and increase of oil flow rate with the reduction of the reciprocating speed (rpm) (for example, well 1017-1, January 2013, change of the reciprocating speed (rpm) from 240 to 180).
4. increase of water cut and reduction of oil flow rate with the reduction of the reciprocating speed (for example, well 1088, March 2013, change of the reciprocating speed (rpm) from 180 to 160).

Thus, the impact of optimization on current well development performance is not unambiguous. Therefore, the optimization efficiency was determined based on the displacement characteristic method [2, 5, 6].

Let us consider how production wells reacted to the optimization of their operation modes. Displacement characteristics were built for all wells. In general, the total effect amounted to 14.517 thousand tons of additional oil produced. However, the curve analysis showed that it is not possible to build a correct displacement characteristic for a single well since there is no straight section. This well was not considered in the final value of the optimization effect. With this in mind, the reasonable effect of well optimization taking into account all losses makes 14.397 thousand tons, which is quite significant for this deposit [8, 7]. Of the wells with a reasonable optimization effect, 29 reacted positively with a significant effect value, 3 wells had a minor positive effect, 9 wells had a significant negative effect, 1 well had a minor negative effect, and 1 well could not have the effect determined due to the absence of a straight-line section. The optimization efficiency analysis of operation modes of production wells regarding the development objects shows that the total effect for the wells of the first object is negative. Specific optimization losses of wells of the first object make 17.3 t/(well*month) (Table 1).

Table 1. Distribution of optimization effect across development objects

Object	Number of analyzed wells	Total effect, t	Average effect, t/μV	Average duration, month	Specific effect, t/(well*month)
Jurassic	24	-2669	-111	6.4	-17.3
Cretaceous	15	15609	1041	5.9	176.4
Combined (CJ)	3	1457	486	6	81.0

The main technological effect of the optimization accounts for wells of the second production facility – the specific effect makes 176.4 t/(well*month) (Table 1). It is interesting that the positive effect of optimization is also obtained for joint wells – the specific effect makes 81.0 t/(well*month).

4. Conclusions

The applied optimization of the operation modes of the production wells of the Northern Buzachi field produced a significant technological effect. According to the 42 wells considered, the technological effect amounted to 14397 tons of additional oil produced.

The total effect of optimization of the wells of the first production facility is negative. At the same time out of 24 wells 9 wells have negative effect. For the main and consolidating well stock, the losses are approximately equal. For the horizontal wells of the first object, the effect of optimization is positive.

All considered wells of the second production facility have a positive effect of optimization. At the same time the wells of the consolidating stock have the maximum specific effect.

The ambiguity of the optimization effect for different wells causes the need for theoretical study of the impact of changing the operation mode of the production well on the efficiency of oil production.

References

[1] Almukhametova E M 2018 Recommendations on thermal non-stationary waterflooding when modeling oil field development *IOP Conference Series: Earth and Environmental Science* **194 (6)** 062002

[2] Beliveau D 2008 Waterflooding Viscous Oil Reservoirs *SPE* 113132

[3] Colonomos P A Feasibility study of Cyclic Steam Injection in a Deep Heavy Oil Reservoir in Western Venezuela *SPE* 15091

[4] Das S K 1997 Vapex: An Efficient Process for the Recovery of Heavy Oil and Bitumen *SPE* 50941

[5] 1994 Foams: fundamentals and Applications in Petroleum Industry (ACS. Washington)

[6] Kotenev Yu Mukhametshin A V Sh and Sultanov Sh Kh 2018 Energy-efficient technology for recovery of oil reserves with gas injection *IOP Conference Series: Earth and Environmental Science* **194 (8)** 82019

[7] Mukhametshin V V 2018 Rationale for trends in increasing oil reserves depletion in western Siberia cretaceous deposits based on targets identification *Bulletin of the Tomsk Polytechnic University, Geo Assets Engineering* **329 (5)** 117-124

[8] 2011 MORE 6.7 Technical Reference (ROXAR)

[9] Mustoni J L, Norman C A and Denyer P 2010 Deep Conformance Control by a Novel Thermally Activated Particle System to Improve Sweep Efficiency in Mature Waterfloods of the San Jorge Basin *SPE* 129732

[10] Ohms D, et al. 2009 Incremental Oil Success From Waterflood Sweep Improvement in Alaska *SPE* 121761

[11] Rocha de Farias M L, et al. 2012 A Comparative Study of Emulsion Flooding and other IOR Methods for Heavy Oil *SPE* 152290

[12] Sadler K W 1995 An EUB Review of In Situ Oil Sands Bitumen Production *SPE* 30240

IOP Conf. Series: Materials Science and Engineering **860** (2020) 012007 doi:10.1088/1757-899X/860/1/012007

Simulation of power supply systems with external faults and methods for increase of residual voltage at terminals of power receivers

K R Bakhteev, A I Fedotov and R Sh Misbakhov

Kazan State Power Engineering University, Engineering center, 51, Krasnoselyskaya str., Kazan, 420066, Russia

E-mail: kam.rav@bk.ru

Abstract. The article discusses the effect of the remoteness of the place of a three-phase short circuit on the depth of the voltage dip depending on the load on the synchronous generators and synchronous motors. A series of experiments was carried out where the length of the 110 kV overhead power line varies, at the end of which a short circuit occurs, as well as the load distribution coefficient of a synchronous machine with higher than normal field-forcing ratio and the influence of these parameters on the level of residual voltage among large industrial consumers. The efficiency of using synchronous machines with a higher than normal field-forcing ratio of the excitation is estimated, and their stability under a sharp increase in loads, which cause voltage changes during the transition process and cause rotor vibrations, accompanied by magnetic losses due to the occurrence of eddy currents in the rotor core, is considered. The results of simulation presented in the Simulink are a MATLAB-based graphical programming environment for modeling, simulating and analyzing multi-domain dynamical systems.

1. Introduction

When a three-phase fault (TPF) occurs, a voltage drop (VD) occurs and the behavior of high-voltage synchronous electric motor (SEM) largely depends on the settings of the protections in the 110 kV grid. As it is known, when a VD occurs, SEM go into the generator mode, turning, in fact, into sources of electricity. If the protection of 110 kV lines does not have the necessary speed for deep VD, then SEM goes out of synchronism and special measures must be taken to resynchronize them.

Modern devices for fast-automatic transfer switch (FATS) with a transformer substation on an overhead power line (OPL) of 110 kV become ineffective [1-3]. The paper suggests ways to increase the residual voltage by increasing the field-forcing ratio of synchronous machines (SM) excitation, as well as using the energy storage system (ESS) in the excitation system.

2. Circuit diagram and simulation model of a typical industrial customer. simulation results

Although the excitation forcing of SM can not always completely remove the problem of the impact of short-term power outages (STPO) on the quality of electricity, it should be noted the effectiveness of this method for prevention STPO: firstly, it requires almost no additional costs, and secondly, significantly reduces the critical length of line (CLL) [4-6], i.e. that zone within which all faults lead to a decrease in voltage at the main step-down substation (MSDS) below the technologically permissible value.

The schematic diagram (Figure 1) shows a diagram where a three-phase fault occurs on a 110 kV OPL in an external power supply system.

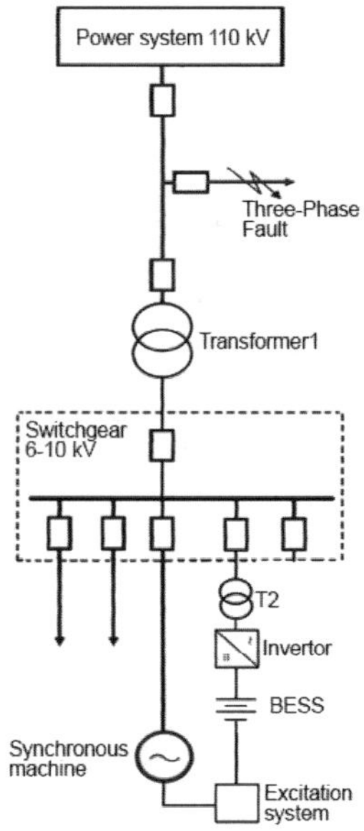

Figure 1. Circuit diagram of the power supply of an industrial customer.

Both SEM and synchronous generator (SG) are considered as SM. In accordance with the circuit diagram of the power supply of an industrial customer, a simulation model of a typical unit of the power supply system of an industrial consumer is developed (Figure 2). This model allows researching the effect of the excitation forcing of SM on the depth of VD caused by a three-phase fault in a 110 kV OPL.

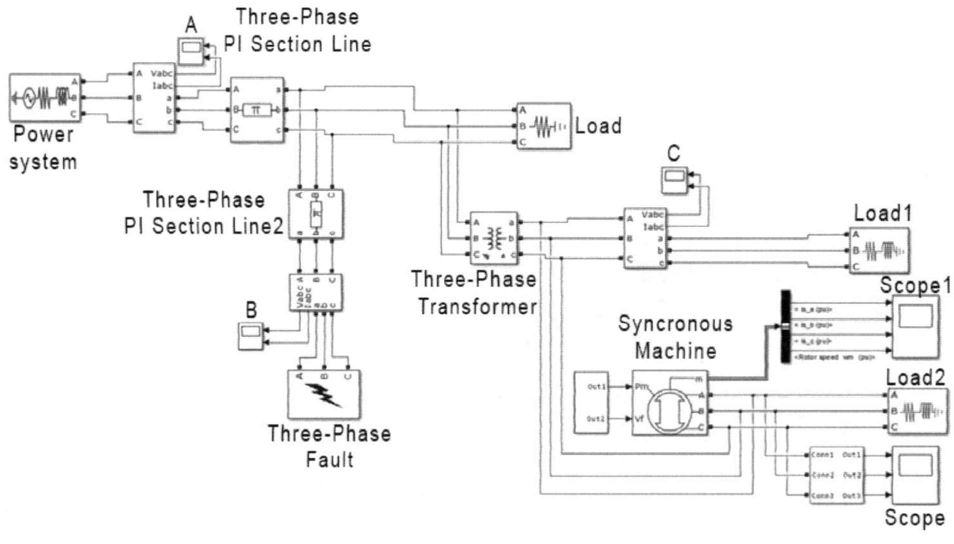

Figure 2. Simulation model of a typical system node.

The parameters for the simulation model blocks are set as follows: the resistive resistance of the power supply is 0.04 ohms, the inductive resistance is 0.016 H. The distance to the TPF varies from 1 to 30 km, which covers the power supply system of the city and its environs. For the 110 kV powerline, reference characteristics of the steel-aluminum wire AC 120/19 and the intermediate support PS 110–9 V were used. SM parameters: Unom=10.5 kV, Xd=1.41 pu, Xq=0.47 pu, Xd'=0.21 pu, Xd"=0.101 pu, Rs=0.003 pu The nominal power of the SM in one considered case was 8 MVA, in the other - 16 MVA. The SMs work with a load distribution coefficient Kdl equal to 1. The coefficient takes into account the load distribution on the SM and is defined as the ratio of the load on the SM terminals to its nominal power. Transformer parameters: the rated power of the transformer is 8 MVA in one considered case, 16 MVA in the other, 110 kV primary winding voltage, and dissipation inductance 0.08 pu, secondary voltage 10.5 kV, and dissipation inductance 0.08 pu.

In the case of STPO, voltage recovery in the 10 kV section of the main power supply occurs due to the action of the Loss of Mains (LoM) protection [1; 4].

At the time of TPF, there is a VD on all sections of buses. The LoM turns off the input switch, ensures the start of the FATS and gives the command to turn on the excitation of the SM, the time of the full transfer cycle is 45 - 60 ms [1; 8; 9].

A three-phase fault occurs at a time of 0,1 s, on a 110 kV overhead line 17 km away from industrial customer buses, and the current amplitude is 8 kA. The resulting short circuit causes VD on the buses of the industrial customer in all 3 phases, which is clearly seen in Figure 3.

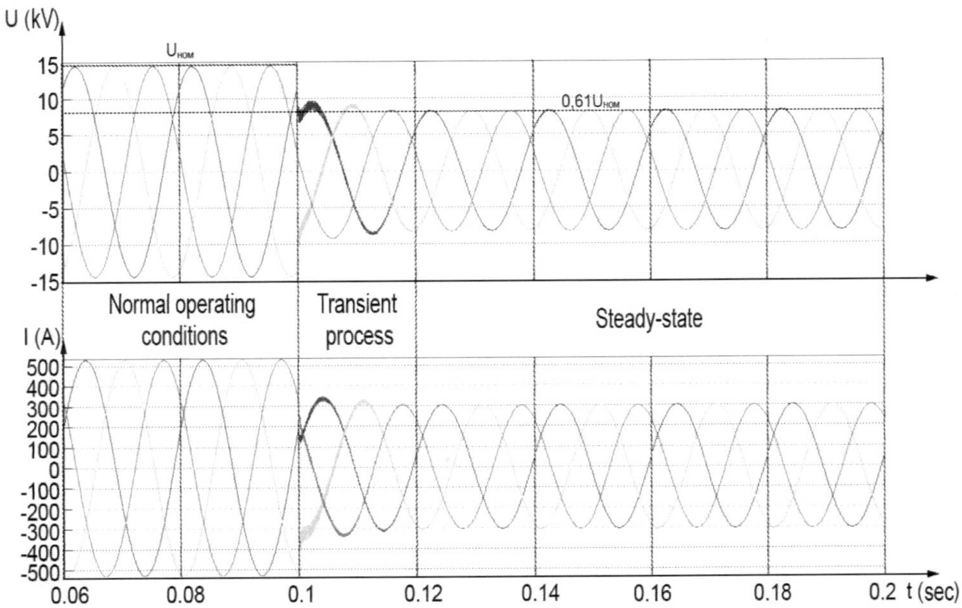

Figure 3. Oscillogram of phase voltages and currents on the buses of an industrial customer without excitation forcing.

The level of residual voltage under these conditions amounted to 61% of the nominal voltage, with an 8 MVA SEM being in operation at an industrial customer.

When switching pathogens to ESS, it is possible to fully use the capabilities of the SM to increase the level of residual voltage. In order to identify the patterns of influence of the field-forcing ration (with independent power supply from the ESS and the power supply of the exciter from its own buses), a series of experiments were carried out with an SG of 8 MVA and the length of the overhead line at which the short circuit occurs. Figure 4 shows a comparison of the graphs of the CLL for the SG.

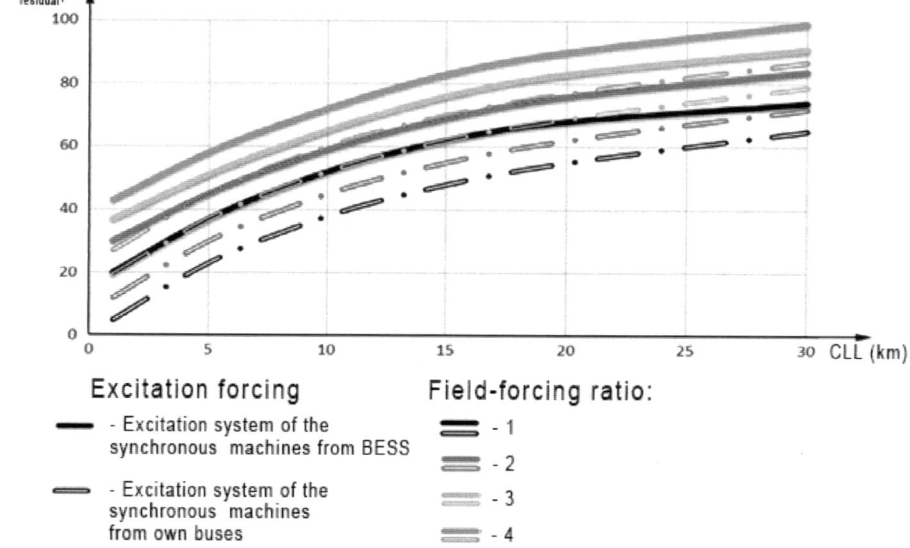

Figure 4. Residual voltage level change

The comparison of the graphs for the SG with a system powered by its own buses and an independent excitation system from the ESS shows that at the level of the specified minimum allowable residual voltage of 80% of the nominal, the CLL for the SG with an independent excitation system from the ESS is less than for the SG with exciter power system from own buses for 18 kilometers. If we compare the efficiency of the field-forcing ratio of the SG with an independent excitation system from the ESS, then four-time field-forcing provides a twofold decrease in the CLL [2; 10]. Fig. 5 shows the moment of switching on the four-time field-forcing ratio of SEM. When we turn on the four-time field-forcing ratio of SEM in 20 ms, we can reach the residual voltage level of 80% of the nominal. By analogy, SG was also considered.

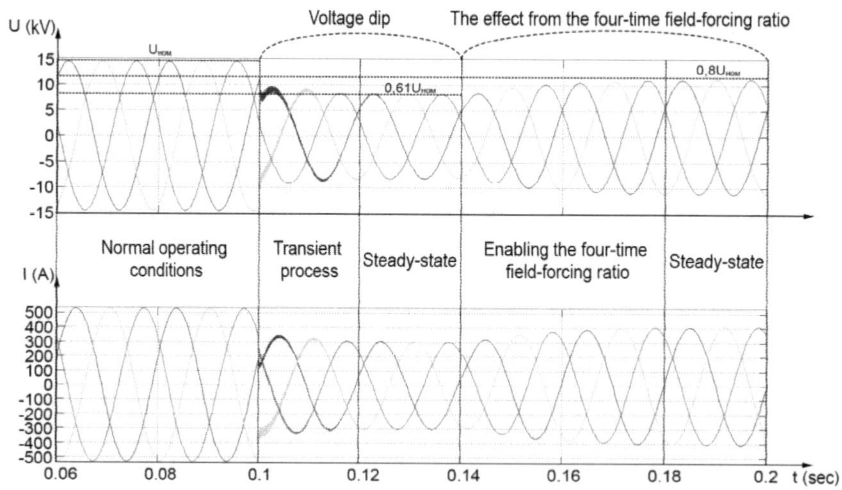

Figure 5. Oscillogram of phase voltages and currents on the buses of an industrial customer at the time of VD with the next turning on four-time field forcing ratio of SM.

However, due to a sudden increase in load on the SM rotor, vibration appears, which causes a voltage change during the transition process. SEM swings are within acceptable limits and SEM continues to work. Figure 6 shows the oscillograms of the phase currents of the SM and the angular velocity of the rotor rotation. Vibrations of the rotor cause energy losses, of which the magnetic losses due to the occurrence of eddy currents in the rotor core are of the greatest importance.

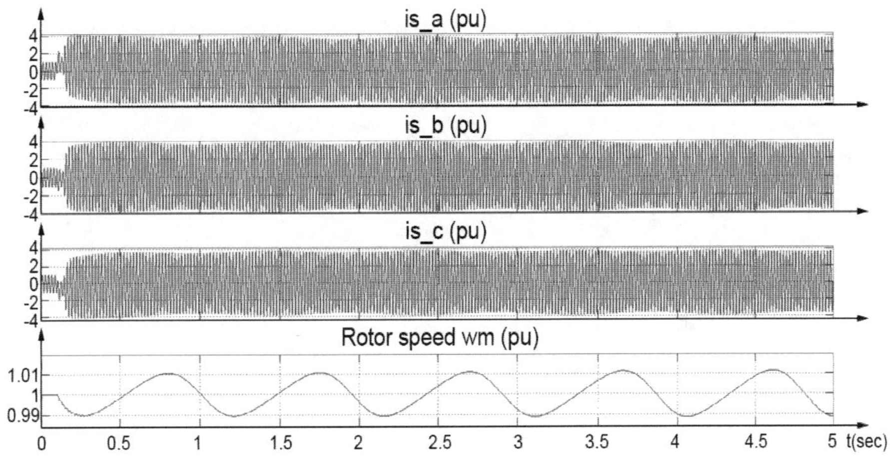

Figure 6. Phase currents in the SM field winding and the angular velocity of the rotor rotation of the SM during the four-time field-forcing ratio

The rotor vibrations have a damped character, and therefore, after some time, the rotor will take a position corresponding to the angle at which the moment balance is established [11].

SM load distribution coefficient is defined as the ratio of the load at the SM terminals to its nominal power. The examples of the influence of various load distribution coefficient on the exciting ability of SM are presented below.

The following experiment shows the dependence of the degree of residual voltage reduction on the Krl voltage at the installed capacity of SM - 8 MVA, Figure 7, and SM - 16 MVA, Figure 8.

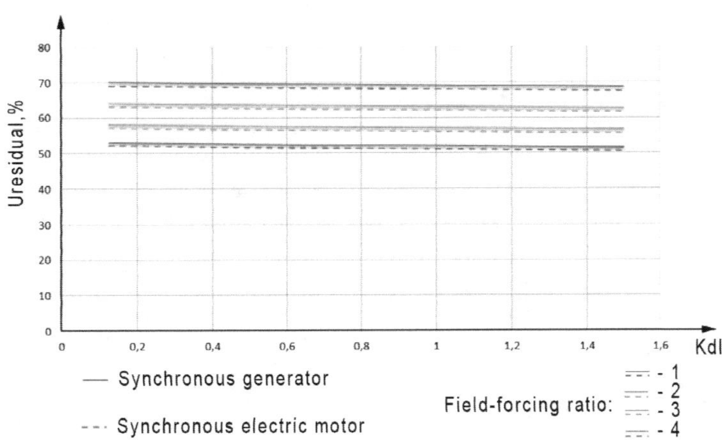

Figure 7. Dependence of the residual voltage on Kdl when installing SM with a capacity of 8 MVA.

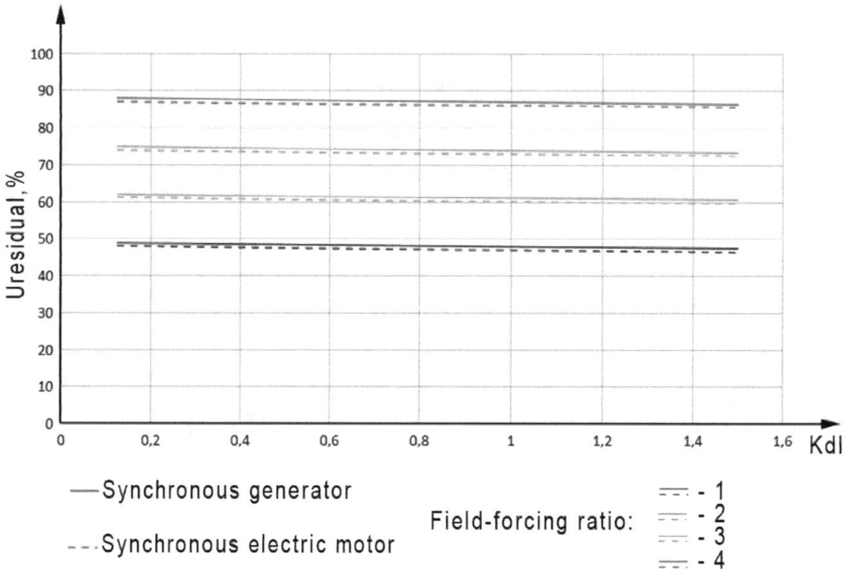

Figure 8. Dependence of the residual voltage on Kdl when installing SM with a capacity of 16 MVA.

3. Conclusion

The experimental results allow making the following conclusion: three-phase fault on 110 kV OPL (average line length is 1-10 km) cause VD sufficient to dysfunction most equipment (residual voltage is less than 60%) at any three-phase fault power.

Faults to the supply and outgoing lines of adjacent substations (the total distance to the fault location is 10-20 km) at short-circuit currents from the system of 15 kA and higher cause VD with a depth of up to 40%. This is enough to maintain the health of most unregulated motors with a voltage of 6-10 kV. With a short-circuit current of 30-40 kA, the residual voltage among consumers approaches 80% of the nominal when the short-circuit point is removed from distribution of busbars of 110 kV substation for 20 km or more. However, a four-time field-forcing ration of the SG excitation makes it possible to reduce the CLL of power line to 10 km.

Fault to remote substations with a voltage of 110 kV (which also corresponds to the electrical distance of the short circuit point from the buses of the distribution substation 110 kV that occurred at nearby district and junction substations with a voltage of 220 and 550 kV) causes voltage dips of up to 30% depth even with a very high short-circuit power. Such devices as industrial controllers, instrumentation circuits, thyristor converters of complex units with stiff Undervoltage protection remain sensitive to VD of this depth. These devices, as a rule, ensure the operation of complex units that perform continuous technological processes in automatic mode. The shutdown of such units due to VD causes defective products and the undersupply of expensive products, but due to the possibility to use a four-time field-forcing ratio, these risks can be minimized.

The presence of SG and SEM on 10 kV buses allows increasing the level of residual voltage up to 25% depending on the power of the SM, for SG the level of residual voltage is 10% higher than for SEM. With close faults, when the level of residual voltage is critically low, using circuitry to increase the short-circuit power on 110 kV buses and the SM ability to excitation forces can significantly reduce the CLL to 3 km.

Thus, to sum up we can conclude that the excitation forcing of SM installed in the internal power supply system of industrial consumers, allows increasing the level of residual voltage during VD caused by a three-phase fault in 110 kV supply grids. The higher than the normal field-forcing ratio of SM contributes to a decrease in the CLL and, as a result, leads to a decrease in the number of process shutdowns at consumers with a continuous production cycle. When choosing a high-voltage current-limiting protector, designed to increase the level of residual voltage at the MSDS at three-phase fault in 110 kV supply grids, it is necessary to take into account the capabilities of SM in terms of excitation forcing. Switching exciters to ESS allows full use of the capabilities of SM to increase the level of residual voltage.

Acknowledgments
This research is performed with financial support of the Ministry of Education and Science of the Russian Federation within implementation of the federal special program «Research and development in the priority directions of scientific and technological complex of Russia for 2014-2020», the agreement on granting a subsidy № 075-15-2019-057, unique identifier of applied scientific research (project) RFMEFI57418X0188. All research articles should have a funding acknowledgment statement included.

References
[1] Belyaev A V *Emergency automation in load nodes with synchronous electric motors of high power. Part 2*, available at: https://www.twirpx.com/file/98009/.
[2] Bakhteev K, Fedotov A and Misbakhov R 2019 The improving efficiency of electric receivers on the industrial enterprises in case of short-term power outages *Proceedings of the 2019 20th International Scientific Conference on Electric Power Engineering (EPE)* Ostrava Czech Republic 347-352 DOI: 10.1109/EPE.2019.8778181
[3] Belyaev A V 2005 Emergency control at load nodes with high power synchronous motors *Library of Electrical Engineering, supplement to the journal Energetik* **88**
[4] Fedorov V K, Leonov E N, Fedorov D V 2016 Influence of distributed generation on losses and quality of electric energy *Omsk Scientific Bulletin* **6 (150)** 72-76
[5] Fedotov A and Bakhteev K 2016 Influence excitation forcing of the synchronous machines on the level of residual voltage in case of short-time violations of electrical power supply *Proceedings of the higher educational institutions. Energy sector problems* **7-8** 64-72
[6] Fedotov A, Bakhteev K and Leonov A 2016 Influence of the small distributed generation on the level of residual voltage at short circuits *Proceedings of the higher educational institutions. Energy sector problems* **3-4** 45-49
[7] Faybisovich D L 2012 Handbook for the design of electrical networks *Rev and add ENAS* **4** 376
[8] Vergutin A A 2017 Improving the reliability of power supply to industrial enterprises through the use of high-speed backup power supply *Scientific and Technical Bulletin of the Bryansk State University* **2**
[9] Gamazin S I, Zhukov V A, Tsyruk S A and Pupin V M Microprocessor-based high-speed ATS as a means of improving the reliability of power supply for responsible consumers *Problems of Energy* **6** 11-12
[10] Bakhteev K, Fedotov A, Chernova N and Misbakhov R 2019 Methodological approaches to the choice of energy storage and optimization of their parameters to improve the electric power quality in various types of electric power systems *Proceedings of the 10th International Scientific Symposium on Electrical Power Engineering, Elektroenergetika 2019* Slovak Republic 488-493
[11] Glebov I A 1988 *Scientific principles of designing excitation systems of powerful synchronous machines* (Moscow: Nauka)

Methods of mathematical statistics application in assessing the density of actual and forecasting distribution density of residual oil reserves

K F Gabdrahmanova, G R Izmailova and L Z Samigullina

Ufa State Petroleum Technological University, Branch of the University in the City of Oktyabrsky, 54a, Devonskaya St., Oktyabrsky, Republic of Bashkortostan, 452607, Russia

E-mail: klara47@mail.ru

Abstract. The authors propose a method for mathematical statistics application for residual reserves in deposits predicting. They propose a method for making residual the current reserves distribution model based on geological and physical characteristics of the reservoir analysis. The article takes the first step to justify the use of statistical methods in the conditions of operational facilities of the Romashkino and Tuymazy oil fields. One of the urgent problems associated with the effective oil fields development is not only a reliable estimate of the distribution density of residual reserves, but their forecast for the subsequent development as well. There are many ways to estimate residual reserves nowadays. All of them are, to one degree or another, based on residual oil saturation ratio estimating, which in the conditions of insufficient accuracy of geophysical methods, as well as their rise in price (carbon-oxygen logging, for example) or their limited application, calls for other approaches in the residual oil reserves value assessing, based, for example, on the field data. Logging methods have their limitations. Their application requires other approaches in residual reserves amount assessing. Having analyzed multiple studies results, such as, for example, field data, implemented in the SPSS program regression analysis, which allows statistical data the processing the authors propose the mathematical statistics methods application in assessing the density of the actual and forecasting the residual oil reserves' distribution density. The mathematical modeling results for oil from a porous medium displacement process are presented for Romashkino and Tuymazy oil deposits.

1. Introduction

According to geological exploration studies in Aktash region, the authors [1-3] carried out a correlation analysis, on the basis of which the regression equations were derived. Considering the structure of the operational object of Aktash region, it can be noted that its geological model is characterized by a significant heterogeneity of the composing collectors, a significant difference in the collector properties of the allocated groups and a limited distribution area of productive sediments, which is a characteristic feature of the territory. Undoubtedly, all this was, on the whole, a complicating factor throughout the entire period of the area developmen.

The researchers [4-6] performed a detailed statistical analysis of the distribution features of the main parameters characterizing the filtration and reservoir propeties of the each productive layer composing the terrigenous Devonian deposits within the Aktash area. The obtained results were tested at the

Romashkino and Tuimazy fields, where the correctness of the derived regression equation was confirmed.

2. Materials and methods

A single-factor correlation analysis and visualization of the various geological, physical, and field parameters influence on the residual *Qres* reserves value and the ultimate oil recovery was performed [6]. Further, the contribution of each of these components to the ultimate oil recovery value and residual reserves was evaluated in order to obtain a predictive regression equation.

For these purposes, the regression analysis implemented in the SPSS statistical data processing program is best suited [7].

To assess the significance (impact) of the various components on the residual reserves density a pairwise correlation analysis was performed first, i.e. the influence of each of the factors on the residual oil reserves value was studied with the aim of their further inclusion in the regression equation for the residual reserves development forecasting. In this case, the Pearson correlation coefficient and the significance coefficient were determined [8–9].

At the same time, the following geological, physical, and fishing factors were taken into account in the correlation analysis:

H_{noth} – net oil thickness, m;
IRpor – porosity fragmentation variability;
IRoil – oil saturation fragmentation variability;
IRperm – permeability fragmentation variability;
K_{cl} – clay mass ratio, %;
N – the interlayers number in a pack;
G – hydraulic conductivity;
K_{fl} – flushing ratio, u.r.
Q_{init} – initial stock density, t/m^2;
K_{perm} – average number of permeable intervals in the section, 1/m;
OWR – oil water ratio, t/t;
N_{cl} – relative clay ratio, u.r.
K_{bwr} – bound water ratio.

Pairwise correlation calculation results are presented in table 1.

From this table it follows that the following factors are the most significant: initial net oil thickness H_{noth}, clay mass ratio K_{cl}, the number of interlayers in pack *N*, hydraulic conductivity *G*, flushing ratio K_{fl}, initial stock density Q_{init}, average number of permeable intervals in the section K_{perm}, relative clay factor N_{cl}, bound water ratio K_{bwr}.

Since K_{cl} and N_{cl} are related quantities, one of them can be neglected, given the different degree of significance (N_{gl}>K_{gl}). The coefficient of bound water K_{bwr} depends on the relative clay content, therefore, it can also be neglected. The number of packs *N* and the average number of permeable intervals in the section K_{perm} are also related quantities that reflect heterogeneity (multilayer), and for the calculations taking into account their significance one can take the average number of permeable intervals in the section coefficient.

So, to obtain a regression equation for the relationship between the residual reserves density and geological field data, the authors used the following factors: hydraulic conductivity *G*, flushing ratio K_{fl}, net oil thickness H_{noth}, relative clay ratio N_{cl}, and the average number of permeable intervals in the section K_{perm}.

Regression analysis to obtain the corresponding equation was carried out in the SPSS program, the regression equation of the dependence of Q_{res} on geological and production factors was derived for each pack individually and for the whole reservoir DI. As a result, the authors obtained the following regression equations.

$$Q_{res}=0.001G-0.028K_{perm}+0.046\,H_{noth}-0.084N_{cl}-0.016\,K_{perm}-0.033, \qquad R^2=0.7 \qquad (1)$$

Table 1. Pairwise Correlation Results

Factor	Correlation coefficient	Q_{res}
H_{noth}	Pearson Correlation	0.772(**)
	Sig. (2-tailed)	0.000
IR_{por}	Pearson Correlation	0.012
	Sig. (2-tailed)	0.851
IR_{oil}	Pearson Correlation	0.063
	Sig. (2-tailed)	0.320
IR_{perm}	Pearson Correlation	0.014
	Sig. (2-tailed)	0.821
K_{cl}	Pearson Correlation	-0.368(**)
	Sig. (2-tailed)	0.00021
N	Pearson Correlation	0.195(**)
	Sig. (2-tailed)	0.002
G	Pearson Correlation	0.732(**)
	Sig. (2-tailed)	0.000
K_{perm}	Pearson Correlation	-0.161(*)
	Sig. (2-tailed)	0.011
Q_{init}	Pearson Correlation	0.805(**)
	Sig. (2-tailed)	0.000
K_{perm}	Pearson Correlation	-0.447(**)
	Sig. (2-tailed)	0.000
OWR	Pearson Correlation	0.079
	Sig. (2-tailed)	0.557
N_{cl}	Pearson Correlation	-0.419(**)
	Sig. (2-tailed)	0.00014
K_{bwr}	Pearson Correlation	-0.419(**)
	Sig. (2-tailed)	0.000

** Correlation is significant at 1% (p=0.01).
* Correlation is significant at 5 % (p=0.05).

In order to verify the calculated (forecasted) Q_{res} value with the actual data, the sum of the residual reserves density for each pack and for the reservoir DI was calculated. The results are summarized in table 2.

Table 2. Comparison table of calculated and actual values of the residual reserves density

Pack	$\Sigma\, Q_{rescalc}$, t/m^2	$\Sigma\, Q_{resact}$, t/m^2	$Q_{res\ calc}$ average, t/m^2	$Q_{res\ act}$ average, t/m^2
4	10.39	10.75	0.179	0.185
7	7.59	6.81	0.194	0.174
9	4.76	4.03	0.198	0.17
11	13.94	11.92	0.244	0.209
13	13.69	11.3	0.240	0.198
23	4.03	4.11	0.287	0.294
DI	49.11	48.94	0.197	0.196

In addition, the obtained regression equations (1) can be used to predict the terrigenous Devonian reserves development at the initial stage of oil field development under given initial conditions: with known hydraulic conductivity, flushing coefficient, oil-saturated thickness, relative clay content and average number of permeable intervals in the section.

In addition to the residual reserves density, the the reservoir depletion degree is evidenced by the oil recovery coefficient. Therefore, it is very important to predict the final oil recovery factor *(ORF)* at the initial stage of field development. For this purpose, the authors derived the appropriate regression equations for each pack and for the DI formation. After a preliminary assessment of the geological and physical factors influence (significance) on this indicator, the most significant (influencing) factors on the oil recovery factor are: net oil thickness H_{noth}, number of packs N, hydraulic conductivity G, amount of water produced per pack $Q_{waterpack}$, flushing factor K_{fl}, initial reserves density Q_{init}, average number of permeable intervals in the section K_{perm}, the recovered oil amount of selected pack $Q_{oilpack}$.

$$ORF = 0.001\ H_{noth} + 0.0008\ G - 0.069\ K_{perm} + 0121 K_{fl} + 0.195 \quad (1)$$

Basing on the results, the comparative curves of the calculated and actual values of the oil recovery factor were compiled (Figure 1).

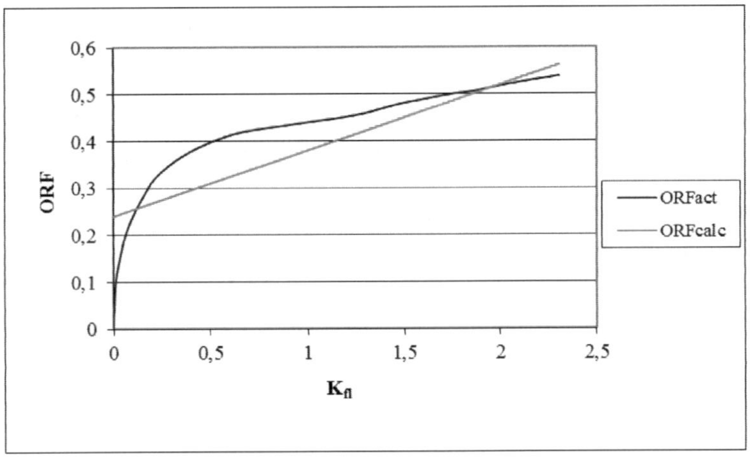

Figure 1. Dependence of the estimated oil recovery factor and the actual oil recovery factor on the flushing ratio

Figure 1 shows graphs reflecting the dynamics of the dependence of the calculated and actual oil recovery factor on the K_{fl} flushing ratio for the "merger site", from which it follows that the maximum oil recovery factor can be achieved with the values K_{fl} = 2.0, which is in good agreement with the available field production data.

To test the universality of the results, it was decided to derive a similar regressive equation for the dependence of Q_{res} on various geological and production factors in the conditions of the Tuymazy oil field (block XII) for the DI layer of the terrigenous Devonian.

3. Results and Discussion

After that, in order to assess the suitability of the equation (1) for calculating the Q_{res} for Devonian terrigenous sediments in the Aktash area, the field data for the Tuymazy oil field were used in the equation (1), which was derived to determine the Q_{res} from the DI reservoir.

According to the reconciliation results table 3 was compiled.

Table 3. Comparison table of the calculated and actual values of the residual reserves density in Tuymazy oil field

Layer	z, t/m^2 (Tuymazy oil field)	$\Sigma\ Q_{resact}$, t/m^2	$\Sigma\ Q_{rescalc}$, t/m^2 (Aktash oil field)
DI amount	38,59	37,37	50,239
The average deviation,%	3,0	-	34,4

The data in table 3 shows hat the average deviation of the calculated Q_{res} from the actual Q_{res} obtained using the regression equation derived for the Tuymazy oil field (1) is 3.0%.
The obtained result allows concluding that the obtained regression equations are inapplicable outside the relevant conditions.

4. Conclusion

Basing on the research results the following conclusions can be drawn.

1. In the conditions of the Devonian terrigenous sediments the regression equations were obtained that can be used to calculate the current residual reserves density, as well as to make their forecast;
2. The equation obtained for the terrigenous deposits of the Romashkino oil field can be used to calculate the residual reserves density for similar deposits of other areas or the Ural-Volga region deposits, taking into account the geological, physical and field factors characteristic of these conditions.

References

[1] Gabdrakhmanova K F, Izmailova G R, Larin P A, Vasilyeva E R, Madjidov M A and Marupov S R 2018 Nomogram method as means for resource potential efficiency predicative aid of petrothermal energy *Journal of Physics*: *Conf. Series* **1015 (3)** 032036

[2] Yakupov R F, Mukhametshin V Sh and Tyncherov K T 2018 Filtration model of oil coning in a bottom water-drive reservoir *Periodico Tche Quimica* **15 (30)** 725-733

[3] Gabdrakhmanova K F, Izmaylova G R and Larin P A 2018 The way of using geothermal resources for generating electric energy in wells at a late stage of operation *IOP Conf. Series: Earth and Environmental Science* **194 (8)** 082012 1-9

[4] Khuzina L B, Mukhametshin V Sh, Tyncherov K T and Shaikhutdinova A F 2018 On the choice of the oscillators' installation site International *Journal of Civil Engineering and Technology* **9(9)** 1952-1959

[5] Igtisamova G R and Kovalev N O 2017 Reliability Evaluation for Improved Screw Dies of Coiled Tubing unit Injector. *Tribology in industry* **39 (1)**

[6] Tyncherov K T and Selivanova M V 2017 Modular wavelet approach for reinforced plastic pipelines operation quality evaluation. In *2017 International Conf. on Industrial Engineering, Applications and Manufacturing (ICIEAM)* pp. 1-4

[7] Zeygman Y V, Mukhametshin V S, Khafizov A R, Kharina S B, Abutalipova E M and Avrenyuk A N 2017 Peculiarities of selecting well-killng fluids composition for difficult conditions (Russian). *Oil Industry Journal* **01** 66-69

[8] Mukhametshin V V 2017 The need for creation of a unified comprehensive method of geological and field analysis and integration of data on effective influence on the bottom-hole formation zone (Russian) *Oil Industry Journal* **04** 80-84

[9] Almukhametova E M, Akimov A V, Kalinina S V, Fatkullin I F and Gizetdinov I A 2017 Efficiency of preliminary discharge of stratum water in Tuymazinskoe oil field In *IOP Conf. Series: Earth and Environmental Science* **87 (6)** 062001

[10] Yakupov R F, Mukhametshin V S, Zejgman Y V, Chervyakova A N and Valeev M D 2017 Metamorphic aureole development technique in terms of Tuymazinskoye oil field (Russian) *Oil Industry Journal* **10** 36-40

Efficient operation of oil producing wells in non-stationary fluid withdrawal mode

K L Gorshkova, N N Alaeva and I P Sitdikova

Almetyevsk State Oil Institute, 2A Lenin Str., Almetyevsk, 423452, Russia

E-mail: chri_leon@mail.ru

Abstract. The article deals with the efficient non-stationary well operating mode during fluid withdrawal. In this well operating mode, the optimal pump switching periods are identified to increase the oil flow rate and reduce the water content of the fluid. The purpose of the study is to improve the efficiency of pilot production wells operation. A descriptive method was used during the study, which involved analysis of the reservoir-well-pump system transient phenomena on a simulation stand, interpretation of transient graphs, comparison and synthesis of the results. The paper presents the results of experimental studies of transient pressure changes on the simulation stand for different pump operating modes. The results of the research have shown that the pressure change depends on the pump operating mode, i.e. reservoir characteristics depend on the change in the hydraulic resistance of the reservoir-well-pump system.

1. Introduction

One of the most important challenges in oil industry is the enhancement of oil recovery from developed fields. Most of the oil fields of the Republic of Tatarstan are quite mature and oil remaining in place is classified as hard to recover, its share constantly growing.

Today, the focus is on the implementation of low-cost oil recovery enhancement methods. One of these methods is the non-stationary fluid withdrawal, which is used for efficient recovery of oil remaining in place [1-3].

The analysis of the cyclic impact (non-stationary mode) on oil recovery in a fractured-porous reservoir is provided in the works of the researchers Yu. M. Molokovich, A. I. Markov, E. I. Suleymanov, R. G. Farkhullin, N. N. Neprimerov, R. S. Khisamov [1-11]. The works [12, 13] show that the cyclic effect on a fractured-porous reservoir leads to its enhanced oil recovery in comparison with the stationary fluid withdrawal mode. The work of V. G. Belov, A. Yu. Gorshenin, V. A. Ivanov et al. covers a non-stationary oil recovery method [14], which involves exposure to periodic depression without stopping the submersible well pumping equipment. Based on this, a method aimed to put wells in an efficient operating mode [15] was developed, where a well is put in a stationary operating mode after non-stationary operation.

2. Problem statement

The main objective in respect of fractured-porous reservoirs is to enhance the oil flow rate and reduce the water content of the fluid in the non-stationary well operating mode. In a non-stationary mode, pressure filtration waves occur and propagate in the reservoir [16,17], which leads to creation of the conditions for oil involvement in the filtration flows of the well's bottom-hole zone and allows enhancing the oil recovery factor. The impulse effect of changing pump performance is transmitted from the pump along the wellbore to the bottom and then to the exploited reservoir.

Laboratory and field experiments are required to investigate into the effect of a production well non-stationary operation with the bottom-hole pressure fluctuating over time on the reservoir oil recovery and reduction of the water content of hydrocarbons produced.

3. Theory

Let us suppose a central production well with a radius r_c is drilled in the reservoir. From the time $t = 0$, the bottom-hole pressure begins to fluctuate during well operation. A constant pressure is maintained at the outer reservoir boundary of the radius R. If we neglect capillary and gravitational forces at the scale approximation, then the non-stationary filtration across the averaged reservoir thickness h is described by a system of equations [5]:

$$\frac{1}{r}\frac{\partial}{\partial r}(\sigma_l r\frac{\partial p_l}{\partial r}) = \beta_l h\frac{\partial p_l}{\partial t} + A\sigma_2(p_l - p_{3-l}), \tag{1}$$

$$\frac{1}{r}\frac{\partial}{\partial r}(f_l\sigma_l r\frac{\partial p_l}{\partial r}) = \beta_{b,l} h S_2 l\frac{\partial p_l}{\partial t} + m_l h\frac{\partial S_l}{\partial t} + \bar{f}A\sigma_2(p_l - p_{3-l}), \tag{2}$$

where the indices $l = 1,2$ refer to the cracks and blocks in the reservoir, S_l - water saturation, $\sigma_l = k_l \cdot k_l^* \cdot h$, $k_l^* = \frac{k_{bl}^*}{\mu_b} + \frac{k_{Hl}^*}{\mu_H}$, $\bar{f} = \frac{k_{bl}^*}{\mu_b k_l^*}$, $\bar{f} = \begin{cases} \bar{f}_1, & p_l \geq p_2, \\ \bar{f}_2, & p_l < p_2; \end{cases}$, k_l - absolute permeability, k_{bl}^*, k_{Hl}^* - water-oil relative permeabilities, $\bar{f}_l$ - water cut in the flow; $\beta_l = \beta_{b,l}S_l + \beta_{H,l}(1 - S_l)$, $\beta_{b,l} = \frac{m_l}{m}(\beta_l + \beta_b m)$, $\beta_{H,l} = (\beta_l + \beta_H m)$ - porous medium, water and oil compression factors, $m = m_1 + m_2$; and A - fluid exchange ratio.

The solution of the system (1)-(2) is defined on an interval $r \in (r_c, R)$ subject to the following conditions:

1. initial

$$p_l(r,0) = p_0,\ S_l(r,0) = S_{l,0} \geq S_{l,*}, \tag{3}$$

2. borderline

$$\begin{gathered} p_l(r_c,t) = p_0 - p_{l,c}\cos(\frac{2\pi}{T}t) < p_R, \\ p_l(R,t) = p_R \geq p_0, \\ S_l(R,t) = S_l^*; \end{gathered} \tag{4}$$

where $S_{l,*}$ and S_l^* are the connate and cutoff water saturation values, and T and $p_{l,c}$ are the period and amplitude of the reservoir exposure.

Dependence of bottom-hole pressure on impulse impact when pump performance changes [18]:

$$P_{pl} = P_0 + P(1 - e^{-\frac{t}{\tau_{pl}}}), \tag{5}$$

where P_{pl} is bottom-hole pressure, *atm*; P_0 is initial pressure, *atm*; P is pulse pressure change, *atm; t* is transient period, s; τ_{pl} is the time of reservoir response to the change in pump performance, calculated on the basis of transient time for each well.

In case of non-stationary fluid withdrawal, we shall consider a single interconnected reservoir-well-pump system for the purpose of effective oil recovery management.

The main purpose of the experimental studies is to identify conditions for supporting efficient well operation in a non-stationary fluid withdrawal mode.

A simulation stand was used to study the transient processes in the reservoir-well-pump system [19, 20] (Fig.1). Data was collected and transmitted from the simulation stand via the RS-485 interface to the controller and via the RS-232 interface to the engineer's automated workstation with the Master SCADA3 software installed.

The functional diagram of the stand (Fig. 1) consists of: A−1− hydraulic accumulator simulating reservoir capacity; A−2 - hydraulic accumulator simulating well filling; *E*−1 - fluid collection tank; K−1,2,3,4,56,7- ball valves; F−1,2- filters; P- pressure reducer; FE1,FE2- fluid meters; *PT* - pressure gauge; *H* −1 - centrifugal pump.

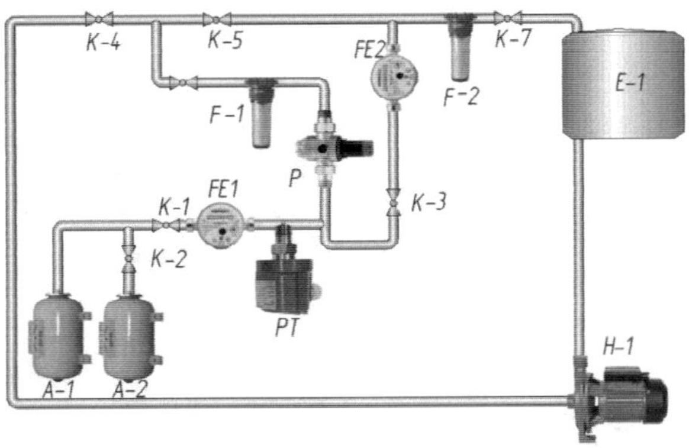

Figure 1. The simulation stand functional diagram

Experimental studies of the transient processes occurring in the reservoir-well-pump system were conducted on this simulation stand, and their results were used to analyze pressure transient processes, compare and summarize the results of experiments. The downhole pump operating modes were simulated by changing the position of the ball valves (*KSH*) on the simulation stand.

To perform laboratory tests, the following activities were carried out:

1.The experiment conditions were defined and the experiment matrix was developed: changing the degree, to which the KSH-1, 2 ball valves open(%). Dependence of the number of experiments on the number of factor levels $N = p^k$, where N — the number of experiments, p— the number of factor levels, and k — the number of factors. The initial state corresponds to the maximum fluid withdrawal, while all ball valves are fully open and the fluid is pumped from the tank to the hydraulic accumulators. Changing the degree of the ball valves (KSH-1,2) opening, we can simulate a change in the downhole pump operating mode. In this case, the level and factor variability interval is 10. With the number of factors being $k \geq 7$, we have to reduce, i.e., eliminate them, because otherwise a large number of experiments will be required.

2. The real-time transient pressure change graphs are built in the *MasterSCADA3*softwareand data arrays are uploaded to*MicrosoftExcel*.

3. Qualitative characteristics are identified based on the transient pressure change graphs, such as reservoir response time, transient period, and maximum pressure deviation during the transient period.

IOP Conf. Series: Materials Science and Engineering **860** (2020) 012009 doi:10.1088/1757-899X/860/1/012009

4. Experimental results

In each experiment, pressure change graphs were obtained in a real-time mode with visualization in the *MasterSCADA 3* software and data arrays were uploaded to *Excel* (Figure2-15).

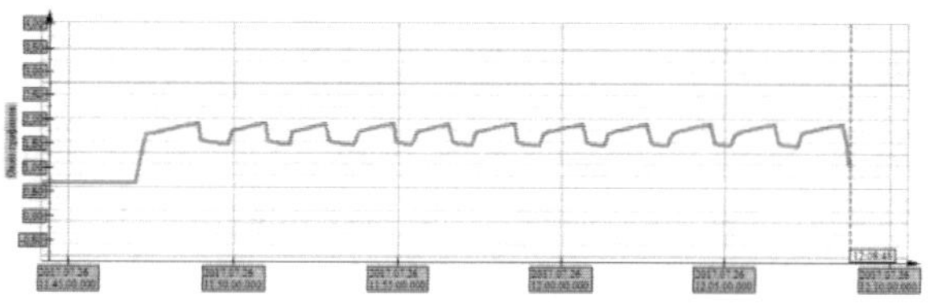

Figure 2. Transient pressure change graph for experiment No.1 in MasterSCADA 3

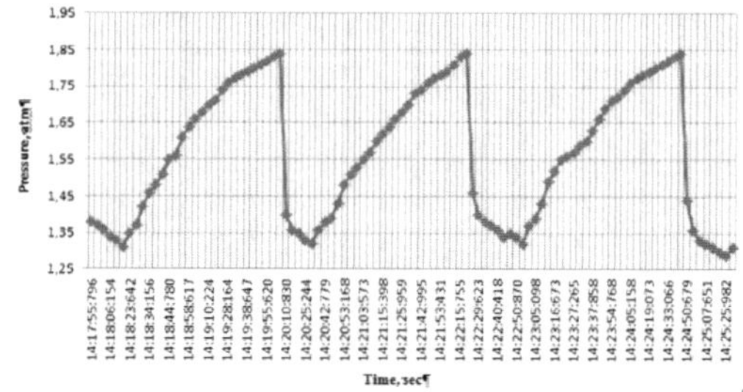

Figure 3. Piece of transient pressure change graph for experiment No.1, exported to Excel

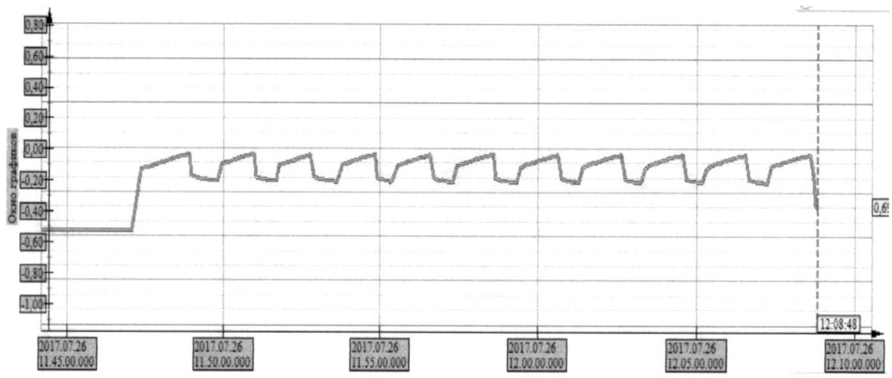

Figure 4. Transient pressure change graph for experiment No.2 in *MasterSCADA 3*

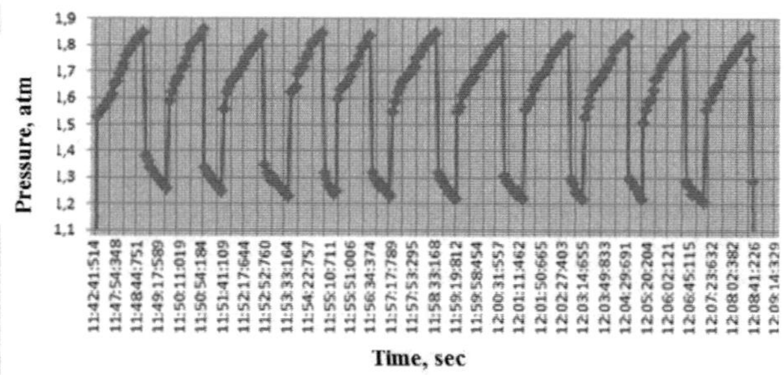

Figure 5. Transient pressure change graph for experiment No.2, exported to Excel

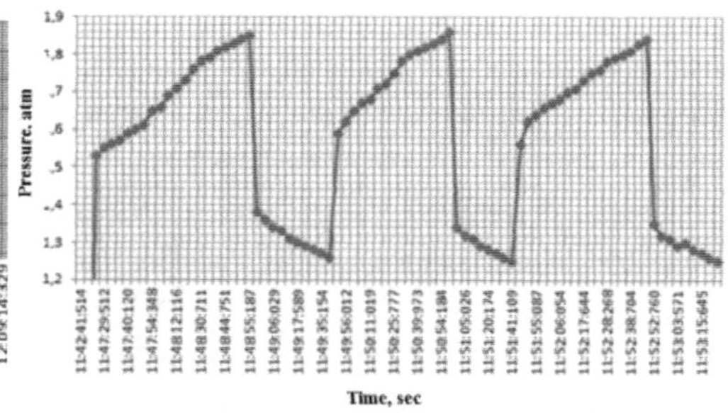

Figure 6. Piece of transient pressure change graph for experiment No.2, exported to Excel

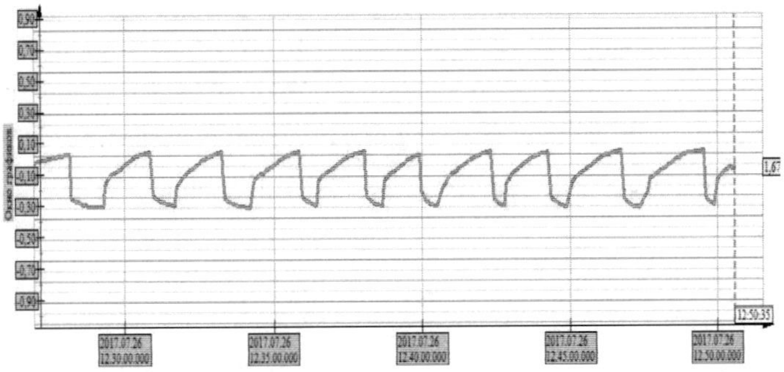

Figure 7. Transient pressure change graph for experiment No.3 inMasterSCADA 3

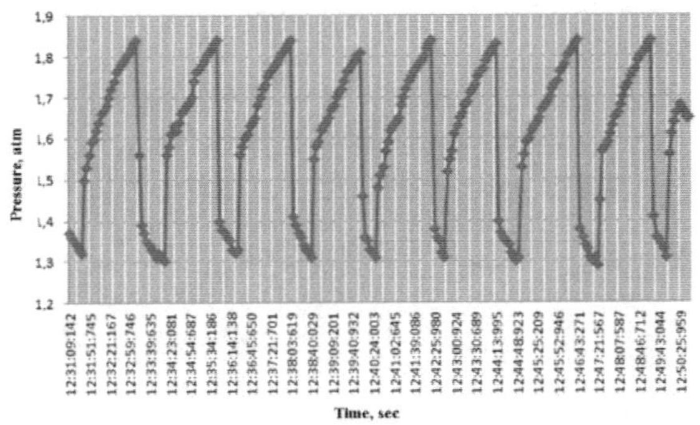

Figure 8. Transient pressure change graph for experiment No.3, exported to Excel

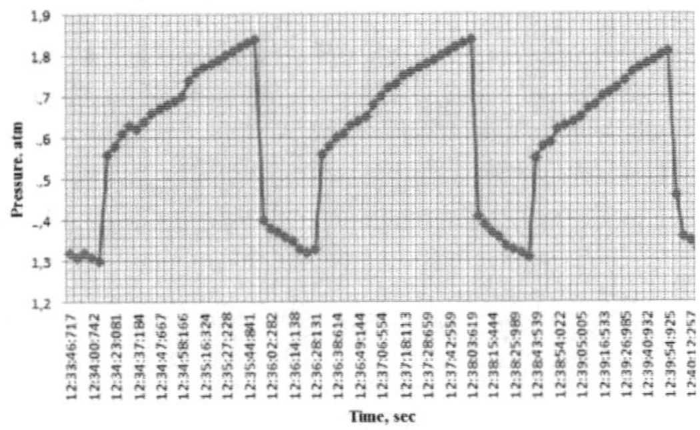

Figure 9. Piece of transient pressure change graph for experiment No.3, exported to Excel

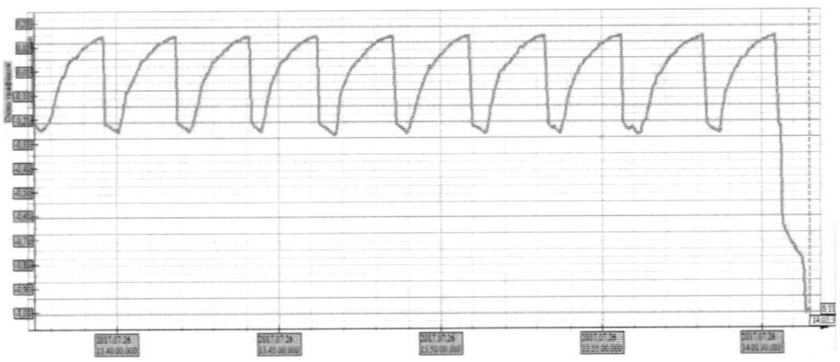

Figure 10. Transient pressure change graph for experiment No.4 in *MasterSCADA 3*

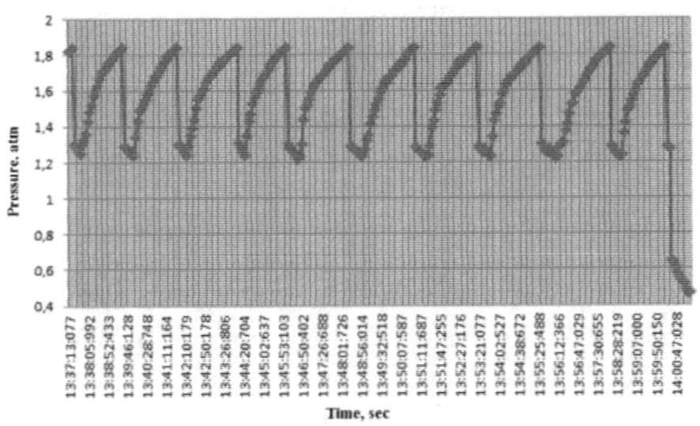

Figure 11. Transient pressure change graph for experiment No.4, exported to Excel

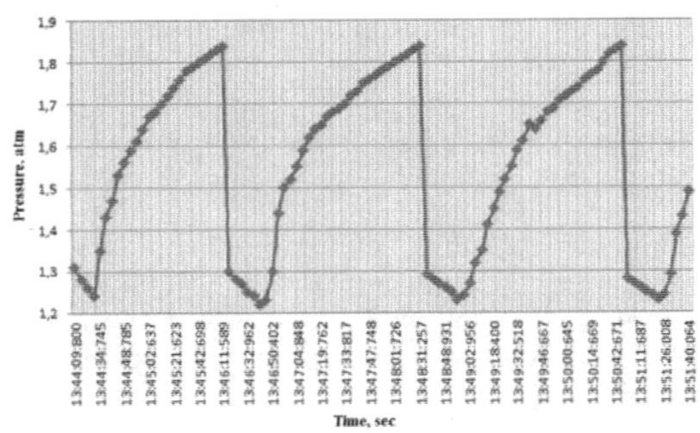

Figure 12. Piece of transient pressure change graph for experiment No.4, exported to Excel

IOP Conf. Series: Materials Science and Engineering **860** (2020) 012009 doi:10.1088/1757-899X/860/1/012009

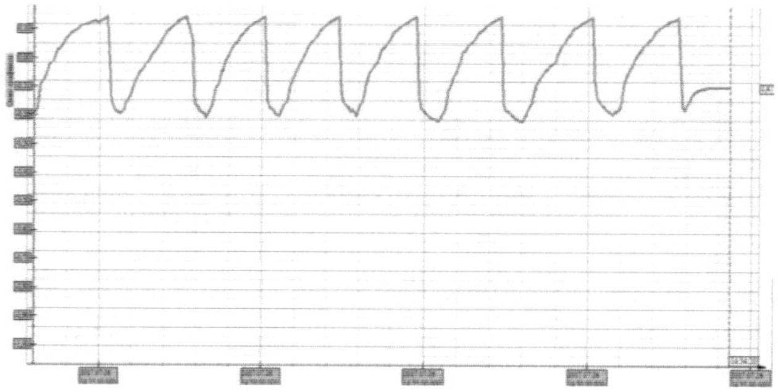

Figure 13. Transient pressure change graph for experiment No.5 in *MasterSCADA 3*

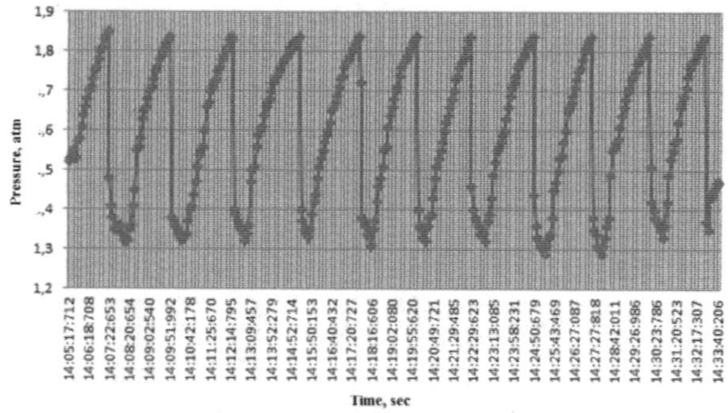

Figure 14. Transient pressure change graph for experiment No.5, exported to Excel

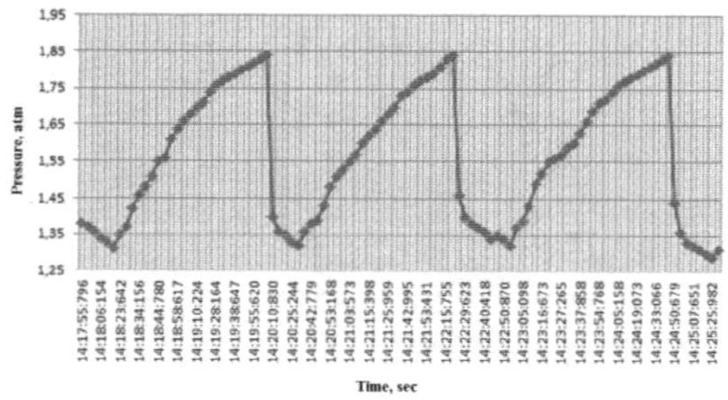

Figure 15. Piece of transient pressure change graph for experiment No.5, exported to Excel

The following qualitative indicators were calculated based on the transient pressure change graphs obtained: reservoir response time, transient period, and maximum pressure deviation during the transient period. For experiments, the following is defined in Table 1: reservoir response time $\tau_{pl\,i}$, (i=1...5),transient period per one pressure change pulse jump t, and the maximum pressure deviation during the transient period p.

Table 1. Transient process qualitative indicators for the conducted experiments

No.	$\tau_{pl\,i}$, (i=1...5)					t, sec	p, atm
1	50	25	19	4	3	136	0.52
2	16	79	12	8	5	192	0.6
3	24	39	25	7	6	136	0.53
4	29	9	33	26	11	133	0.61
5	50	25	4	19	3	136	0.52

Using the calculatedvalues, field studies were performed with the focus on the operating mode control in respect of the eight pilot production wells that were operated in a non-stationary fluid withdrawal mode.

The non-stationary operating mode assumes that the well requires some additional equipment (controller, control stations with a frequency-controlled electric converter, and deephole telemetry),

IOP Conf. Series: Materials Science and Engineering **860** (2020) 012009 doi:10.1088/1757-899X/860/1/012009

which allows achieving all well operating modes (stationary, periodic and non-stationary) in the borehole.

Putting eight production wells in the non-stationary fluid withdrawal mode, the duration of the non-stationary fluid withdrawal mode should be 20 days. The non-stationary mode parameter is the impact amplitude (change in pump performance) from the max swing/min to min swing/min.

It is recommended to perform non-stationary fluid withdrawal in accordance to the calculated well operating mode switch periods. The water content, flow rate, and dynamic level of the fluid must be registered in the non-stationary fluid withdrawal mode.

Table II presents on-line data from the pilot wells at the deposit studied (n_{max} is the number of maximum rod swings, swing/min; n_{min} is the minimum rod swings, swing/min; L is the rod stroke length, m; H_s, H_d are static and dynamic levels, m; P_{pl} is the downhole pressure, *atm;* Q_z is the fluid rate, m3/day; Q_v is the water cut, %).

Table 2. On-line data from pilot wells

No.	n_{max}, swing/min	n_{min}, swing/min	L, m	H_s, m	H_d, m	P_{pl}, atm	Q_z, m3/day	Q_v, %
1	3.3	1.0	2.40	141	411	39	26.0	85
2	2.9	1.0	2.50	59	275	52	7.0	81
3	5.0	1.0	2.10	45	274	51	7.5	82
4	3.7	1.0	2.10	57	174	62	6.4	89
5	3,5*	1.0*	3.50	70	177	60	7.5	56
6	3.5*	1.0*	3.50	80	121	64	7.1	75
7	3.0*	1.0*	3.00	72	254	54	6.6	60
8	2.4	1.0*	3.00	75	494	39	6.8	84

In case of the deposit studied, the non-stationary operation was possible only on four wells: 1, 2, 3, and 4 due to some technical limitations. At the wells 5 and 6, the beam-pumping unit was underutilized because of jamming; 7 – formation of emulsion, and 8 – low dynamic level.

The dynamic level was calculated for well No.1 using the pressure values obtained by means of a downhole measuring system.

When measurements are performed using a sounder, the well annulus must be pre-discharged, i.e., there is a need to bleed off the excess pressure. These measures often lead to foaming of the fluid level in the annular space and the formation of a "foam cap", which makes a clear signal reflection impossible and does not allow shooting the fluid level.

Therefore, the use of sounder data is limited to approximate estimates. Only in case of favorable conditions the results of such measurements can be used to identify almost the full range of reservoir parameters. In addition, echo sounding is not allowed in oil wells with a high gas factor and in case of the product water cut exceeding 80%. The registration period must be at least $1 \div 2$ days. Short-term dynamic level measurements are uninformative.

Since the "foam cap" formed as a result of the fluid level foaming in the annulus can reach at least 30 m, the convergence criteria (6) in operation may be a fluid level less by $(30 \div 60)$m, than the level measured by sounder.

$$H_{dr} = H_{di} - (30 \div 60), \tag{6}$$

where H_{dr} is a calculated dynamic level, m and H_{di} is a measured dynamic level, m.

In well No.1, fluid pressure and dynamic level are measured. However, the dynamic level measurement may not be always precise. This is due to the influence of gas in the well. The dynamic level value can be obtained by calculating the pressure at the bottomhole.

The bottomhole pressure formula is [21]

$$P_{pl} = \rho \cdot H_d \cdot g, \tag{7}$$

where P_{pl} is the bottomhole pressure,*atm*; ρ is the oil density, *kg/m*^{3}; H_d is the dynamic level, *m*; g is the free fall acceleration, *m/s*^{2}.

From formula (7) we can express the fluid dynamic level in the well:

$$H_d = \frac{P_{pl}}{g \cdot \rho}, \tag{8}$$

Figure 17 shows a graph of calculated and measured dynamic level values for well No.1 (the bottom hole pressure, measured level and the one calculated using the formula (8)).

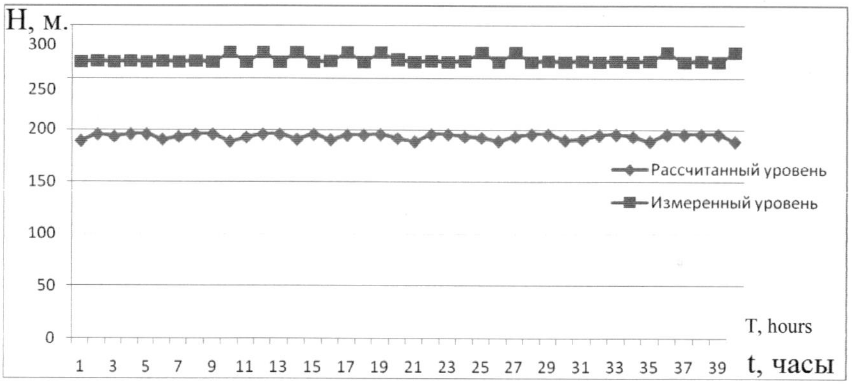

Figure 16. Calculated and measured level graph

As it can be seen from the graph (Figure 16), the difference between the measured and calculated fluid level values is $67 \div 82$m. The average calculated dynamic level value is 193. According to the convergence criterion (6): 193=($193 \div 184$), which attests to the reliability of data obtained by calculation and confirms their applicability in solving problems related to optimal control of well operating modes.

For wells suitable for the non-stationary fluid withdrawal study, the time period for monitoring the water cut in well products was calculated at minimum and maximum number of swings.

The water cut of the extracted fluid is monitored with a glance to this time period, when the reservoir fluid reaches the wellhead flow moisture meter. It is the time interval, which allows a reliable determination of the fluid water content [21, 22].

To ensure the optimal well operating mode in case of a non-stationary fluid withdrawal mode, a consistent pattern is observed manifesting itself as lowering water content of the fluid depending on the depression cycles.

Figure 17 shows graphs of changing average product water content during 2 hours and the 2-hour oil mass at the oil metering unit for the entire periodof field research.

IOP Conf. Series: Materials Science and Engineering **860** (2020) 012009 doi:10.1088/1757-899X/860/1/012009

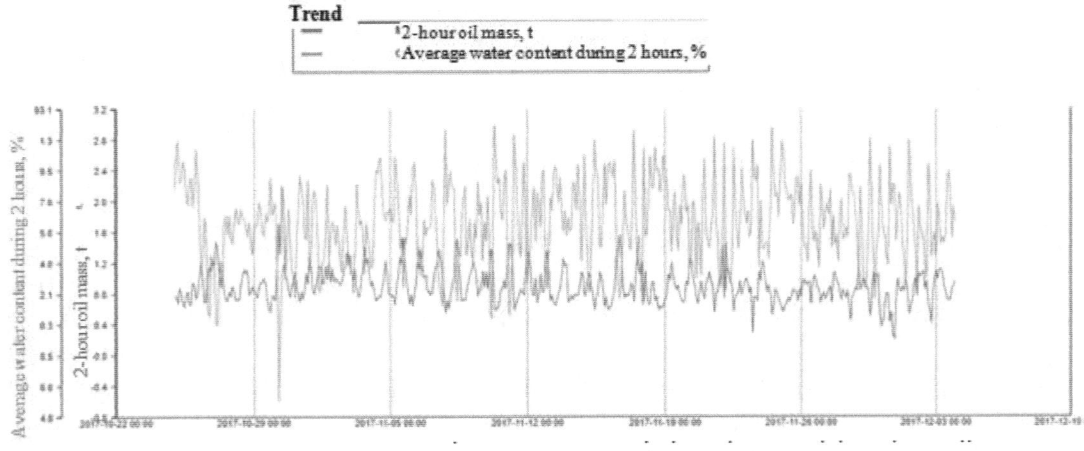

Figure 17. Graphs of changing average product water content during 2 hours and the 2-hour oil mass at the oil metering unit for the entire period of field research

5. Discussion

The laboratory studies of transient processes revealed that the pressure change occurs exponentially over time $5\tau_{\text{пл}}$, during which the pressure changes by 99.3% of the created pulse pressure, while the reservoir response continues. To effectively control the filtration pressure wave, it is advisable to use the pump operating mode change period equal to the reservoir response time $3\tau_{\text{пл}}$, since during this time the pressure changes by 95% of the created pulse pressure. It was found that after a period of time exceeding $3\tau_{\text{пл}}$ the pressure changes slower, while $3\tau_{\text{пл}}$ is different for each experiment, since the reservoir-well-pump simulation system is influenced by the change in the system's hydraulic resistance.

Based on the results of field studies conducted with the use of the appropriate producing well operating mode control tools, there are options for periodic generation of impulses to the reservoir with subsequent pump stop, where it is seen that the change of the pump engine speed from the maximum value to the specified minimum value leads to a change in the pressure on the reservoir, when porous reservoir blocks accumulate pressure energy almost equal to the reservoir pressure during the build-up period, and then, during pumping, oil is squeezed out of the pore space into cracks under the influence of the difference $(P_{pl} - P_1)$. This oil moves towards the bottomhole through the cracks. Therefore, we can state that when stopping the well, the fluid water content decreased somewhat before the start of non-stationary treatment. However, the reliability of this fact requires multiple verifications. For this purpose, the procedure for evaluating the water content reduction effect has been changed. Figure 18 shows graphs of changing average product water content during 2 hours and the 2-hour oil mass at the oil metering unit for the entire period of field research. After obtaining the time constants τ for the build-up and pumping sub-steps and 95% durations of the amplitude of each of them, equal to 3τ, the time of fluid portion movement from the pump intake to the moisture meter and the time (1hour) required to count reliable moisture meter values was added thereto, since the fluid flow through the pump is stationary only at this time interval, i.e. the pump flow rate is equal to the inflow from the reservoir. The time from the start of pumping to the first vertical dotted line is 95% of the disturbance amplitude. The adjustable parameters are the amplitude and time of impact on the reservoir. The amplitude can be adjusted by the difference between the maximum and minimum quantities, i.e. the flow rate of the fluid pumped. The exposure time is tuned by the software using a controller that controls the frequency control electrical converters. This is done to be able to watch the process of water content reduction in the fluid and identify the time to stop the non-stationary withdrawal based on a certain criterion (minimum water content of the fluid).

6. Summary and conclusions

It is confirmed that the pressure change occurs exponentially over time $5\tau_{пл}$, during which the pressure changes by 99.3% of the created pulse pressure;

It is revealed that, to effectively control the filtration pressure wave, it is advisable to use the pump operating mode change period equal to $3\tau_{пл}$, since during this time the pressure changes by 95% of the created pulse pressure;

It was found that the reservoir response time $3\tau_{пл}$ for each transient process is different, since the reservoir-well-pump simulation system is influenced by the change in the system's hydraulic resistance;

To identify the optimal period of non-stationary impact on the reservoir from the producing well, it is recommended to obtain the reservoir response time $\tau_{пл}$ at the maximum and minimum fluid withdrawal.

In case of non-stationary well operation, the well flow rate increases and its water cut decreases in comparison with stationary and periodic operating modes. During the stationary mode, the average oil flow rate at the metering unit was 10.245 m^3/day with an 80.1% water content. In case of the non-stationary mode, the average oil flow rate at the metering unit was 15.2 m^3/day with a 76.28% water content.

7. Acknowledgement

The research was performed under the Service Contract No. 290/0002/2115/48 dated 01.05.2017 entered into with PJSC TATNEFT named after V.D. Shashin on the topic: Development of Automatic Control Algorithms for Producing Well Operating Modes and Selection of Automation Tools to Support a Non-Stationary Fluid Withdrawal Method.

References

[1] Molokovich Yu M, Markov A I, Davletshin A A and Kushtanova G G 2000 *Piezometry of the borehole vicinity. Theoretical background* (Kazan: Publ. house DAS)

[2] Farkhullin R G, Molokovich Yu M and Markov A I 2000 *Development of a fractured-porous reservoir by non-stationary drainage* (Kazan: Regent)

[3] Farkhullin R G 2002 *A set of field studies aimed at oil reserves recovery monitoring* (Kazan: Publ. house TATPOLIGRAF)

[4] Molokovich Yu M et al. 2000 *Development of a fractured-porous reservoir by non-stationary drainage* (Kazan: Regent)

[5] Molokovich Yu M 2006 *Nonequilibrium Filtration and Its Application in Oil Industry* (Izhevsk: Regular and Chaotic Dynamics R&D Centre; Institute of Computer Research)

[6] Neprimerov N N 1987 Influence of field experiment on the filtration theory development *Collection: Problems of filtration theory and enhanced oil recovery process mechanics* (Moscow: Nauka) pp 153-162

[7] Neprimerov N N 1986*Technology of Optimal Formation Development* (Kazan)

[8] Khisamov R S, Khamidullin M M, Shaidullin R G, Galimov I F, Vanyurikhin I S and Galiev F A 2012 RF, Patent No. 2 453 689 *Method for Oil Deposit Development*

[9] Gavrilov A G, Ovchinnikov M N and Shtanin A V 1999 Geological structures recognition and evaluation of water saturation in oil fields by the hydrodynamical methods *Proc.Int.Conf. Geometrization of Physics IV* (Kazan) pp 208-210

[10] James P Brill and Hemanta Mukherjee 1999 Multiphase Flow in Wells *First Printing Henry L., Doherty Memorial Fund of AIME Society of Petroleum Engineers Inc* (Texas, Richardson)

[11] Kitsios V 2010 *Recovery of fluid mechanical modes in unsteady separated flows* PhD thesis, Department of Mechanical Engineering, The University of Melbourne and Institute PPRIME, Universite de Poitiers

[12] Chekalin A N, Konyukhov V M and Volkov Yu A 1996Analysis of the cyclic impact influence on fractured-porous formation oil recovery *Materials of the workshop "The concept of oil recovery enhancing method development"*(Bugulma) pp 79-83

[13] Samoilov V V, Frolov S A, Samoilov D Yu and Usov V V 2014Results of research on the sucker rod pump well control methods based on the non-stationary well fluid withdrawal method*Automation, telemetry and communication in the oil industry* **8** 26-30

[14] Belov V G, Gorshenin A Yu, Ivanov V A, Kozlovsky V S, Musaev Kh Ts and Fedoseev A I and Shelekhov A L 2006 RF, Patent No. 2228352 *The method of non-stationary oil extraction from a reservoir*

[15] Belov V G, Gorshenin A Yu, Ivanov V A, Kozlovsky V S, Musaev Kh Ts and Fedoseev A I 2006 RF, Patent No. 2 289 019 *The method for putting wells in the most efficient operating mode*

[16] Hollaender F, Hammond PS, Gringarten AC 2002 Harmonic testing for continuous well and reservoir monitoring *SPE Annual Technical Conference and Exhibition*(San Antonio) pp 1–12

[17] Kuo CH 1972Determination of reservoir properties from sinusoidal and multirate flow test in one or more wells *SPE* **12 (6)** pp 499–507

[18] Samoilov V V and Frolov S A 2014An unconventional approach to implementing EOR methods. *LLC Tatintek* (Almetyevsk:printing house of OOO Tatavtomatizatsiya)

[19] Tomus Yu B, Gorshkova Ch L, Alaeva N N and Orekhova L G 2018 Studying transient processes in non-stationary fluid withdrawal modes on the simulation stand*Collection of works:Achievements, Challenges and Prospects of the Oil and Gas Industry Development, materials of the International Scientific and Practical Conference*(Almetyevsk State Oil Institute) pp 199-203

[20] Gorshkova Ch L, Orekhova L G and Alaeva N N 2018To the issue of managing the producing well operating modes when using a non-stationary fluid withdrawal method*Collection of works: Achievements, Challenges and Prospects of the Oil and Gas Industry Development, materials of the International Scientific and Practical Conference*. (Almetyevsk State Oil Institute) pp 203-205

[21] Gorshkova Ch L, Alaeva N N and Orekhova L G 2019To the issue of choosing the well pump operating mode in case of non-stationary fluid withdrawal *Bulletin of Higher Educational Institutions. Oil and Gas* **5** 145-152

[22] Alaeva N N, Gorshkova Ch L, Barankov E Yu 2019 RF, Patent No. 2700738 *A method for improving the reliability of the fluid water content monitoring for oil producing wells equipped with sucker rod pumps*

IOP Conf. Series: Materials Science and Engineering **860** (2020) 012010 doi:10.1088/1757-899X/860/1/012010

Optimization of structure of control system with fuzzy controller

E A Muravyova[1] and E R Enikeeva[2]

[1] Ufa State Petroleum Technological University, Branch in Sterlitamak, 2 October av., Sterlitamak, 453118, Russia
[2] Almetyevsk State Oil Institute, 2 Lenin street, Almetyevsk, 423452, Russia

E-mail: enikeeva@mail.ru

Abstract. The article gives a brief description of the three-phase separator of the "Heather-Triter" type. A three-phase separator of the "Heather-Triter" type is used to obtain commercial oil from well products, and to separate well products for gas, oil, and water. To regulate the oil level and phase separation, the separator is equipped with direct-acting mechanical control valves on the oil outlet line and on the water outlet lines. The issues of modeling the level control system in a three-phase separator of the "Heather-Triter" type, and criteria for the quality of production management are considered. The description of the model of a three-phase separator of the "Heather-Triter" type is given. It works with various methods of fuzzy controllers. A method for improving the quality of level control in the Troika is described by the Heather-Triter phase separator using a fuzzy controller. The issues of optimizing the structure of a control system with a fuzzy controller are considered. A method for improving the quality of level control in a three-phase Heather-Triter separator is described.

1. Introduction

Development and introduction of automated process control systems is mostly trendy in the development of up-to-date industrial production. In this regard, optimization of industrial processes is becoming more widely used. Of great importance in the optimization is a properly developed and most realistic mathematical model of the process control system [8]. The mathematical model allows simulating the processes occurring within the system. By studying the simulation model, it is possible to obtain the data and information that can be used to control a real object [5].

Fuzzy sets are a tool enabling to develop models for controlling complex objects which do not always make it possible to develop precise mathematical models using integro-differential equations[7].
In this article, we will describe mathematical model of a three-phase separator "Hiter-Triter" type, we will consider methods for optimization of the control system structure using the fuzzy controller.

2. Brief characteristics of the three-phase separator of "Hiter-Triter" type

A three-phase separator of the Heather-Triter type is used to produce marketable oil from well products, to separate well products into gas, oil and water.

The Heater-Triter separator (hereinafter referred to as the "separator") is a horizontal cylindrical apparatus containing an emulsion input unit, an emulsion heating unit by a flame tube, a coalescence unit, a water, oil and gas separation unit. To improve the separation of water and oil, it is possible to wash the emulsion with hot water and add a demulsifier. On the cylindrical part of the body and on the bottoms of the devices are technological fittings, a fitting for the installation of instrumentation and automation equipment and hatches. On the bottom (from the inlet side of the mixture), a flange connection of the heater's flame tubes to the housing is provided.

To regulate the oil level and phase separation, the separator is equipped with direct-acting mechanical control valves on the oil exit line and on the water exit lines.

The problems of separator control are as follows:

1. There are cases when mechanical control valves do not cope with their task, because of which the phase separation level is not maintained at a given value, and water begins to flow to the oil outlet line.
2. Systems of regulation of oil and water levels are interconnected, which negatively affects the quality of transients. The consequence of this is water pollution with oil and high water cut after the separator.

It seems advisable to combine the control of the separator taking into account its state in all of the above parameters on the basis of fuzzy logic methods.

3. The level control system modeling in a three-phase separator «Hiter-Triter» type

3.1.Production control quality criteria

The quality of the products of this process is determined by the water content of the oil, which should be less than 10%. This parameter is constantly monitored at the exit of the BCS. This parameter should not exceed the established norm. Oil quality is monitored at the top level of PCS. This division employs specialists responsible for monitoring the output parameters of production as a whole. They receive product data automatically.

The quality of an automated control system is determined by a combination of properties that ensure the effective functioning of both the control object itself and the control device, i.e. the entire management system as a whole.

The properties that make up this set and have quantitative meters are called quality criteria of the control system.

The quality of an automated system can be assessed by criteria such as the complexity of the synthesis of the system and the required computing power (processor time and memory). The complexity of the synthesis of the system in our case implies a waste of time on the development of a fuzzy controller.

To reduce the complexity, it is proposed to use the following methods to optimize the structure of the control system with a fuzzy controller:

1. Transition from one fuzzy controller to series-connected fuzzy controllers;
2. Transition from one fuzzy controller to parallel fuzzy controllers;
3. Feedback elimination method.

The use of these methods leads to a decrease in the number of rules drawn up during the synthesis of a fuzzy controller, which reduces the time required to develop a fuzzy controller. Also, by reducing the number of rules, the requirement for computing power is reduced, that is, less processor time and memory are required for data processing.

3.2. Description of the three-phase separator model of "Hiter-Triter" type

We will create a Heather-Triter separator model, which will consider the dependence of the liquid level and water level on the flow rate of the emulsion, its emulsion and the degree of opening of the valves for oil and water output. The conceptual model of the Heather-Triter separator is shown in Figure1.

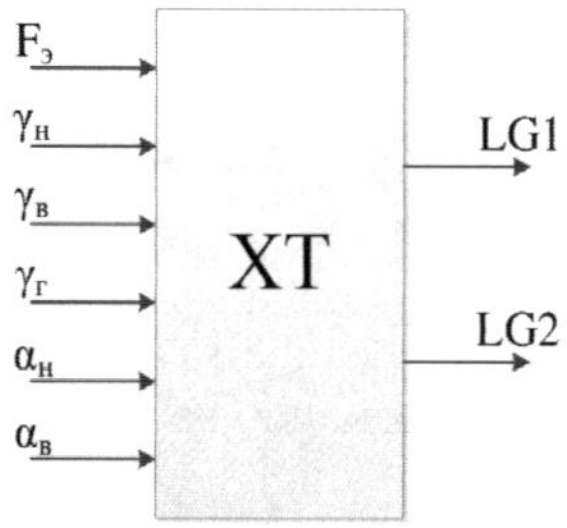

Figure1. Conceptual model of the Heather-Triter separator

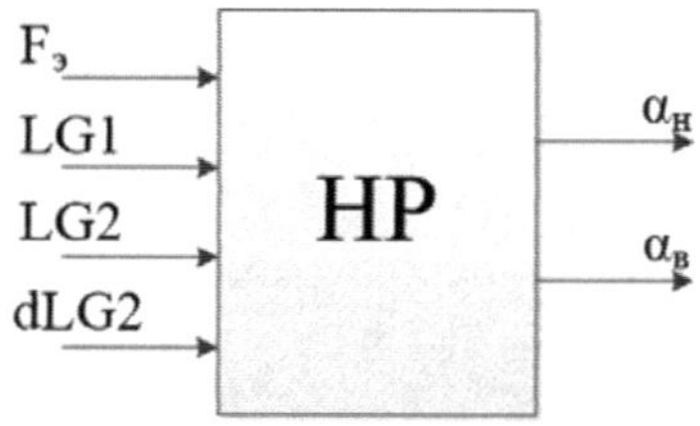

Figure 2. Conceptual model of a fuzzy controller

Designation of model parameters:
F_e - water-gas-oil emulsion flow rate, m^3/h;
γ_o – volume fraction of oil in the emulsion, %;
γ_w – volume fraction of water in the emulsion, %;
γ_g – volume fraction of gas in the emulsion, %;
α_o – oil outlet valve opening degree, %;
α_w – the degree of opening of the valve at the outlet of water, %;
LG1 – fluid level, m;
LG2 – water level, m.

The productivity of the oil and gas separator with direct heating (Heather-Triter) in terms of liquid by technological parameters is 6848 m^3/day, and the maximum allowable 11810 m^3/day.

We assume that in the water-gas-oil emulsion of gas there will be 5%, after the first stage of separation, the volume fraction of water in relation to the liquid can be from 60 to 98%, therefore, the volume fraction of oil from 2 to 40%. The inflow of water-gas-oil emulsion will be equal to 7208.4 m3/day (300.35 m^3/h), and the maximum allowable inflow of 12431.6 m^3/day (517.98 m^3/h).

The liquid level in the apparatus should be maintained within 0.4-0.8 m, and the water level in the range of 0.6-0.8 m.

Create a level management model:

$$\begin{cases} \dfrac{dLG2}{dt} = k_1 \times (F_{w1} - F_{w2}); \\ \dfrac{dLG1}{dt} = k_2 \times (F_{o1} - F_{o2} + F_{w1} - F_{w2}) \end{cases} \quad (1)$$

where, $\dfrac{dLG2}{dt}$ – water level change, %/h;

$\dfrac{dLG1}{dt}$ – fluid level change, %/h;

k_1, k_2 – scaling factors, %/m^3;

F_{w1} – inlet water flow, m^3/h;

F_{w2} – outlet water flow, m^3/h;

F_{o1} – input oil consumption, m^3/h;

F_{o2} – output oil consumption, m^3/h.

Calculation of scaling factors.

The volume of the three-phase Heather-Triter device is 80 m^3. The volume occupied by the liquid level in the range of 0.4-0.8 m will be considered equal to 10 m^3.

Therefore

$$k_1 = \frac{100}{10} = 10\% / m^3 \tag{2}$$

The volume occupied by the phase separation level in the range of 0.6-0.8 m will be considered equal to 20 m^3. Therefore

$$k_2 = \frac{100}{20} = 5\% / m^3 \tag{3}$$

The inlet water flow rate is calculated by the formula:

$$F_{w1} = F_э \times \gamma_w, \tag{4}$$

Inlet oil consumption is calculated by the formula:

$$F_{o1} = F_э \times \gamma_o, \tag{5}$$

The flow rate of water and oil at the outlet depends on the degree of opening of the valves. Therefore, we obtain the functions:

$$F_{w2} = f(\alpha_w), \tag{6}$$

$$F_{o2} = f(\alpha_o), \tag{7}$$

The synthesis of this fuzzy controller is complicated by the fact that it is necessary to draw up a large number of rules for it, namely 54 = 625 rules.

4. Optimization of the control system structure using the fuzzy controller

In control theory, special attention has always been paid to the synthesis of mathematical models and control algorithms with insufficient information about the control object and the useful signals acting on it, and interference.

Over time, it became clear that the use of classical methods of control theory is no longer sufficient to control such systems, and new methods and approaches need to be developed. One of these approaches is based on fuzzy sets and fuzzy logic. Initially, this approach was applied and proved to be effective in creating expert systems. Somewhat later, it began to be used to create expert management systems and more recently, for the synthesis of regulators and control systems for technological systems.

In this article, we consider methods for optimizing the structure of a control system with a fuzzy controller:

1. Transition from one fuzzy controller to series-connected fuzzy controllers;
2. Transition from one fuzzy controller to parallel fuzzy controllers;
3. Feedback elimination method.

Optimization criteria will be:

The complexity of the synthesis;

required computing power (processor time and memory).

4.1. Method of consecutively connected fuzzy controllers

This method is perfect in cases where the system is better divided into blocks (links). The scheme of this method is presented in Figure 3.

IOP Conf. Series: Materials Science and Engineering **860** (2020) 012010 doi:10.1088/1757-899X/860/1/012010

The application of this method allows reducing the complexity of the synthesis process of a fuzzy controller, in the event that it is easier to calculate small blocks and conduct experiments with them than to calculate large blocks and conduct experiments with it.

A fuzzy controller is a static link, therefore, a series connection of fuzzy regulators will also be a static link. This statement reduces the computational process, because the internal connections of the fuzzy controller are necessary only at the stage of its synthesis and are not calculated during its operation.

Another advantage of this method is the ability to replace one of the blocks when changing the system, i.e. there is no need to make calculations for the entire controller, but it is enough to calculate a new block.

Consider an example of calculating the number of rules. For the structure shown in Figure 3, suppose that n1=5 terms are described for input x1, n2=6 terms for x2, n3=7 terms for x3, and n4=8 terms for y1.

Then the number of rules for HT (Figure 3) will be equal to where - the number of rules; rules.

Upon transition to the series connection of fuzzy regulators (HT1 and HT2), the number of rules will be equal where - the number of rules; rules.

This method should be used when the condition is met.

This method is applicable in cases where the output parameter of the fuzzy controller is independent of all input parameters. The scheme of this method is presented in Figure 4.

The transition from one fuzzy controller to two parallel fuzzy controllers can help reduce the number of rules in the synthesis, which, in turn, can reduce the complexity of developing a fuzzy controller.

The requirement for a microprocessor that implements a fuzzy controller is also reduced, namely, less processor time and memory are required for data processing.

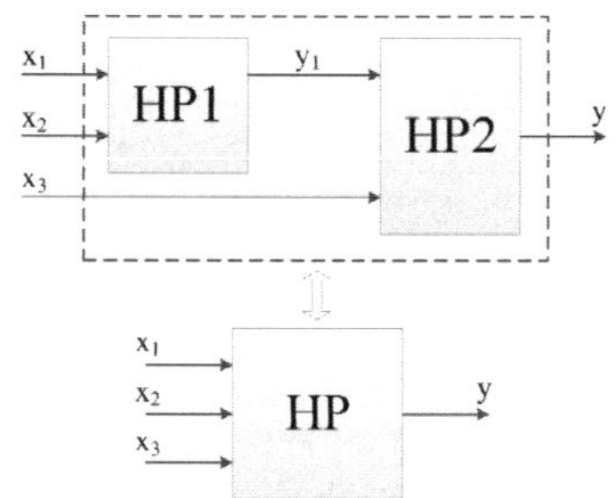

Figure 3. Structure of the serial connection of fuzzy regulators

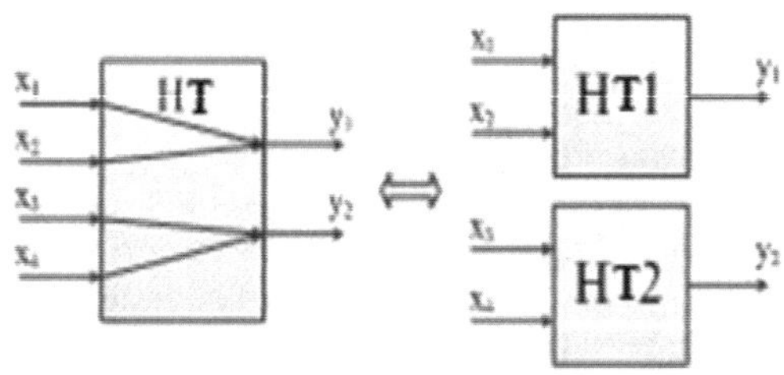

Figure 4. Structure of parallel fuzzy controllers

Consider an example of calculating the number of rules. For the structure shown in Figure 4, suppose that n_1=5 terms are described for input x_1, n_2=6 terms for x_2, n_3=7 terms for x_3, and n_4=5 terms for x_4.

Then the number of rules for HT (Figure 4) will be equal

$n_1 \times n_2 \times n_3 \times n_4 = M$, where M - the number of rules;

$M = n_1 \times n_2 \times n_3 \times n_4 = 5 \times 6 \times 7 \times 5 = 1075$ rules.

In the case when two parallel fuzzy controllers are used (HT1 and HT2)

$n_1 \times n_2 + n_3 \times n_4 = M'$, where M' is the number of rules; $M' = n_1 \times n_2 + n_3 \times n_4 = 5 \times 6 + 7 \times 5 = 65$ rules.

This method should be used when the condition $M' \langle M$ is met.

IOP Conf. Series: Materials Science and Engineering **860** (2020) 012010 doi:10.1088/1757-899X/860/1/012010

4.2. Method of the fuzzy controllers parallel connection

The essence of this method is to switch from a fuzzy controller with feedback to a fuzzy controller without feedback, which reduces the number of input parameters, as a result of which the number of rules drawn up during the synthesis and operation of the fuzzy controller is reduced.

In practice, to compensate for the disturbance, it is necessary to control the current control action, i.e. to reduce the disturbance one must either increase or decrease the control action. Therefore, it is necessary to apply a control action to the input of the fuzzy controller.

4.3. Method of feedback exception

If the change in the control action does not depend on the current value, you can reduce the number of inputs, which will lead to a decrease in the rules. In our case, the control action will be a change in the degree of valve opening, which does not depend on the current value.

Let us consider an example of calculating the number of rules. For the structures shown in Figure 5, suppose that n_1=6 terms are described for input x_1, n_2=5 terms for x_2, and n_3=7 terms for y.

We calculate the number of rules for a fuzzy controller with feedback in the Figure 5 rules.

Then for the fuzzy controller shown in Figure 5, the number of rules will be equal to: rules.

5. Optimization of the fuzzy controller for control of the separator

To implement optimal control, we divide the conceptual model of HT presented in Figure 2 into two parallel ones (Figure 6).

Optimization is carried out by two methods:

1. Division of one fuzzy controller into two parallel fuzzy controllers. As a result, the number of rules is reduced, 52+52=50 rules, which reduces the time to develop a fuzzy controller, reduces the requirement for a fuzzy controller system (requires less CPU time and memory to process data).
2. Obtaining at the output of HT2 the change in valve opening ($\Delta\alpha_w$), instead of the new value of the degree of opening of the valve of the water flow at the outlet (α_w). This is implemented to reduce the input parameters for HT2, and therefore, to reduce the rules drawn up.

Designations for model with parallel HT (Figure 6):

$\sum F$ – total oil and water consumption at the outlet, m^3/h;

$\Delta\alpha_w$ – change in the opening of the outlet water flow valve, %;

1/p – an integrating unit in which the calculation of a new values α_w:

$$\alpha_w = \alpha_w^{/} + \Delta\alpha_w \tag{8}$$

where, $\alpha_w^{/}$ – the degree of opening of the outlet water flow valve until the water level changes;

$f(\alpha_w)$ – structural unit in which the discharge water flow is calculated, depending on the degree of valve opening.

$$F_w = f(\alpha_w) \tag{9}$$

where, F_w – outlet water flow, m^3/h;

F_o – output oil consumption, m^3/h;

Oil consumption at the outlet is calculated by the formula:

$$F_o = \sum F - F_w \tag{10}$$

where, F_w – output oil consumption, m^3/h;

$f(F_o)$ – a structural unit in which the degree of opening of the valve at the oil outlet is calculated, depending on the oil consumption at the outlet.

$$\alpha_o = f(F_o) \tag{11}$$

The calculation of a fuzzy controller consists of the following steps: fuzzification; aggregation; activation; accumulation; defazzification.

Initially, expert information was collected. As a result of the collection of expert information, reference point tables for HT1, HT2 are derived. The reference point tables describe the entire area of definition of inputs and outputs of HT [1-3].

Table1. Table of reference points for HT1

LG1, m / Fe, m^3/hour	0,4	0,5	0,6	0,7	0,8
100	70	90	100	110	130
150	120	140	150	160	180
200	170	190	200	210	230
250	220	240	250	260	280
300	270	290	300	310	330

Next, linguistic variables were developed for the input and output parameters of the fuzzy controller.

Fuzzification is considered completed if correspondence is established between the numerical values of the input variables of the fuzzy inference system and the values of the membership function of the corresponding terms of the linguistic variables μ(Fe), μ(LG1), μ(LG2), μ(dLG2).

For example, when Fe = 160 m^3/h, the second and third terms are used, the membership function $\mu(Fe_2)$ is 0.8, and $\mu(Fe_3)$ is 0.2.

At the stage of aggregation, we calculate the degree of truth of the conditions for some rules of the fuzzy inference system. If Fe = Fe_2 ($\mu(Fe_2)$ = 0.8) and LG1 = LG12 (μ(LG12) = 0.55) then ;

where, z – truth conditions for rules.

At the activation stage, we calculate the degree of truth of additional consequents. For HT1 we will use production rules of the form:

If LG1 = LG1i и Fe = Fei, then ∑F = ∑Fi и ∑F = ∑FjC

where C – degree of belonging of a variable to a term ∑F [1-3].

If LG1 = LG11 и Fe = Fe1, then ∑F = ∑F1, for this consequent, the degree of truth will be equal to 1, since the rule corresponds to the extreme reference point (C=1). There is no additional consequent. Similarly, we calculate the degrees of truth of other consequents.

Production rules for HT1, HT2 were drawn up. For example, see Table II shows the production rules for HT1.

Table2. The system of production rules for HT1 in tabular form.

LG1 / Fe	LG11	LG12	LG13	LG14	LG15
Fe1	$\sum F_1$	$\sum F_1, \sum F_2^{0,46}$	$\sum F_1^{0,46}, \sum F_2$	$\sum F_1^{0,18}, \sum F_2$	$\sum F_2, \sum F_3^{0,18}$
Fe2	$\sum F_2$	$\sum F_2, \sum F_3^{0,46}$	$\sum F_2^{0,46}, \sum F_3$	$\sum F_2^{0,18}, \sum F_3$	$\sum F_3, \sum F_4^{0,18}$
Fe3	$\sum F_3$	$\sum F_3, \sum F_4^{0,41}$	$\sum F_3^{0,46}, \sum F_4$	$\sum F_3^{0,18}, \sum F_4$	$\sum F_4, \sum F_5^{0,18}$
Fe4	$\sum F_4$	$\sum F_4, \sum F_5^{0,46}$	$\sum F_4^{0,46}, \sum F_5$	$\sum F_4^{0,18}, \sum F_5$	$\sum F_4, \sum F_5^{0,12}$
Fe5	$\sum F_5$	$\sum F_5, \sum F_6^{0,34}$	$\sum F_5, \sum F_6$	$\sum F_5^{0,34}, \sum F_6$	$\sum F_6$

We calculate the accumulation for Fe = 160 m^3/h, LG1 = 0.54 m, these values correspond to the subclauses $\mu(\sum F_2)$ = 0.44 and $\mu(\sum F_3)$ = 0.2024. We apply the method of finding the maximum value from the corresponding subclauses, therefore μmax = 0.44.

Defuzzification is carried out by the method of center of gravity:

$$\sum F = \frac{\int_{F_{min}}^{F_{max}} F \cdot \mu\left(\sum F\right) \cdot d\sum F}{\int_{F_{min}}^{F_{max}} \mu\left(\sum F\right) \cdot d\sum F} \quad (12)$$

where, $\sum F$ – total oil and water consumption at the outlet, m^3/h;
$\sum F_{max}$ – maximum total oil and water consumption at the outlet;
$\sum F_{min}$ – minimum total oil and water consumption at the outlet.

6. Conclusion

The article describes a way to improve the quality of level control in a three-phase Heather-Triter separator through the use of a fuzzy controller and its further optimization. Application of optimization to HT will allow obtaining greater accuracy of regulation, as well as leading to a decrease in computing capacity. In turn, increasing the accuracy of the regulation of technological parameters will make it possible to improve the quality of the produced marketable oil.

References

[1] Muravyova E A, Enikeeva E R and Nurgaliev R R 2017 Automatic system for maintaining the optimum liquid level and the development of a liquid level sensor *Neftegazovoye delo* **2** 171-176

[2] Muravyova E A, Enikeeva E R and Nurgaliev R R 2017 Development of an automated system for monitoring and regulating fluid levels *Proceedings of the 5th International Conference on Industrial Engineering (ICIE 2019)(Electronic Materials* vol II) ed AA Radionov, OA Kravchenko, VI Guzeev, YV Rozhdestvenskiy (Kazan: Publishing House "Buk") pp 151-153

[3] Solovev K A, Nadyrshin A S, Bashirov M G, Sultanov R G, Sololoveva O I, Krishko K A and Gareyev I M 2017 RF Patent No. 167890 *Fuzzy controller with double base rules*

[4] Savolainen R S and Karhela T 2017 A framework for runtime verification of industrial process control systems *IEEE 15th International Conference on Industrial Informatics (INDIN 2017)* ed S Maltsev (Emden: Univ Appl Sci Emden Leer) pp 687-694

[5] Bhattacharya A K and Sambasivam D 2016 Multi-objective Differential Evolution for automated control of Industrial Process input material composition incorporating supply-side fluctuations *2016 IEEE Congress on Evolutionary Computation, CEC 2016* ed G Yen, K Tan (Vancouver, CANADA) pp 3003-3010

[6] Shahzad A, Lee M and Kim H D 2015 New Security Development and Trends to Secure the SCADA Sensors Automated Transmission during Critical Sessions *SYMMETRY-BASEL* **4** 1945-1980

[7] Ghosh A A and Hepsiba D 2015 Model Predictive Control Of Temperature Process In Pasteurization *ICIIECS 2015 - 2015 IEEE International Conference on Innovations in Information, Embedded and Communication Systems* ed S Palehal (Coimbatore: Karpagam Coll Engn) pp245-258

[8] Stepanov F and Stepanov A M 2017 Mathematical modelling of intellectual self-organizing automatic control system: action planning research *Procedia Engineering* Volume **2** 617-622

[9] Ritu R and Mohan B M 2020 General structure of Interval Type-2 fuzzy PI/PD controller of Takagi–Sugeno type *Engineering Applications of Artificial Intelligence* **1** 103- 110

[10] Aruna R and Jaya Christa S T Modeling, system identification and design of fuzzy PID controller for discharge dynamics of metal hydride hydrogen storage bed *International Journal of Hydrogen Energy* **2** 473-479

Application of analytic hierarchy process for selection of heat transfer meters

N V Abdulkina and I P Sitdikova

Almetyevsk State Oil Institute, 2A Lenin Str., Almetyevsk, 423452, Russia

Abstract. At the present time, the process of super-viscous oil deposits development is of current interest among the oil companies. The production of super-viscous oil in the Republic of Tatarstan is carried out using the steam-assisted gravity method. The article proposes a method for solving the problem of selecting a heat meter devices on sites of the Ashalchi field for the production of super-viscous oil. Analytic hierarchy process was used and the parameters required for the optimal selection of heat meters have been determined in the course of investigations. Based on these parameters, a hierarchy for the heat metering unit and a pairwise comparison matrix for the main criteria of the flowmeters have been made. As a result of the analysis, the optimal flowmeter in terms of the chosen criteria has been selected.

1. Introduction

In order to make management decisions and predict possible results, the decision-maker is faced with a complex system of mutually dependent components (resources, desired goals or source material, devices or groups of devices, etc.) that need to be analyzed.

To solve decision-making problems, various mathematical tools of the systems approach are used in various fields. In particular, the universal analytic hierarchy process is now widely used, which interactively allows finding such an alternative that predicts the solution of the problem in the best way. This method (Analytic Hierarchy Process) was proposed by the American scientist Thomas Saaty in 1970, which is rapidly-growing and widely used for comparing objects and solving practical problems of complex control systems. [1]. In order to use this method, it is necessary to systematize large amounts of information; namely, it is necessary to consider several solutions, select a suitable criterion, and determine conditions for problem solving.

The analytic hierarchy process can be presented as the most reasonable solution in a challenging environment of multi-criteria problems with hierarchical structures that include tangible and intangible factors, in contrast to the approach based on linear logic elements. The principle of the method is to divide the problem into simple components by building multi-level hierarchy that combines all the required components, to analyze source data at each level of processing and to compare it. As a result, we get numerical values of the various hierarchy elements interaction, based on which the alternative option is preferred [2, 3].

IOP Conf. Series: Materials Science and Engineering **860** (2020) 012011 doi:10.1088/1757-899X/860/1/012011

The main objective of the hierarchy is to evaluate the higher levels at the stage of interaction between different levels of the hierarchy under consideration, and to exclude direct dependence on elements at these levels.

2. Problem statement

Let us review the analytic hierarchy process application when selecting heat meter devices used in the production of super-viscous oil at the Ashalchi field.

The Ashalchi field experimental program has been developed since 2006. The analysis of various thermal treatments for the development of super-viscous oil deposits shows that the most reliable results are obtained when using steam-assisted gravity technology, which includes processes: combination of gravity drainage and steam drive. Based on foreign experience of horizontal wells used for the production of heavy high-viscosity oil, TATNEFT has developed its own technology with the use of paired horizontal wells at a distance of 5-6 m, strictly one above one another. The steam-assisted gravity effect is achieved through a pair of horizontal wells, the upper of which is designed for injecting steam, and the lower for selecting super-viscous oil. Paired horizontal wells are used to drill areas of deposits with an oil-saturated thickness of more than 10-12 meters. Areas of deposits with an oil-saturated thickness of less than 10 meters are drilled with horizontal wells to have a cyclic steam effect.

Saturated vapor is used as a heat-transfer medium on exposure to the steam-assisted gravity. When injecting into the reservoir, the vapor goes up to the roof, making a high-temperature "chamber" and heating the viscous oil. The mechanism of steam drive is that due to the low density of vapor compared to other phases, the "steam chamber" expands upwards and sideward. Discharge vapor runs to the upper part of the reservoir. At the boundary of the "steam chamber", the vapor is condensed in the process of thermal energy transfer to the environment, and the heated oil is driven out by the condensed vapor and flows down under its own weight in the top-down direction. The "steam chamber" space expands as long as the movable oil and condensed vapor are taken from its base [4].

Thus, the increase in reservoir recovery in the process of steam injection is achieved by reducing the oil viscosity, which helps to improve the reservoir coverage.

Oil is always in contact with a high-temperature "steam chamber", i.e. heat losses are minimal, which makes this method profitable from an economic point of view [5].

The steam used for injection is generated by several boiler plants. The original working substance for steam generation in boiler plants is water, and the original energy carrier is natural gas. The steam is the main heat-transfer medium in this production technology, a special place is taken by means of its metering and flow rate control of the steam injected into injection wells. The energy efficiency of this technology is largely determined by the measurement accuracy, which depends on both the metering principle and the quality of the steam flowmeter.

Measuring steam flow rate is a very specific task. It is primarily determined by the parameters of the heat-transfer medium in the steam pipelines: by high temperatures and pressure. The vortex flow meters are the most common among the available devices for measuring the flow rate of heat-transfer fluid.

3. Theory

Suppose we have a number of alternatives (solutions): B_1, B_2, ... B_k. Each of the alternatives is evaluated by a list of criteria: K_1, K_2, ... K_n. The criteria are compared in pairs. The result can be represented as an antisymmetric matrix. The matrix element a (i,j) is the intensity of hierarchy element i relative to the hierarchy element j. For pairwise comparison, the author of the Saaty method proposed a special rating scale consisting of five main and four intermediate judgments:

- equal superiority – 1;
- moderate superiority – 3;

- a significant superiority – 5;
- strong superiority – 7;
- very strong superiority-9;
- Even-numbered scores are given in intermediate cases: 2, 4, 6, 8.

Then a matrix is made. In the process of filling the matrix, if element *i* is more important than element *j*, then the cell (*i, j*) corresponding to row *i* and column *j* is filled with an integer, and the cell (*j, i*) corresponding to row *j* and column *i* is filled with a reciprocal number (fraction).

In addition, pairwise comparison of options for each criterion is performed in the same way as it was done for the criteria, and the corresponding tables are filled in. For each table, the consistency of local priorities is checked by measuring three characteristics [6, 7].

Let us review three types of vortex flow meters that are most frequently used in the Ashalchi field. It is necessary to select the optimal flowmeter in the cost-efficiency ratio and to determine the parameters for the most appropriate option.

4. Experimental results

Let us build a hierarchy for the heat metering unit from the top through intermediate criteria to the lower level (Figure 1).

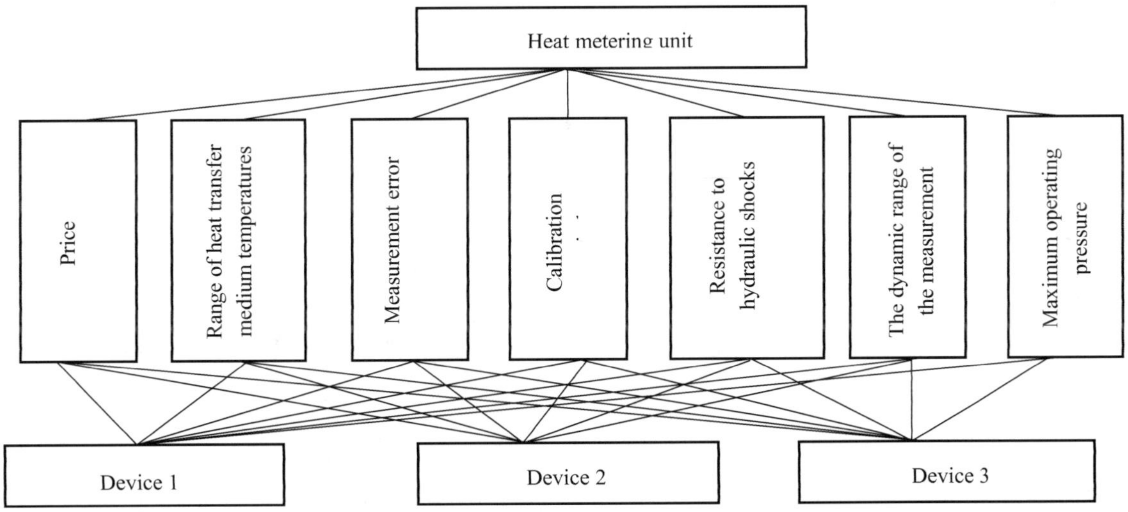

Figure 1. Hierarchy for the heat metering unit

We perform a comparative analysis of each criterion in pairs, using a special rating scale consisting of five main definitions and four intermediate definitions (Table 1).

The relative superiority scale contains the reciprocal values 1/3, 1/5, 1/7, 1/9 and intermediate values 1/8, 1/6, 1/4, 1/2. Since the factors are compared twice with each other in a sequential search of all possible pairs, only the part lying below or above the diagonal is filled in to make up the matrix [8, 9].

IOP Conf. Series: Materials Science and Engineering **860** (2020) 012011 doi:10.1088/1757-899X/860/1/012011

Table 1. Scale of relative superiority

Importance	Definition	Comment
1	Equal superiority	Two options have identical superiority
3	Moderate superiority	Moderate superiority of option A over option B
5	Strong superiority	Strong superiority of option A over option B
7	A significant superiority	Significant superiority of option A over option B
9	Very strong superiority	Very strong superiority of option A over option B
2,4,6,8	Intermediate decisions between two definitions	They are used as an intermediary solution

We build a symmetric pairwise comparison matrix (Table 2).

Table 2. Symmetric pairwise comparison matrix

Criteria	Price	Range of heat transfer medium temperatures	Measurement error	Calibration period	Resistance to hydraulic shocks	The dynamic range of the measurement	Maximum operating pressure	w^*	w^*_{norm}
Price	1	2	3	4	5	7	3	3.06	0.306
Range of heat transfer medium temperatures	1/2	1	5	5	5	7	5	3.00	0.300
Measurement error	1/3	1/5	1	5	5	7	5	1.78	0.178
Calibration period	1/4	1/5	1/5	1	2	5	2	0.79	0.079
Resistance to hydraulic shocks	1/5	1/5	1/5	1/2	1	5	4	0.7	0.070
The dynamic range of the measurement	1/7	1/7	1/7	1/5	1/5	1	2	0.30	0.030
Maximum operating pressure	1/3	1/5	1/5	1/2	1/4	1/2	1	0.36	0.036
Total	2.75	3.94	9.74	16.20	18.45	32.50	22	9.99	0.999

Let`s measure parameters: standardized priority vector, maximum Eigen value, consistency index. The value of the native factor $w^*$is determined by the formula:

$$w^* = \sqrt[n]{a_{i1} \cdot a_{i2} \cdot a_{i3} \cdot \ldots a_{in}}, \quad (1)$$

where a_{ij} - coefficients of the symmetric pairwise comparison matrix *j=0...n*; *n* - dimension of the symmetric matrix.

$w_1^* = \sqrt[7]{1\cdot 2\cdot 3\cdot 4\cdot 5\cdot 7\cdot 3} = 3.06$, $w_2^* = \sqrt[7]{1\cdot 0.5\cdot 5\cdot 5\cdot 5\cdot 7\cdot 5} = 3.00$,

$w_3^* = \sqrt[7]{0.33\cdot 0.2\cdot 1\cdot 5\cdot 5\cdot 7\cdot 5} = 1.78$, $w_4^* = \sqrt[7]{0.25\cdot 0.2\cdot 0.2\cdot 1\cdot 2\cdot 5\cdot 2} = 0.79$,

$w_5^* = \sqrt[7]{0.2\cdot 0.2\cdot 0.2\cdot 0.5\cdot 1\cdot 5\cdot 4} = 0.70$, $w_6^* = \sqrt[7]{0.14\cdot 0.14\cdot 0.14\cdot 0.2\cdot 0.2\cdot 1\cdot 2} = 0.30$,

$w_7^* = \sqrt[7]{0.33\cdot 0.2\cdot 0.2\cdot 0.5\cdot 0.25\cdot 0.5\cdot 1} = 0.36$.

The standardized priority vector is determined by the formula:

$$w^*_{норм} = w_1^* / \sum_{i=1}^{n} w_i^* \quad . \tag{2}$$

The maximum eigenvalue of the considered symmetric matrix (Table 2): $\lambda^*_{\max} = 2.75\cdot 0.306 + 3.94\cdot 0.3 + 9.74\cdot 0.178 + 16.2\cdot 0.079 + 18.45\cdot 0.07 + 32.5\cdot 0.03 + 22\cdot 0.036 = 8.08$

We check if the condition is met:

$$\lambda^*_{\max} \geq n, \tag{3}$$

This condition is met: 8.08>7.

The consistency index is calculated using the formula:

$$I_c = \frac{\lambda^*_{\max} - n}{n-1}, \tag{4}$$

$$I_c = \frac{8-7}{6} = 0.18$$

Given that $Ic<0.2$, the discrepancy between the real and ideal comparison schemes is within the allowed range, so the consistency condition is met.

At the next stage of solving the problem of choosing a flowmeter, we make up pairwise comparison matrix of parameters according to the reviewed criterion (Table 3-9).

Table 3. "Price" comparison matrix

Device	1	2	3	$w*$	$w*_{norm}$
1	1	5	7	3.27	0.739
2	1/5	1	2	0.74	0.167
3	1/7	1/2	1	0.41	0.093
Total	1.34	6.50	10	4.42	0.999

The maximum eigenvalue of the reviewed symmetric matrix (Table 3): $\lambda^*_{\max} = 1.34\cdot 0.739 + 6.5\cdot 0.167 + 10\cdot 0.093 = 3.01$.

IOP Conf. Series: Materials Science and Engineering **860** (2020) 012011 doi:10.1088/1757-899X/860/1/012011

We check if condition (3) is met. Given that 3.01>3, the condition is met.
The consistency index is determined by the formula (4): $I_c = 0.005 < 0.2$.

Table 4. "Range of heat transfer medium temperatures" comparison matrix

Device	3	1	2	w^*	w^*_{norm}
3	1	5	7	3.27	0.700
1	1/5	1	7	1.12	0.240
2	1/7	1/7	1	0.27	0.058
Total	1.34	6.14	15	4.66	0.998

According to Table 4: $\lambda^*_{\max} = 1.34 \cdot 0.700 + 6.14 \cdot 0.240 + 15 \cdot 0.058 = 3.28 > 3$. $I_c = 0.14 < 0.2$.

Table 5. "Measurement error" comparison matrix

Device	3	1	2	w^*	w^*_{norm}
3	1	2	2	1.59	0.494
1	1/2	1	2	1	0.311
2	1/2	1/2	1	0.63	0.196
Total	2	3.50	5	3.22	1.000

According to Table 5: $\lambda^*_{\max} = 2 \cdot 0.494 + 3.5 \cdot 0.311 + 5 \cdot 0.196 = 3.06 > 3$. $I_c = 0.03 < 0.2$

Table 6. "Calibration period" comparison matrix

Device	3	2	1	w^*	w^*_{norm}
3	1	2	2	1.59	0.485
2	1/2	1	3	1.14	0.348
1	1/2	1/3	1	0.55	0.168
Total	2	3.33	6	3.28	1.000

According to Table 6: $\lambda^*_{\max} = 2 \cdot 0.485 + 3.33 \cdot 0.348 + 6 \cdot 0.168 = 3.14 > 3$. $I_c = 0.07 < 0.2$.

Table 7. "Resistance to hydraulic shocks" comparison matrix

Device	1	2	3	w^*	w^*_{norm}
1	1	2	2	1.59	0.494
2	1/2	1	2	1	0.311
3	1/2	1/2	1	0.63	0.196
Total	2	3.5	5	3.22	1.000

According to Table 7: $\lambda^*_{\max} = 2 \cdot 0.494 + 3.5 \cdot 0.311 + 5 \cdot 0.196 = 3.06 > 3$. $I_c = 0.03 < 0.2$.

Table 8. "Dynamic range of measurement" comparison matrix

Device	1	2	3	w^*	w^*_{norm}
1	1	7	3	2.76	0.683
2	1/7	1	1/2	0.41	0.101
3	1/3	2	1	0.87	0.220
Total	1.47	10	4.5	4.04	0.943

According to Table 8: $\lambda^*_{\max} = 1.47 \cdot 0.683 + 10 \cdot 0.101 + 4.5 \cdot 0.220 = 3 = 3$. $I_c = 0 < 0.2$.

Table 9. "Maximum operating pressure" comparison matrix

Device	1	2	3	w^*	w^*_{norm}
1	1	7	5	3.27	0.766
2	1/7	1	1/5	0.3	0.070
3	1/5	5	1	1	0.234
Total	1.34	13	6.20	4.27	1.070

According to Table 9: $\lambda^*_{\max} = 1.34 \cdot 0.766 + 13 \cdot 0.070 + 6.2 \cdot 0.234 = 3.38 > 3$. $I_c = 0.19 < 0.2$.

5. Discussion of results

We have made pairwise comparison matrices for alternative levels. We have determined native factors, standardized vectors, and Eigen values and confirmed the consistency of the matrices.

Next, we need to synthesize the final solutions. To do this, we evaluate the standardized vectors with the selected weights of the criteria. As a result, we get the expression:

$$\begin{pmatrix} 0.739 & 0.240 & 0.311 & 0.168 & 0.494 & 0.683 & 0.766 \\ 0.167 & 0.058 & 0.196 & 0.348 & 0.311 & 0.101 & 0.070 \\ 0.093 & 0.700 & 0.494 & 0.485 & 0.196 & 0.220 & 0.234 \end{pmatrix} \cdot \begin{pmatrix} 0.005 \\ 0.140 \\ 0.030 \\ 0.070 \\ 0.030 \\ 0.000 \\ 0.190 \end{pmatrix} == \begin{pmatrix} X_1 \\ X_2 \\ X_3 \end{pmatrix} = \begin{pmatrix} 0.219 \\ 0.062 \\ 0.198 \end{pmatrix}$$

The values obtained from the selection of heat flowmeters using the analytic hierarchy process are shown in the Bar chart (Figure 2).

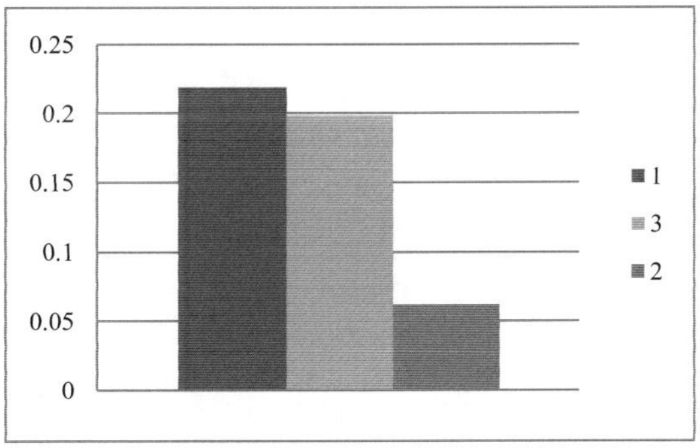

Figure 2. Bar chart of comparison between heat flowmeters

6. Summary and conclusions

Result analysis shows that the element of the standardized vector X_1=0.219 has the maximum score. This means that the flowmeter of the first option is the most appropriate in terms of the set of reviewed criteria.

We suggest the appropriate solution to the problem of choosing flow meters for heat metering, which allows efficient generation of a system of automation tools.

References

[1] Samkov A, Sokolova N, Migovich N and Ryzhikov N 2014 Energy-saving equipment and technologies samples selection methods *Energy: economics, technology, ecology* **36 (2)** 57-62 (in Russian)
[2] Saati T 2015 About an intangible approach measurement to relative measurements based on the main eigenvector of the matrix of pairwise comparisons *Cloudofscience* **2 (1)** 5-39
[3] Korobov V and Tutygin A 2010 Advantages and disadvantages of the hierarchies analysis method *News of the Russian State Pedagogical UniversityA.I. Herzen* **122** 108-115 (in Russian)
[4] Khisamov R, Musin M, Musin K, Fayzullin I and Zaripov A 2013 *Laboratory and pilot works on the extraction of super-viscous oil from the mine results Generalization* (Kazan "Fan" Academy of Sciences of the Republic of Tatarstan)
[5] Muslimov R 2013 Increasing the role of unconventional types of hydrocarbon raw materials for long-term sustainable economic development (on example of the Republic of Tatarstan) *Geo-resources* **54 (4)** 45-50
[6] Vertakova Yu 2005 Management decisions, development and selection (*M: KNORUS*)
[7] Ivanco M, Hou G and Michaeli J 2017 Sensitivity analysis method to address user disparities in the analytic hierarchy process *Expert Systems with Applications* **90** 111-126 DOI: 10.1016/j.eswa.2017.08.003.
[8] Pairwise comparison matrices in decision making 2015 *Lecture Notes in Economics and Mathematical Systems* **677** 29-90 DOI: 10.1007/978-3-319-08323-0_2
[9] Mitroshin P 2016 Hierarchy analysis method for selection of expert groups in competence-based model of education *Components of Scientific and Technological Progress* **30 (4)** 32-37

IOP Conf. Series: Materials Science and Engineering **860** (2020) 012012 doi:10.1088/1757-899X/860/1/012012

Four-way shell-and-tube heater with ultrasonic generator

D Sidina[1], R Vakhitova[1] and V Kuneevsky[2]

[1] Almetyevsk State Oil Institute (ASOI), 2a, Lenina st., Almetyevsk, 423450, Russia
[2] Tatar Oil Research and Design Institute (TatNIPIneft) of PJSC TATNEFT, 32, M. Djalil st., Bugulma, 423230, Russia

E-mail: daria-mazankina@yandex.ru

Abstract. This article offers a heat-exchanger of improved design which allows reducing time between overhauls and maintenance. Since to date, the challenge of providing enterprises and public utility installations with heat-transfer media of required parameters is the most pressing. The collection chamber is equipped with a variable frequency ultrasonic radiator. Bays are installed with the possibility of extracting them. The case is made as one-piece and enclosing all bays. Heat exchange sections are made in the form of sectors. The dividers are equipped with windows that provide the flow of the annular medium in the opposite direction to the flow of the pipe medium. Tests were conducted on a four-way shell-and-tube heat exchanger. The heater was disassembled and inspected, after inspection it was found that scale on the heating surface was completely absent. As a result of the analysis, we can conclude that the design efficiency of the four-way shell-and-tube heater. When working on soft water and ultrasonic protection, there is no carbonate scale on the heating surface. When developing a combined vertical heat exchanger, a comparison was made with high-speed water sectional heat exchangers. The task of creating a simple and cheap design of the heat exchanger was performed by simplifying the design of the heat exchange sections and reducing the metal consumption. The stated technical problem was solved by improving the multi-pass heat exchanger with an ultrasonic emitter, which has a casing and heat-exchange sections in the form of tube bundles with tube sheets at the ends.

1. Introduction

Over the past decades, there has been a sustainable growth of energy consumption in many engineering processes of enterprises and a sharp rise in the cost of energy costs and production prime cost. This is due to many factors, which include the following: outperforming growth rates of fuel and electricity tariffs, lack of funds for technical upgrade and retrofit of supply systems for the industry enterprises, a significant reduction in the load on production capacities, etc.

The low efficiency of thermal energy and fuel consumption is largely due to the lack of a comprehensive procedure for heating system research. For this reason, it is impossible to justify the optimal parameters and modes for operation thereof, as well as to find reserves for heat and fuel economy. Reliable and cost-effective provision for enterprises and public utility installations with heat-transfer media of required parameters and introduction of new high-tech solutions for heat generation will guarantee the generation of cheap heat being one of the primary objectives in the energy industry and allows lowering utility tariffs for households.

Nowadays one of the pressing challenges in the energy industry is to provide enterprises and public utility installations with heat-transfer media of required parameters, together with the use of new high-tech solutions [1, 2, 3, 4, 5,6].

2. Problem statement

The primary objective was to develop a simple and low-cost design of a heat exchanger by simplification the bay design and reducing the metal consumption along with overhaul life and maintenance periods [7, 8, 9, 10].

The stated engineering challenge is solved by improving a multi-pass heat exchanger with an ultrasonic radiator with a casing, identical bays in the form of tube banks with tube arrays at the ends. The bays are connected to the common collection chambers of the tube medium and the casing by means of flanges, therewith transversely to the tube arrays in the collection chambers of the tube medium, partitions are installed to form compartments communicating adjacent heat exchange sections (bays).

The novelty is that the collection chamber of the tube medium is additionally equipped with a variable frequency ultrasonic radiator, and the bays are inserted with the possibility of extracting into the casing. The casing is made as one-piece and enclosing all bays, which are made in the form of sectors using separators transversely to the tube arrays. Moreover, the separators are equipped with apertures to ensure the flow of the annular medium in the direction opposite to the tube medium flow.

3. Research results

The multi-pass heat exchanger with the ultrasonic radiator 1 is shown in Figure 1 and contains identical bays 2 (section view C-C) in the form of tube banks 3 (section view B-B) with tube arrays 4 and 4' at the ends connected to common collection chambers 5 of the tube medium using flanges 6, 7. Therewith, transversely to the tube arrays 4 in the collection chambers of the tube medium 5, partitions 8 are installed, forming compartments 9 (section views A-A and B-B) containing adjacent bays 2 (section view C-C). The collection chamber of the tube medium 5 is supplemented with an ultrasonic radiator 1 of variable frequency. The heat bays 2 (section CC) are inserted with the possibility of extraction into a single casing 10 with flanges 7 and are made in the form of sectors using perpendicularly installed tube sheets 4 dividers 11. The separators 12 (Figure 4.2) are equipped with apertures 11 to ensure the flow of the annular medium in the direction opposite to the tube medium flow. The ultrasonic radiator 1 (section AA) frequency change depends on the amount and type of salts dissolved in the tube medium.

The multi-pass heat exchanger with an ultrasonic radiator (Figure 1) works as follows. The tube medium flow (for example, cold water) passes through the collection chambers of the tube medium 5, tube banks 3 with tube arrays 4 and 4', the upper 4 of which is installed between the flanges 6 and 7, and the lower array 4' is hermetically inserted owing to the seal 13 in a one-piece casing 10, based on its annular limiter 14. In collection chambers of the tube medium 5 (section views C - C and B - B), partitions 8 are installed, forming compartments 9 connecting the hydraulically adjacent bays 1 in a way that they direct the flow of the tube medium through banks 3 of bays 2 (section C - C) only in one direction (clockwise or counterclockwise).

Therewith, the partitions 8 absorb axial loads during thermal compression occurring in the heat exchange sections 2 during heat exchange. At the same time, the annular medium flow in one of the collection chambers 5 (Figure 1) is processed by ultrasonic waves generated by ultrasonic radiator 1 to slow down the process of scale and sediment deposition by 8-10 times in pipe bundles 2. Therewith, ultrasonic waves act directly on the tube medium resulting in an intense deposition effect in this collector chamber 5 (the greatest positive effect is achieved with installation of an ultrasonic radiator in the lower collector chamber of the tube medium 5), the passage cross-section of which greatly exceeds the pipe clear opening 3, which increases the overhaul life.

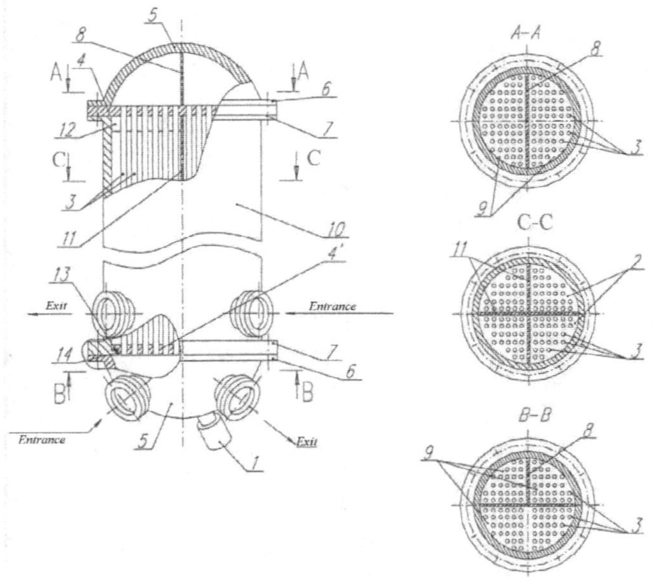

Figure 1. Four-way shell-and-tube heat-exchanger

In laboratory conditions, the tube medium with a different salt content was examined by treating with ultrasonic waves and identifying which frequency provides the maximum deposition effect. Then, after analyzing the annular medium passing through the multi-pass heat exchanger, the ultrasonic radiator 1 frequency was set in the most effective range for salt deposition in the collection chamber 5, which increases the overhaul life and maintenance period by 1.2 -1.5 times with regard to the uncontrolled ultrasonic radiator, i.e., in general, 10 - 15 times. Therewith, annular medium flow (for example, hot water) passes through a one-piece casing 10 through the windows 12 (Figure 2) of the separators 11 along the bays 2 (section view C - C) in the direction opposite to the tube medium flow. The results are that heat exchange takes place, i.e. one of the media is heated, and the other is cooled. Therewith, the number of bays 2 required for heat exchange is determined by the number of sectors, whereon the one-piece casing 10 is divided by separators11 (Figure 1).

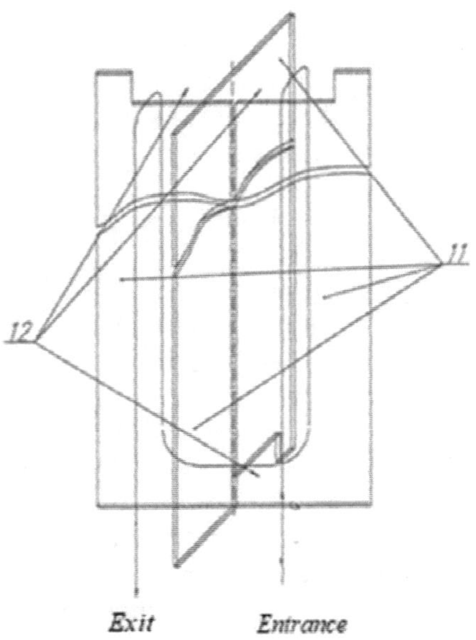

Figure 2. Water motion pattern in the annular medium

When carrying out the necessary maintenance, the upper collection chamber 5, tube banks 3 with tube arrays 4 and 4' and separators 11 are removed and the collection chamber with an ultrasonic radiator is descaled. When inspecting the tubes 3 the decision is taken on the following:

first, with an insignificant clogging of the tube banks 3 the washing and cleaning thereof is to be performed;

second, with a significant clogging of tube banks 3, the tube banks 3 with the tube arrays 4 and 4' and the separators 11 are to be replaced with new or previously cleaned tube banks 3 with the tube arrays 4 and 4' with the separators 11, which significantly reduces the downtime of the device and equipment technologically connected with this device [11, 12].

Where after, the multi-pass heat exchanger is assembled in reverse order and put into operation.

Easy-to-assemble design of the heat exchanger is cost-effective and simple to manufacture and maintain with a longer overhaul life and maintenance period as compared to counterparts due to treatment of the tube medium by ultrasonic waves (in practice, 10 ... 15 times).

To examine the effectiveness of the ultrasonic protection of the compound shell-and-tube heater manufactured by the PSP Teplogaz LLC, an ultrasonic single-channel generator of the Agropribor research and development enterprise was used.

For over 20 years, the Agropribor company has been analyzing ultrasonic methods, as well as mass-producing the "Volna" ultrasonic unit for these purposes. In boiler facilities of Moscow, Volgograd, Saratov, Orenburg, Vologda, Voronezh, Sverdlovsk regions and in other Russian regions, as well as in neighboring countries (Baltic States, Ukraine, Belarus), these units are successfully operated. With high efficiency, the "Volna" ultrasonic units are used to prevent scale formation in boilers of various capacities, as well as in waste-heat exchangers, in high-speed heat exchangers, in compressor coolers and other equipment.

The use of one unit provides an economic effect of 6-19 thousand rubles.

To date, a huge number of "Volna" units are in operation (more than 6 thousand units). Monitoring of operation of the "Volna" ultrasonic units made it clear that after the introduction of ultrasound, the scale deposition, as previously formed in boilers and heat exchangers, begins to peel off and then settle like flakes or loose mud. There is no further forming of scale. The mud produced in the course of operation is removed by periodic purges. During a preventive inspection, the boiler can be flushed with a jet of water. The water jet pressure shall be 0.2-0.3 MPa. Thereby, there is no longer any necessity in time-consuming and expensive equipment cleaning and all environmental issues associated with this, are resolved as well. If ultrasonic units run continuously, this allows operating the heat-power equipment in a nonscale-deposition mode. This allows keeping the efficiency factor at a level close to an optimal one, and this leads to significant fuel savings. The use of ultrasonic units allows increasing the service life of the equipment and reduces the number of possible emergencies because when exposed to ultrasound, the heated surfaces do not overheat to a great extent, the heat transfer increases and metal corrosion decreases.

When using the ultrasound method, environmental pollution by harmful wastewaters from water treatment plants is minimized.

After a few months of operation of the plants, all capital investments (acquisition, installation, and adjustment) will pay off. The units are easy to operate and maintain. After getting an insight into the design of the installation and the rules for operation thereof, the boiler room personnel can easily enter into operation.

To prevent scale formation in heat exchange equipment and boiler units, it is recommended to use "Volna" ultrasonic units.

By design, the "Volna" ultrasonic unit is presented as a pulse generator complete with two magnetostrictive converters [13]. The generator is needed to create shock electrical pulses, converters turn shocks into mechanical vibrations of ultrasonic frequency. To provide vibration transmission to the boiler unit, converters are welded to the boiler external surfaces. Using electric welding, these are

IOP Conf. Series: Materials Science and Engineering **860** (2020) 012012 doi:10.1088/1757-899X/860/1/012012

welded at certain points. The number of converters and their location points are selected depending on the design and the boiler unit capacity, and on the average amount of scale deposited during service.

4. Experimental results

The testing of the four-way shell-and-tube heat exchanger was conducted in the building No. 4 of the Perm State Technical University, wherein it was put into operation to heat the water fed to the hot water supply system.

Before putting into operation the heater was visually inspected. There was no scale on the heating surface.

The heater was heated by the system water of the Perm hydro-technical utilities. The subject for heating was the tap water of the Chusovsky water-intake facility of the Perm municipal water system.

The testing was conducted on November 1, 2002. The mode of operation of the heater during the test is described in Table 1.

Table 1. The mode of operation of the heater during the test

Month	Calcium content, mg-eq/l	Hydrocarbonate alkalinity, mg-eq/l	pH value	Average heating temperature, ^0C
November	5.5	4.06	7.0	70
December	10.0	7.01	7.0	70
January	9.7	7.32	7.0	70
February	11.3	8.44	7.0	70
March	12.2	8.45	7.0	70
April	11.2	7.93	7.0	70
May	1.25	0.61	7.0	70
June	1.9	1.62	7.0	70

Table 2 provides the calculation data for the intensity of carbonate scale formation in the absence of ultrasonic protection.

Table 2. Estimated amount of carbonate scale without ultrasonic protection

Month	Intensity of scale formation, mg/h.m^2)	weight of scale per month, kg/m^3
November	384.17	0.277
December	1206.00	0.897
January	1221.55	0.909
February	1640.78	1.103
March	1773.56	1.320
April	1527.99	1.100
May	13.12	0.010
June	52.35	0.038

The total weight of carbonate deposits, calculated according to Table 2, amounted to 5.654 kg/m $^{1\ 2}$. Which corresponds to the estimated total thickness of the carbonate scale layer at the critical point - 2.262 mm.

On July 1, 2003, the heater was turned off and disassembled. When inspecting the disassembled heater, it was found that there is no scale on the heating surface (the inner surface of the brass tubes) whatsoever. As a result of the work performed, the conclusion on the design effectiveness of the four-way shell-and-tube heater can be made. When operating on untreated water and ultrasonic protection, there is no carbonate scale on the heating surface.

5. Conclusion

When developing a joint vertical heat exchanger, comparisons were made with water-to-water high-speed sectional heat exchangers according to industry standards, and the following objectives were reached:

1. The scale formation and deposition of pipeline corrosion products on both heat exchange surfaces when using acoustic descaling devices, in particular in hot-water supply systems.
2. Facilitated clean-up of heating surfaces from impurities.
3. A compact heat exchanger was developed with overall dimensions in plan view with consideration of service passages, and similar to the overall dimensions of plate heat exchangers.

The developed joint vertical heat exchangers with the same heating surface with water-to-water high-speed sectional heat exchangers have approximately equal weight, metal consumption, and cost, but have a variety of advantages.

Based on the experience of State Unitary Enterprise Mosteploenergo, it is known that acoustic descaling devices generally protect the heating surface of conventional water-to-water high-speed sectional heat exchangers against scale and deposits in heat conductor facilities in Moscow better than plate heat exchangers.

References

[1] Shkulova N, Ivanova and Gusein Ya 2016 Method for increasing the efficiency of the long-running heat-exchanging equipment of the TVG-8M hot water boiler *Journal of Resources of the European North. Technology and development economics* **3** 59 – 69 (in Russian)

[2] Fireushin A and Romanchuk A 2018 Improving the efficiency of heat transfer in apparatuses such as "pipe in pipe" *Oil and gas business***16** **(6)** 65-71 (in Russian)

[3] Galkin P 2015 Application of ultrasound to intensify the heat transfer process in a shell-and-tube heat exchanger *Materials of the VIII International Scientific and Practical Conference of Young Scientists "Actual Problems of Science and Technology - 2015"* pp 140-141 (in Russian)

[4] Sharipov T, Chetvertkova O and Sharafiev R 2015 Study of the influence of structural parameters of transverse partitions on the thermal efficiency of a shell-and-tube heat exchanger *Materials of the VIII International Scientific and Practical Conference of Young Scientists "Actual Problems of Science and Technology - 2015"* pp 155-158 (in Russian)

[5] Novik A and Nezamaev N 2019 Intensification of the heat transfer process in shell-and-tube heat exchangers *Abstracts of the IX scientific and technical conference "Science Week - 2019" St. Petersburg State Technological Institute (Technical University)* pp 27 (in Russian)

[6] Pyazhskiy V, Konovalov D, Dakhin S, Bulygin Yu and Shatskiy V 2018 Mathematical model of heat surface cooling system in the starting mode. *Bulletin of the South Ural State University. Series: Mathematical Modeling. Publisher: South Ural State University (National Research University) (Chelyabinsk)* **11** **(4)** 136-145

[7] Wang Wei, Makeev A and Povorov S 2018 Thermal - hydraulic performance analysis cold side of the plate heat changer using water-waters *Science and Practice Bulletin* **4** **(4)** 263-276

[8] Blagin E, Shimanov A, Anisimov M and Uglanov D 2019 Study of the influence of the scale formation process in heat exchangers for preheating a distillation desalination plant on the efficiency of their work. *Bulletin of the International Academy of Refrigeration. Publisher: Public Organization "International Academy of Refrigeration" (St. Petersburg)* **2** 37-42

[9] Qian Hao, Kudashev S and Plotnikov V 2019 Experimental study on heat transfer of pulsating flow enhanced the plate heat exchanger *Science and Practice Bulletin* **5** **(8)** 81-92

[10] Proskurin Yu, Kazantsev M, Grisha B and Dubrovkin M 2018 RU, Patent No. 177038 2018 (in Russian)

[11] Gnedochkin Yu, Kuneevsky V and Rudenko G 2005 RU, Patent No. 47504 "Heat exchanger" (in Russian)
[12] Kuneevsky V, Gnedochkin Yu, Rudenko G and Dunaev A 2005 RU, Patent No. 44375 2005 "Heat exchanger" (in Russian)
[13] Nikolaevsky N 2002 Ultrasonic method of preventing scale formation *Heat supply magazine* **10** (in Russian)

IOP Conf. Series: Materials Science and Engineering **860** (2020) 012013 doi:10.1088/1757-899X/860/1/012013

The mathematical description of the part as the source data in the computer-aided design process

T G Makarova, A R Minulina, M Yu Filimonova, T A Brodskaya

Almetyevsk State Oil Institute, 2A Lenin street, Almetyevsk, 423452, Russia

E-mail: tngm417@mail.ru

Abstract. The paper reviews the issues of source data generation for automated design of the technological process of machining and geometric information transfer from one design stage to another. When dealing with design issues, the input of source data related to the work-piece and manufacturing conditions plays an important role. The formalized description of these objects is carried out using mathematical models. A drawing of a machine-building part or work-piece is a geometric model of the manufacturing item. However, this model is inconvenient when establishing links in the design of its manufacturing processes. Drawing models are the basis of graphic modules of computer-aided design systems. However, they are inconvenient for achieving process design goals, since the object is known to be surface in machining or assembly process, but not zero-dimensional objects on the plane. It represents a mathematical model of a machine-building part consisting of the basic elements of the form, which can be used in the automated development of the technological process of machining.

1. Introduction

Improving labour efficiency in the production of work-pieces is associated with the introduction of integrated computer-aided design systems. When designing in CAD/CAM/CAE systems, it is very important to automate the process of transferring geometric information from one design stage to another, as well as to establish feedbacks from later design stages to earlier ones.

In the work-piece production process, starting from the structural design and ending with the manufacturing stage, a special role is played by the drawing as the main model. When dealing with design issues, the input of source data related to the work-piece and manufacturing conditions plays an important role. A computer-aided design system efficiency depends on the degree of its automation and software.A subsection

2. Problem statement

Characterization of the manufacturing object (part, assembly unit, or design process) is a challenge and it has not been fully solved yet. The formalized description of these objects is carried out using mathematical models. The challenge is that when building models, their elements should be chosen based on the laws of designing technological processes. A drawing of a machine-building part or work-piece is a geometric model of the manufacturing item. However, this model is inconvenient when establishing links in the design of its manufacturing processes, since the drawing contains the initial elements (segments, arcs, points, etc.), i.e. one-dimensional manifolds on the plane that make up the projection of the part, cross-sections, and sections. Using these manifolds, recognition of the geometric structure of a part is a difficult task to formulate. Also, the drawing implicitly reflects the information required to build the technological process. For example, surface machinability for a given basing structure, the length of the surface, etc. Thus, one and the same part can have an infinite number of mathematical models, depending on the selection of the multiple source elements.

3. Theory

A machine-building part is defined as a material bounded by a number of surfaces or a combination of them, one located against the other, based on the intended service and the most economical manufacturing process.

A solid mathematical model of a part can be represented as:

$$M=(M, R), \tag{1}$$

where M is the manifold surfaces bounding the part;

R – manifold relations between them (surface parameters, quality indicators, shape errors, etc.) [1,2,3].

The geometric model of the part specified in the drawing can be represented as follows:

$$M_{draw}=(M,R), \tag{2}$$

where M is the manifold of initial primitive elements (segments, arc lines, points, etc.);

R – a manifold of relations determining the properties of primitives (type and thickness of lines, segment point coordinates, etc.). For example, you can set a circle with three parameters: coordinates of the midpoint and radius. These relationships define the drawing model of the part. The manifold of relations, in this case, is massive and difficult to formalize.

Solid and drawing models are different models of the same part.

Drawing models are the basis of graphic modules of computer-aided design systems. However, they are inconvenient for achieving process design goals, since the object is known to be surface in machining or assembly process, but not zero-dimensional objects on the plane.

Therefore, when designing the assembly and machining process, such models are not used, since the objects of such processes are parts and surfaces, but not line segments.

Geometric models also differ depending on the goals and stages of design. The solution involves the development of geometric information converters for the transition from one design stage to another. These converters are developed based on relational database management systems, where the source data related to the part is entered in the conversational mode for each surface. In this case, the automated transfer of geometric information from the graphic module to the process design module consists of automatic recognition of surfaces and their geometric properties in the source file made in the CAD module. Currently, the issue of automated information transfer is partially solved[4].

Mathematical models and their components can be presented in various forms: linear models, in the form of differential and finite-difference equations, elements of mathematical statistics and logic, in the form of graphs, simulation, etc. And they can also be divided into analytical and simulation.

Numerous mathematical methods have been developed to solve problems using models. The variety of these methods is caused by requirements applied to the method of solving a specific task. Therefore, the selection of the method comes to an optimization goal: to find a given task solution method that would meet the requirements of the ECM and have the highest accuracy [5].

Many of the methods for finding optimal options are iterative and based on continual improvement of options. Finite methods leading to an extremum in a finite number of steps are used. Deterministic methods include methods with a rigid algorithm. Random search methods or static methods are those in which trial steps or working steps are performed with random elements. In addition to these examples, there are many others.

The logical-algebraic language for describing machine-building parts is universal and allows you to identify all the geometric properties of the part elements that are necessary for the design of the

IOP Conf. Series: Materials Science and Engineering **860** (2020) 012013 doi:10.1088/1757-899X/860/1/012013

technological process. As well as its primitives, which describe the part, coincide with the objects of processing during machining.

Let`s present the geometric model of a part using the logical-algebraic language in the following form:

$$M=(M, R), \tag{3}$$

where M is the manifold surfaces bounding the part and which can be used to "design" the part.;

R is the manifold of relations that we divide into two groups:

1) relations that define the properties of the surface (geometrical parameters and the location parameters);

2) relationships that are formed using logical operations & (and), V (or), - (not).

Any quite simple part can be represented as a combination of elementary elements that are easy to describe mathematically. In the formation of machine-building parts, the following surfaces are most common: plane, cylindrical, conical, spherical, toroidal, composite surface.

A plane in three-dimensional space is defined by three points that do not lie on a single line. Based on the information from the drawing, there are various ways to encode the plane.

The cylindrical surface is defined by seven parameters: coordinates of two mismatched points lying on the axis of rotation; the radius of the circle of the plane intersection that is perpendicular to the axis of rotation, and the cylindrical surface.

A spherical surface is defined by four parameters: coordinates of the sphere midpoint and the radius value.

The toroidal surface is defined by eight parameters: coordinates of the great circle center passing through the small circle centers; the radius of a great circle; the radius of the small circle, the normal vector to the plane of the centers.

The conical surface is defined by eight parameters: coordinates of mismatched points A_1 and A_2 belonging to the rotation axis of a conical surface; radii of circles of planes intersection that are perpendicular to the axis of rotation and passing through the points A_1 and A_2.

A surface of standard form can be obtained by approximating it with splines.

For the analytical equation of a surface based on the parameters of its external representation, the matrix M is used for converting coordinates from a relative system to an absolute one.

$$M = \begin{vmatrix} m_{11} & m_{12} & m_{13} & X_1 \\ m_{21} & m_{22} & m_{23} & X_2 \\ m_{31} & m_{32} & m_{33} & X_3 \\ 0 & 0 & 0 & 1 \end{vmatrix} \tag{4}$$

where $m_{11}= \cos\beta\cos\Upsilon$; $m_{12}= -\cos\beta\sin\Upsilon$; $m_{13}=0$;

$m_{21}= - \sin\alpha \sin\beta \cos\Upsilon+\cos\alpha \sin\beta$; $m_{22}= \sin\alpha \sin\beta \sin\Upsilon - \cos\alpha \cos\beta$, $m_{23}=\sin\alpha \cos\beta$;

$m_{31}= -\cos\alpha \sin\beta \cos\Upsilon - \sin\alpha \sin\Upsilon$; $m_{32}=\cos\alpha \sin\beta \sin\Upsilon - \sin\alpha \cos\Upsilon$; $m_{33}=\cos\alpha \cos\beta$.

Then the inverse matrix M^{-1} is calculated to transform coordinates from the external representation to the internal one:

$$M^{-1} = M_{\Upsilon}^{-1} \cdot M_{\beta}^{-1} \cdot M_{\alpha}^{-1} \cdot M_0^{-1} \tag{5}$$

Let`s take a cylindrical surface (Figure 1) as an example of such a transformation.

External representation:

$A_1 = [x_1^1, x_2^1, x_3^1]$ – the first point on the surface axis;

$A_2 = [x_1^2, x_2^2, x_3^2$- the second point on the surface axis;

R is the radius of the perpendicular section circle.

In its coordinate system, the equation of a cylindrical surface has the form:

IOP Conf. Series: Materials Science and Engineering **860** (2020) 012013 doi:10.1088/1757-899X/860/1/012013

$$\begin{cases} x_1 = t \\ x_2 = R cos\alpha \\ x_3 = R sin\alpha \end{cases} \quad (6)$$

where $0 < \alpha < 2\pi$.

The coordinate transformation matrix M is obtained using sequential movements: parallel translation of the $00^1 = [c_1, c_2, c_3]$ system, consecutive rotations of the absolute coordinate system around the O_{x3} and $O_{\text{ч}2}$ axes to the Υ and β angles.

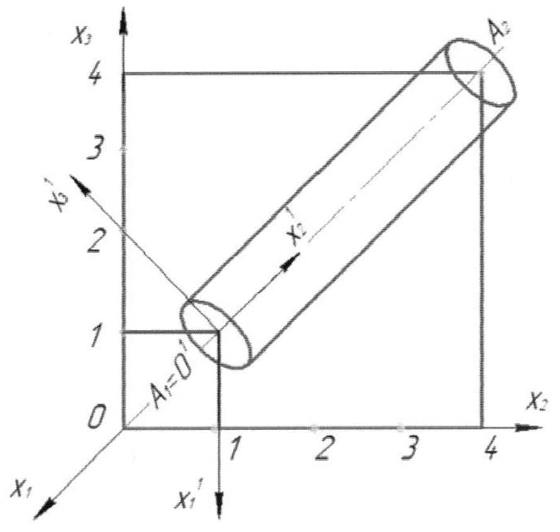

Figure 1. The coordinate transformation when moving from the external representation of a cylindrical surface to the internal one.

As for geometrically complicated parts, having more than ten bounding surfaces, their description becomes massive. Therefore, intermediate spatial form elements are introduced, which are formed from the original set of surfaces. In this case, the part is designed not from the original surfaces, but larger spatial elements: parallelepipeds, cylinders, truncated cones, spheres, tori, pyramids, prisms, etc. Each of the basic elements is characterized by geometric parameters that determine the length of the elements in space and the parameters of their location.

To get the part using these elements, you need to perform the following operations: stretching or compressing along the coordinate axes; parallel translation along coordinate axes; rotation about the coordinate axes.

4. Experimental results

Let's describe a part of the flange type (Figure 2), for example.

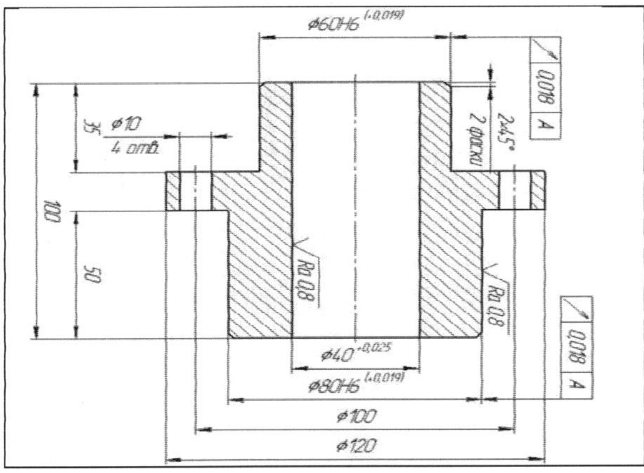

Figure. 2. The flange type part drawing detail

This part consists of several form elements: C01, C02, C03, C04, C05 cylinders (Figure 3).

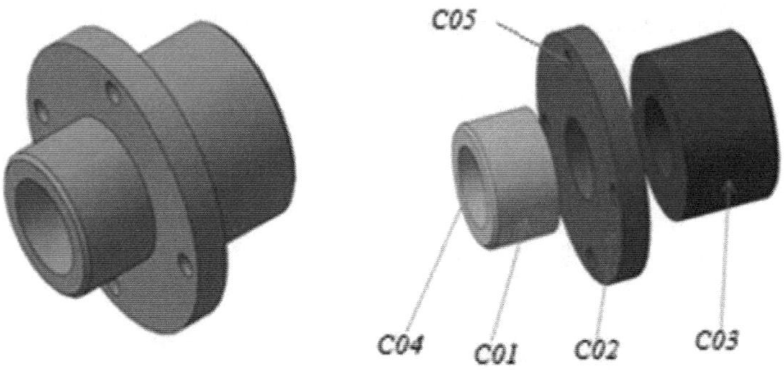

Figure 3. Model of the flange type part

The first letter is the name of the element, and the digits are the number of the element.
You can record the relative location of the part's basic elements:

$$D = C01VC02VC03 - (C04\&C05)$$

Let`s present the basic elements that form the part and their geometric parameters and location parameters in Table 1.

Table 1. The basic elements of the part and their geometrical parameters.

Part number	Geometrical parameters			Location parameters					
	P1	P2	P3	x	y	z	α	β	ϒ
C01	35	60	-	0	0	0	0	0	0
C02	15	120	-	0	0	0	0	0	0
C03	50	80	-	0	0	0	0	0	0
C04	100	40	-	0	0	0	0	0	0
C05	15	10	-	0	50	0	0	0	0

You can build a coordinate transformation matrix M for a cylindrical surface (spatial case). We assume that the absolute coordinate system is obtained as a result of sequential movements:
- parallel translation along the OO'=(C_1; C2; C_3) vector - M_0 matrix;
- rotation of the OX_3^0; OX_2^0coordinate axes, respectively on the M_γ and M_β angles of the matrix.
M_0=M_0 M_γ and M_β ,
where

$$M_0=\begin{pmatrix}1&0&0&C1\\0&1&0&C2\\0&0&1&C3\\0&0&0&1\end{pmatrix};\ M_\gamma=\begin{pmatrix}\cos\gamma&\sin\gamma&0&0\\-\sin\gamma&\cos\gamma&0&0\\0&0&1&0\\0&0&0&1\end{pmatrix};\ M_\beta=\begin{pmatrix}\cos\beta&0&-\sin\beta&0\\0&1&0&0\\\sin\beta&0&\cos\beta&0\\0&0&0&1\end{pmatrix} \quad (7)$$

Let us find the equation of a cylindrical surface for a special case:

$$\begin{aligned}&C01\quad A_1(2;\ 0;\ 1),\ A_2(37;0;36), R=30\\&C02\quad A_3=A_2(37;0;36),\ A_4(52;0;51),\ R=60\\&C03\quad A_4=A_5(52;0;51),\ A_6(102;0;101),\ R=40\\&C04\quad A_6=A_1(2;\ 0;\ 1), A_7(102;0;101),\ R=20\end{aligned} \quad (8)$$

C05 $A_1(2;\ 0;\ 1)$, $A_2(37;0;36)$, $R=30$

The equation of the C01 cylindrical surface in its own coordinate system looks like:

$$X_1 = t\ ;\ X_2 = R\cos\alpha;\ \ X_3 = R\sin\alpha\ ,\ R=30$$

$$\begin{pmatrix} X_1 \\ X_2 \\ X_3 \\ 1 \end{pmatrix} = \begin{pmatrix} \frac{\sqrt{2}}{2} & 0 & -\frac{\sqrt{2}}{2} & 2 \\ 0 & 1 & 0 & 0 \\ \frac{\sqrt{2}}{2} & 0 & \frac{\sqrt{2}}{2} & 1 \\ 0 & 0 & 0 & 1 \end{pmatrix} \times \begin{pmatrix} t \\ 30\cos a \\ 30 \sin a \\ 1 \end{pmatrix} \Rightarrow \begin{cases} x_1 = \frac{\sqrt{2}}{2}t - 15\sqrt{2}\,\sin a + 2 \\ x_2 = 30\cos a \\ x_3 = \frac{\sqrt{2}}{2}t + 15\sqrt{2}\,\sin a + 1 \end{cases} \tag{9}$$

The equation for the first cylinder:

$$\frac{1}{1800}(X_3 - X_1 + 1)^2 + \frac{1}{900}X_2^2 = 1 \tag{10}$$

We find the equation for the C02 element:

$$\begin{pmatrix} X_1 \\ X_2 \\ X_3 \\ 1 \end{pmatrix} = \begin{pmatrix} \frac{\sqrt{2}}{2} & 0 & -\frac{\sqrt{2}}{2} & 37 \\ 0 & 1 & 0 & 0 \\ \frac{\sqrt{2}}{2} & 0 & \frac{\sqrt{2}}{2} & 36 \\ 0 & 0 & 0 & 1 \end{pmatrix} \times \begin{pmatrix} t \\ 60\cos a \\ 60 \sin a \\ 1 \end{pmatrix} \Rightarrow \begin{cases} x_1 = \frac{\sqrt{2}}{2}t - 30\sqrt{2}\,\cos a + 37 \\ x_2 = 60\cos a \\ x_3 = \frac{\sqrt{2}}{2}t + 30\sqrt{2}\,\sin a + 36 \end{cases} \tag{11}$$

Then we get the equation for the second cylinder:

$$\frac{1}{7200}(X_3 - X_1 + 1)^2 + \frac{1}{3600}X_2^2 = 1 \tag{12}$$

In a similar manner we find the equation for the third element C03:

$$\begin{pmatrix} X_1 \\ X_2 \\ X_3 \\ 1 \end{pmatrix} = \begin{pmatrix} \frac{\sqrt{2}}{2} & 0 & -\frac{\sqrt{2}}{2} & 52 \\ 0 & 1 & 0 & 0 \\ \frac{\sqrt{2}}{2} & 0 & \frac{\sqrt{2}}{2} & 51 \\ 0 & 0 & 0 & 1 \end{pmatrix} \times \begin{pmatrix} t \\ 40\cos a \\ 40 \sin a \\ 1 \end{pmatrix} \Rightarrow \begin{cases} x_1 = \frac{\sqrt{2}}{2}t - 20\sqrt{2}\,\sin a + 52 \\ x_2 = 40\cos a \\ x_3 = \frac{\sqrt{2}}{2}t + 20\sqrt{2}\,\sin a + 51 \end{cases} \tag{13}$$

So we obtain the equation for the cylinder C03:

$$\frac{1}{3200}(X_3 - X_1 + 1)^2 + \frac{1}{1600}X_2^2 = 1 \tag{14}$$

We find the equation for the element C04:

$$\begin{pmatrix} X_1 \\ X_2 \\ X_3 \\ 1 \end{pmatrix} = \begin{pmatrix} \frac{\sqrt{2}}{2} & 0 & -\frac{\sqrt{2}}{2} & 2 \\ 0 & 1 & 0 & 0 \\ \frac{\sqrt{2}}{2} & 0 & \frac{\sqrt{2}}{2} & 1 \\ 0 & 0 & 0 & 1 \end{pmatrix} \times \begin{pmatrix} t \\ 20\cos a \\ 20 \sin a \\ 1 \end{pmatrix} \Rightarrow \begin{cases} x_1 = \frac{\sqrt{2}}{2}t - 10\sqrt{2}\,\sin a + 2 \\ x_2 = 20\cos a \\ x_3 = \frac{\sqrt{2}}{2}t + 10\sqrt{2}\,\sin a + 1 \end{cases} \tag{15}$$

We obtain the equation for the fourth cylinder:

$$\frac{1}{800}(X_3 - X_1 + 1)^2 + \frac{1}{400}X_2^2 = 1 \tag{16}$$

$$\begin{cases} \frac{1}{1800}(x_3 - x_1 + 1)^2 + \frac{1}{900}X_2^2 = 1 \\ \frac{1}{7200}(x_3 - x_1 + 1)^2 + \frac{1}{3600}X_2^2 = 1 \\ \frac{1}{3200}(x_3 - x_1 + 1)^2 + \frac{1}{1600}X_2^2 = 1 \\ \frac{1}{800}(x_3 - x_1 + 1)^2 + \frac{1}{400}X_2^2 = 1 \end{cases} \tag{17}$$

5. Discussion of results

The logical-algebraic model of the part (17) allows you to determine the necessary properties of surfaces required for the design of the technological process of machining. Based on the obtained universal model, it is possible to change the parameters of the studied part in a wide range quickly, without significant costs and to automate the processes of transferring geometric information at different design stages using computer technologies. This allows working out the design of the part for producibility at the early stages of design.

6. Summary and conclusions

Achieved part model is required for solving numerous issues of computer-aided design of the machining process. For example, when preparing initial information for design, generating a database for technological purposes, keeping it up-to-date and searching for information, selecting the structure and calculating process parameters, as well as optimizing technological processes.

References

[1] Kuzmin V V and Skhirtladze A G 2008 *Mathematical modeling of technological processes of assembly and machining of machine-building work-pieces* (Moscow: High school)

[2] Calyculin S Yu, Kuzmin V V et al 2019 Development of a mathematical model of a part based on graphic system models *Engineering technologies and systems* **29 (1)** 67-76

[3] Calyculin S Yu and Turbeyeva T B 2018 Methods of converting two-dimensional drawings of parts to a table view *Automation and control in mechanical engineering* **3 (32)** 16-20

[4] T. B. Turbeyeva T B and Volkova G D et al. 2018 Conceptual modeling of design processes for geometrically complicated machine-building work-pieces *Automation and control in mechanical engineering* **2 (31)** 11-22

[5] Minulina A R and T. G. Makarova T G 2018 M odeling the reliability of the failure control system and recovery of oil pumping stations *Scientific life* **11** 24-30

IOP Conf. Series: Materials Science and Engineering **860** (2020) 012014 doi:10.1088/1757-899X/860/1/012014

Forecasting technical state and efficiency of electrical switching devices at electric complexes in oil and gas industry

T V Tabachnikova [1], E I Gracheva [2], O V Naumov[3] and A N Gorlov[3]

[1] Almetyevsk State Oil Institute, 2 Lenin Street, Almetyevsk, 423452, Russia
[2] Kazan State Power Engineering University, 51 Krasnodelskaya Street, Kazan, 420034, Russia
[3] South-West State University, 94, 50 old October Street, Kursk, 305040, Russia

E-mail: tvtab@mail.ru

Abstract. This work is topical, since it allows predicting the physical state of the contact joints of low-voltage switching devices and to avoid their destruction, which occurs due to mechanical and chemical effects on the contacts during operation, leading to an increase in the transition resistance and overheating of the contacts. The following criteria were adopted as serviceability criteria for the contact connections of switching devices: the maximum temperature of their heating, the maximum value of the resistance of the pole of the device and its parts, as well as the excess of the temperature of the contact heated part over the ambient temperature. The authors obtained graphs of the dependence of the resistance value of the contact connections of magnetic starters and circuit breakers on the ambient temperature and load factor. The article determines the analytical dependence of the resistance of the magnetic starter contact connections on the ambient temperature. It also features mathematical models that take into account various operating conditions of switching devices, the input values of which are refined using regression analysis.

1. Introduction

During the operation contact connections of low-voltage switching devices are exposed to mechanical and chemical influences which lead to an increase in their transient resistance, and as a result, by the action of the load current, to their overheating and subsequent destruction. The development rate of defects depends on the design of the contact connection, its location and the intensity of external influences. The time-lag between defect occurrence and the emergency failure of contact is from several months to several years. The main control of contacts is the measurement of their transient resistance to direct current. Contact resistance is measured using M-246 micrometers. The ammeter-voltmeter method measures resistance indirectly. The single bridge (Wheatstone bridge) and double bridge (Thomson bridge) methods are widely used [formula 1, 2].

Electric thermometers, thermal candles, thermal stencils, and thermal indicators were used in the energy industry to control the contact connections of conducting parts before. Nowadays, thermal imagers are being successfully used for these purposes. Now there is a possibility not only to measure the temperature of individual points but also to observe the thermal conditions of electrical installations.

The heat amount released on defective contact depends on the square, the current flowing through it, the transient resistance, and the time. The thermal energy released when the current flows through the contact's transient resistance is transmitted as thermal radiation to the environment, to the conjugated conducting parts and insulating devices. The temperature of contact connections depends on many factors including the area of their surfaces, the heat transfer coefficients of conjugated conducting parts, and the environmental parameters (temperature). In the formula [2] heating temperature limit (excess of heat temperature above the ambient temperature) is adopted as the main criterion of contact connection's intactness at nominal current I_{nom} and in formula [3] the value limit of phase (pole) resistance of the device and its parts. These two intactness criteria of switching devices are equivalent since the heating of contact connections occur before mentioned in formula [2] temperature by the flow of current load $I_l=I_{nom}$ and energy released on the device resistance R_d, a power source of which is proportional to the resistance of conducting system.

The exceed of contact's heated part above the ambient temperature is acceptable as a criterion for evaluating the contact group's state.

2. Materials and methods

The estimated temperature excess is calculated from the current load I_l and the nominal current I_{nom} of the device:

$$t_{norm} = t_{nom}\left(\frac{I_1}{I_{nom}}\right)^2, \tag{1}$$

where t_{norm} is the normalized value of heating temperature excess for the controlled object (contact connection).

The estimated calculated value of the contact resistance at the normal conversion temperature is calculated using this expression:

$$R_{norm} = R_m \frac{K + t_{nom}}{K + t_m}, \tag{2}$$

where R_m is the measured resistance value at t_m, [Om]; K is copper coefficient equal to 235; t_{nom} is the nominal temperature equal to 40 °C [2]; t_m is the ambient temperature at which the resistance measurements were made R_m, °C.

The electrical resistance of the contacts is calculated by the expression:

$$R = R_{nom} + \alpha t_{nom} k_l^2, \tag{3}$$

where α is the temperature coefficient depending on conductor material; k_l is the load factor of the switching device; Rnom is the nominal resistance of the contact groups [3].

The calculation is performed in the following order:

- calculating the contact resistance R_m at ambient temperature t_m=50 °C according to the expression (2):

$$R_m = R_{norm} \frac{K + t_m}{K + t_{norm}} = 275 \frac{235 + 5}{235 + 40} = 259 \text{ mOm}$$

- contact resistances for other ambient temperature values are determined the same way;
- for study the dependence of resistances on the load we determine the contacts resistance at k_l=0.1 using the expression (3):

$$R = R_{nom} + \alpha t_{nom} k_l^2 = 275 + 4.3 \cdot 40 \cdot 0.1^2 = 227 \text{ mOm}$$

- the resistance of the magnetic starter contacts at other values of the load factor is determined.

For magnetic starters having the following nominal resistances of contact groups: 82.5; 33.0; 13.1

mOm; as well as for circuit breakers with the following nominal resistances 34.9; 27.9; 17.5 mOm; the calculation is made the same way [2].

Dependency graphs of contact connection's resistance value (magnetic starters and circuit breakers) on the ambient temperature and the load factor are shown in Fig. 1-4.

Predicting the development of a random process that reflects the functioning of a complex system should be preceded by statistical processing of experiment results to build a correlation field. Then, by using the correlation field we can find an empirical regression, which means to establish a quantitative relationship between the characteristics of the process.

The next step is to approximate the ultimate empirical curve of regression.

The simplest way of approximation of this curve is linear regression:

$$\gamma = \alpha + \beta x \quad (4)$$

The change γ is related to some change in the parameter x, but does not depend on the quantity of "parameter x has already accumulated". Using the principle of least squares it is easy to create a normal equation of linear regression [2]:

$$\begin{cases} \sum y_i - \sum(\alpha + \beta x_i) = 0 \\ \sum y_i x_i - \sum(\alpha + \beta x_i)x_i = 0 \end{cases} \quad (5)$$

Making simple transformations, we bring this system to this form

$$\begin{cases} m\alpha + \beta \sum x_i = \sum y_i \\ \alpha \sum x_i + \beta \sum x_i^2 = \sum y_i x_i \end{cases} \quad (6)$$

β is a regression coefficient that can be easily calculated using determinants:

$$\beta = \frac{m\sum x_i y_i - \sum x_i \sum y_i}{m\sum x_i^2 - (\sum x_i)^2} \quad (7)$$

α is a free regression term:

$$\alpha = \frac{\sum y_i - \beta \sum x_i}{m} \quad (8)$$

The resulting expressions completely determine linear regression for a given sample.

Equation (5) for the free term of the regression can be rewritten in the following way:

$$\alpha = \frac{1}{m}\sum y_i - \beta\frac{1}{m}\sum x_i = \overline{y} - \beta\overline{x} \qquad \text{when} \qquad \overline{y} = \alpha + \beta\overline{x},$$

The average point of the joint distribution of the quantities learnt always lies on the regression line [6].

For the determination of the regression line, it is enough to know only angular coefficient β.

The fact that dependence studied is assumed to be linear allows using a sample correlation coefficient r for estimating the bond strength:

$$r = \beta\sqrt{\frac{m\sum x_i^2 - (\sum x_i)^2}{m\sum y_i^2 - (\sum y_i)^2}} \quad (9)$$

The initial calculation is based on the dependence of the contact connections resistance of a magnetic starter from the ambient temperature (Table 1).

According to the expression (7), we calculate the regression coefficient β:

$$\beta = \frac{m\sum x_i y_i - \sum x_i \sum y_i}{m\sum x_i^2 - (\sum x_i)^2} = \frac{9\cdot 60000 - 225\cdot 2340}{9\cdot 7125 - 225^2} = 1$$

thereat coefficient α is determined by the expression (8)

$$\alpha = \frac{\sum y_i - \beta \sum x_i}{m} = \frac{2340 - 1\cdot 225}{9} = 235$$

- we get the correlation coefficient r according to expression (9):

$$r = \beta\sqrt{\frac{m\sum x_i^2 - (\sum x_i)^2}{m\sum y_i^2 - (\sum y_i)^2}} = 1\sqrt{\frac{9\cdot 7125 - 225^2}{9\cdot 609900 - 2340^2}} = 0.99$$

- the correlation coefficient is very close to 1.0 (0.99) which means that the relationship between x and y is almost linear, and the final equation must be recognized as equality.

Table 1. Dependence of contact connections resistance of magnetic starters on the ambient temperature (R_{nom} = 275mOm)

t, °C	5	10	15	20	25	30	35	40	45
R, mOm	240	245	250	255	260	265	270	275	280

The calculation is made in the following order:
- we take $x_i = t_i$, °C; $y_i = R_i$, mom;
- table 2 is made to simplify the calculation.

Table 2. The results of the intermediate calculations

No.	*X*	*Y*	X^2	Y^2	*X*·*Y*	*X*+*Y*	$(X+Y)^2$
1	5	240	25	57600	1200	245	60025
2	10	245	100	60025	2450	255	65025
3	15	250	225	62500	3750	265	70225
4	20	255	400	65025	5100	275	75625
5	25	260	625	67600	6500	285	81225
6	30	265	900	70225	7950	295	87025
7	35	270	1225	72900	9450	305	93025
8	40	275	1600	75625	11000	315	99225
9	45	280	2025	78400	12600	325	105625
Σ	225	2340	7125	609900	60000	2565	737025

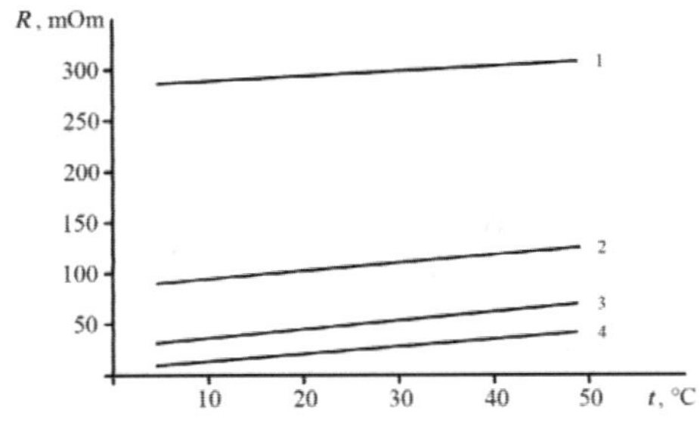

Figure 1. Dependence of contact connection resistance of magnetic starters on the ambient temperature: 1-Rnom=275 mOm; 2-Rnom=82.5 mOm; 3 – Rnom=33 mOm; 4 – Rnom=13.1 mOm

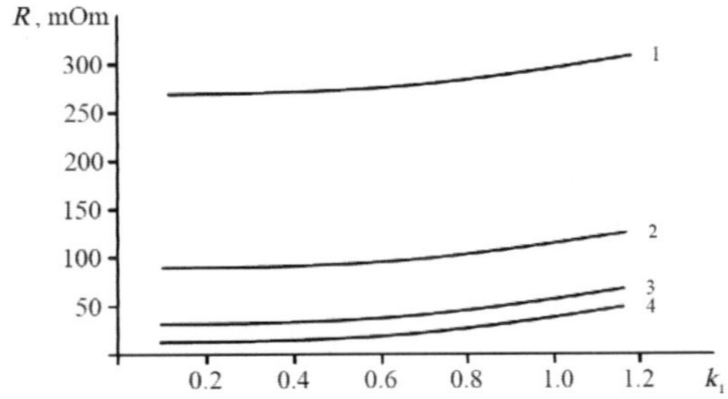

Figure 2. Dependence of contact connection resistance of magnetic starters on the load factor: 1-Rnom=275 mOm; 2-Rnom=82.5 mOm; 3 – Rnom=33 mOm; 4 – Rnom=13.1 mOm

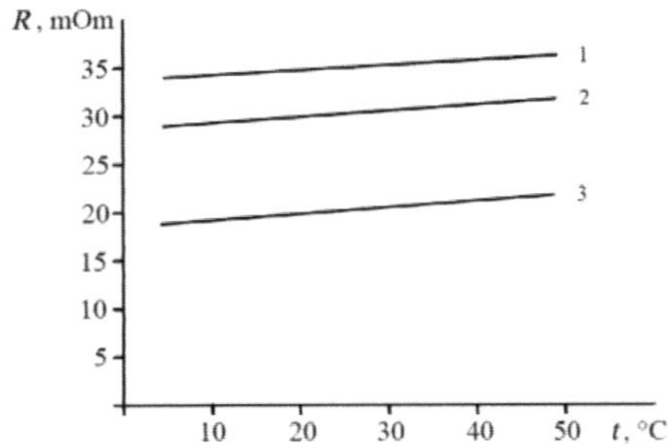

Figure 3. Dependence of contact connection resistance of circuit breaker on the ambient temperature: 1- Rnom =37.4 mOm; 2- Rnom =27.9 mOm; 3- Rnom =17.45 mOm

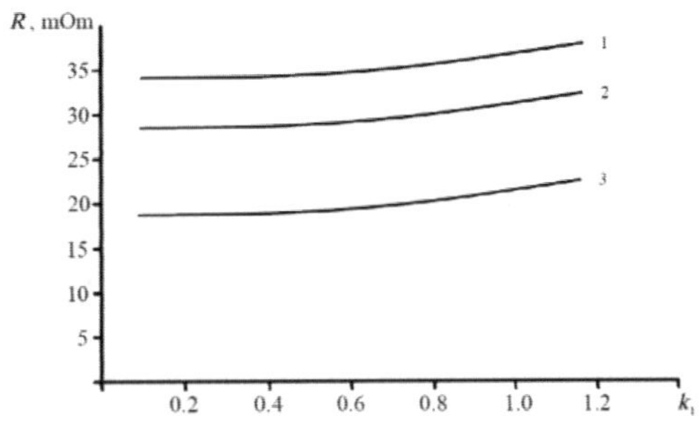

Figure 4. Dependence of the contact connection resistance of circuit breaker on the load factor: 1- Rnom =37.4 mOm; 2- Rnom =27.9 mOm; 3- Rnom =17.45 mOm

3. Results

The obtained expression is consistent with the analytical dependence of contact connections resistance of magnetic starter on the ambient temperature (Figures 1 and 5).

Analytical forecasting of changes in contact resistance of switching devices during operation. The work capacity of the switching device is determined by the resistance of its contacts R_c. We consider the function $R_c(T)$ value of which changes continuously in the time interval $T_1=[t_0, t_n]$. As a result, there are values of this function $R_0, R_1,..., R_i, R_n$ on the interval T_1 [2].

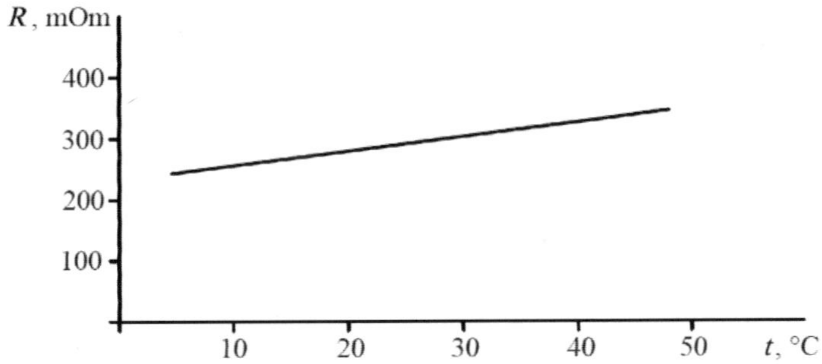

Figure 5. Dependence of contact connection resistance of magnetic starters on the ambient temperature

It is necessary to determine the value of the R(t) function from the known R_i values: $R_{n+1},\ldots, R_{n+i},\ldots, R_i,\ldots, R_{n+m}$ at future times $t_{n+1},\ldots, t_{n+i},\ldots, t_{n+m} \in T_2$, or find out after what time the values of R_{n+i}, $t_{n+i} \in T_2$ will reach the acceptable level R_{adm}.

We can use the following expression to describe how the R_c parameter changes in this interval

$$R_c = R_{nom} - kT, \tag{10}$$

Where $k = (R_{nom} - R_p)/T$; T – operating time, years; $R_p = 5R_{nom}$ – permissible contact resistance [5].

The calculation is made in the following order [2]:

- according to expressions (10, 11), we determine the character of the $R_c(T)$ change

$$R_c = 275 + 36T$$

Where $k = (R_{nom} - R_p)/T = (275 - 1375)/30 = -36$

- we study the change in contact connections resistance during operation, i.e. T = 1,..., 30 years:

$$R_c = 275 + 36 \cdot 1 = 311, \text{ mOm}$$

The calculation for other values of the operating time is made the same way.

The regression coefficient that characterizes the obliquity of the straight line is determined:

$$\alpha = \frac{T}{R_p - R_{nom}} = \frac{1}{1375 - 275} = 0.001$$

where T = 1 year, the operating time of the magnetic starter.

The calculation for other values of the operating time is made the same way.

The results of the calculations are presented in Table 3.

Table 3. Regression coefficient

T, year	1	3	5	7	10	15	17	20	23	25	30
$\alpha \cdot 10^{-3}$	1	3	5	6	9	14	15	18	21	23	27

Dependency graphs of the contact resistance of switching devices on service life are shown in figures 6 and 7.

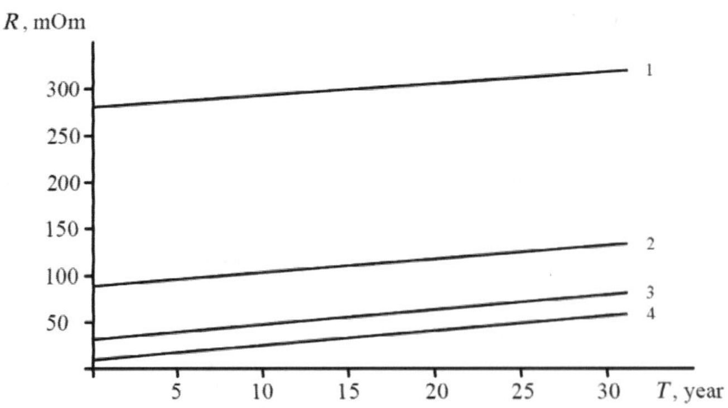

Figure 6. Change in contact connection resistance of magnetic starters during operation: 1 – R_{nom}=275 mOm; 2 – R_{nom}=82.5 mOm; 3 – R_{nom}=33 mOm; 4 – R_{nom}=13.1 mOm.

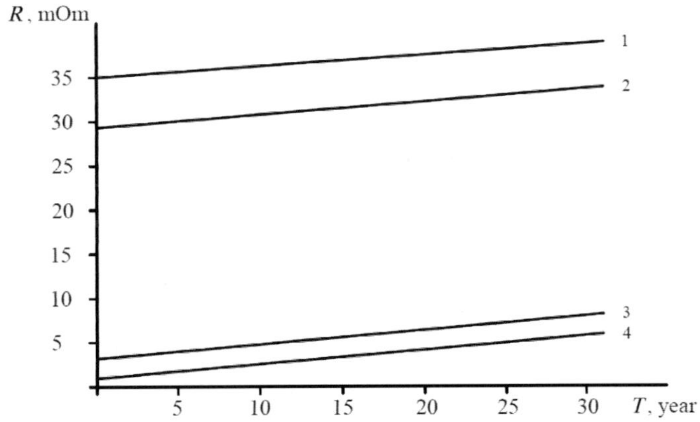

Figure 7. Dependence of contact connection resistance of circuit breakers on service life:
1- R_{nom} =34.9 mOm; 2- R_{nom} =27.9 mOm; 3- R_{nom} =17.45 mOm; 4 – R_{nom} =0.512 mOm.

4. Conclusions

During the analytical process of forecasting changes in contact groups resistance of low-voltage switching devices (magnetic starters and circuit breakers) under different operating conditions, mathematical models are obtained, the input values of which are refined using regression analysis. The change in contact connections resistance of the low-voltage devices is one of their technical condition characteristics and allows evaluating the efficiency of oil and gas industry equipment.

References

[1] Kalyavin V P and Rybakov L M 2000 *Reliability, and diagnostics of electric installations*: teaching guide (Yoshkar-Ola: Mari State University)

[2] Gracheva E I 2008 Prediction of changes in the resistance of contact connections of low-voltage switching devices by various methods *Proceedings of the higher educational institutions. Energy Sector Problems.* **7(8)** 44-52

[3] GOST 2933-83 2002 *Low-voltage electrical devices. Testing methods* (Moscow: Standard publ)

[4] 2002 *Volumes and standards for testing electrical equipment* (Moscow: Energy)

[5] Shevchenko V V and Gracheva E I 2002 Determination of contact connections resistances of low-voltage switching devices *Industrial power engineering* **2** 42-43

[6] Voschinin A P and Akmatbekov R A 1992 *Optimization of regression models and design of experiments* (Bishkek: Ilim)

IOP Conf. Series: Materials Science and Engineering **860** (2020) 012015 doi:10.1088/1757-899X/860/1/012015

Results of mathematical modeling of the mud pump electric drive operation with the compensation unit in the circuit

T V Tabachnikova and L V Shvetskova

Almetyevsk State Oil Institute, 2 Lenin Street, Almetyevsk, 423452, Russia

E-mail: tvtab@mail.ru

Abstract. The mathematical model was developed for the mud pump electric drive with an individual compensation unit in a bypass circuit, which allows simulating the electric drive operation during the start and in steady-state modes with and without regard to the individual compensation unit. The mathematical model of the electric drive of the mud pump was developed with the aim of increasing the reliability and continuity of the drilling rig technological process in case of failure of the frequency-controlled converter of the electric drive. The use of the electric drive parameters given in relative units in a mathematical model allows replicating the obtained simulation results for other modifications of the pump electric drive. It is proved that when an individual compensating unit is included in the electric drive circuit, it reduces the loss of active power in the cable line and overhead line, and helps to increase the efficiency of the mud pump. The simulation results showed the boundary parameters of the dependent and independent variable parameters of the distribution electric network, at which the start-up of the electric drive of the mud pump is completed successfully.

1. Introduction

This article discusses the MDU (mobile drilling unit) 3000, which is used for rotary and downhole motor drilling. To achieve optimal mud pump operation, the developed power must remain constant and equal to the nominal power. The authors have developed a mathematical model of the mud pump electric drive to analyze the steady-state operation modes and start-up modes. To check the validity of the results obtained using the proposed mathematical model, the most commonly used pump locking device with a nominal power of 1000 kW is assumed as an electric drive. Given that the per-unit parameters of pump locking devices with different capacities are close to each other, the simulation results obtained can be valid for other modifications of this series.

Figure 1 shows a schematic diagram of the drilling rig electrical complex with the following reference designations: L – the length of a 6 kV outgoing circuit up to the rig, l_1 – a section of the aerial line from the supply feeder to the rig, T_1 – a step-down transformer in the package transformer substation, $P + jQ$ – the drilling rig equivalent load, l_2 – a high voltage cable line supplying the electric drive of the drilling rig electric pump unit, a shunt compensation unit (SCU 1) is an individual unit designed for compensating reactive power of a 1000 kW induction motor (IM).

The mud pump electric drive is connected to the power supply through a high-voltage frequency converter, however, Figure 1 shows a schematic diagram of the mud pump electrical complex for the scenario, in which the drive is connected directly to the mains.

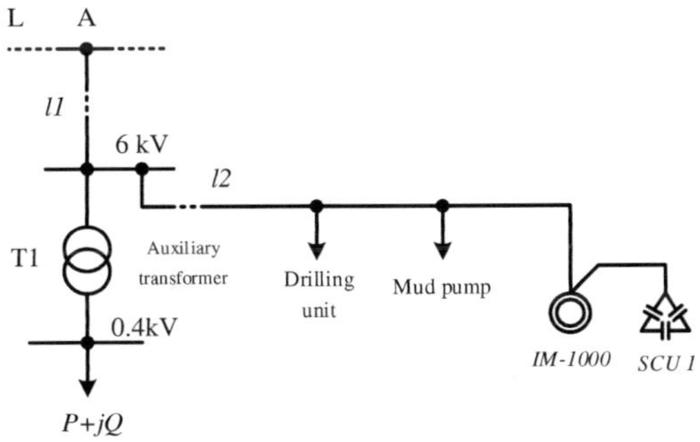

Figure 1. Schematic diagram of the drilling rig electrical complex

2. Problem statement

Currently, a variable frequency drive for mud pumps is used, which allows reducing the number of piston replacements in the course of drilling. However, frequency-controlled converter failures occur from time to time, which entails losses. To avoid frequent stops of the mud pump occurring due to the failure of the high-voltage frequency-controlled converter and in order to improve the operation process reliability and continuity, it is proposed to introduce a bypass (Figure 2), that is, to arrange a direct connection of the pump unit electric drive to the power source. We propose a configuration assuming an electric drive with an individual compensation unit, which will reduce the loss of true power in the cable line and aerial line, and also contribute to the mud pump efficiency improvement. When introducing an individual compensation unit, there will be a need to connect a discharge device in order to discharge the reactive power in the system.

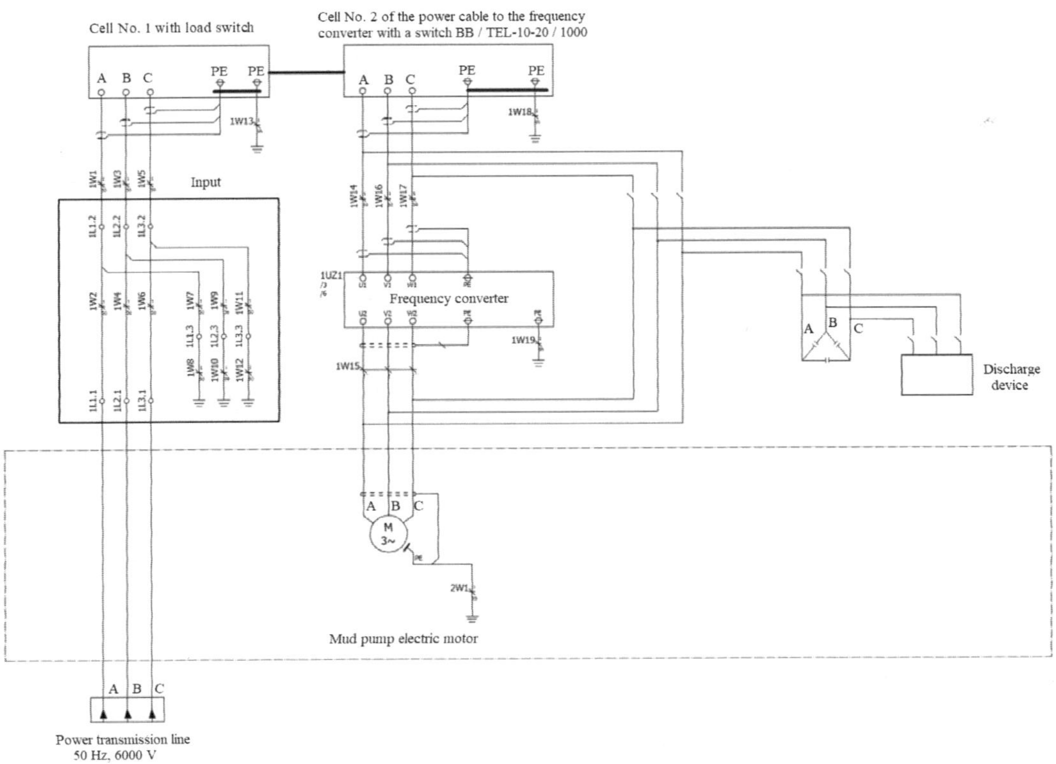

Figure 2. Mud pump electric drive power supply diagram

IOP Conf. Series: Materials Science and Engineering **860** (2020) 012015 doi:10.1088/1757-899X/860/1/012015

3. Theory

Based on the schematic diagram (Figure 1) and the equivalent circuit of the drilling rig electrical complex (Figure 3) connected at the node (A) at the end of the outgoing line, mathematical and simulation models of this complex were developed. The mathematical model includes the following parameters: a l_1 long aerial line section, the node compensation unit SCU2, a step-down transformer T_1, equivalent load P+jQ, a step-up transformer T_2, a mud pump unit, and an individual compensation unit SCU1. The mathematical model of the mud pump electrical complex allows simultaneous simulation of the operation in steady-state and transient modes [1, 2, 3].

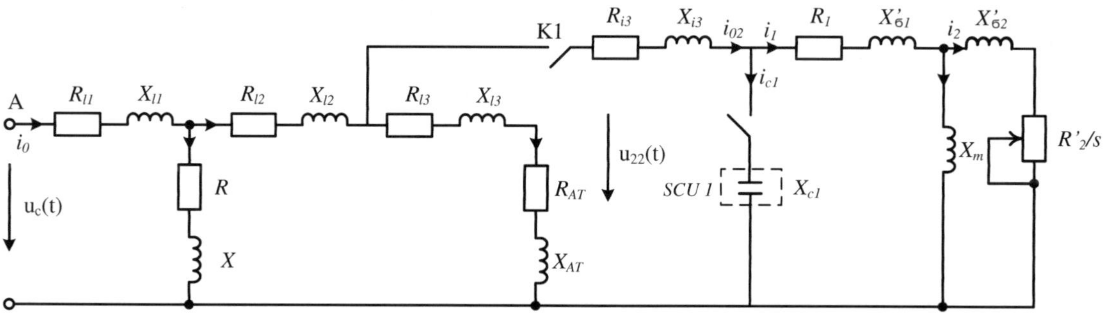

Figure 3. Equivalent circuit of the mud pump electrical complex connected at the node (A) at the end of the outgoing line

A system of analytical dependencies (1 – 5) has been obtained based on the equivalent circuit:

$$u_{22}(t) = R_{l_3} i_{02} + L_{l_3} \frac{di_{02}}{dt} + u_{c1}; \quad (1)$$

$$i_{02} = i_1 + i_{c1}; \quad (2)$$

$$i_1 = i_{02} - C_{c1} \frac{du_{c1}}{dt}; \quad (3)$$

$$u_{c1}(t) = R_1 i_1 + \frac{d\psi_1}{dt}; \quad (4)$$

$$0 = i_2 R'_2 + \frac{d\psi_2}{dt}, \quad (5)$$

where u_{22} and i_{02} are the input voltage and current in the downhole motor power cable line at the well entry; $\boldsymbol{R}_{l_3}$ and $\boldsymbol{L}_{l_3}$ – true resistance and inductance of the downhole motor power cable line; i_{C1}– current at SCU1 (condenser C_1) terminals; u_{c1}– voltage at SCU1 terminals; $\boldsymbol{i}_1$ – stator current, $\boldsymbol{i}_2$ – rotor current; $\boldsymbol{R}_1$ – true stator winding resistance; R'_2 – true stator winding resistance normalised to the stator winding [4, 5].

Let us express the stator and rotor winding currents $\boldsymbol{i}_1$ $\boldsymbol{i}_2$ and through the stator $\psi_1 = i_1 L_1 + i_2 L_m$ and rotor $\psi_2 = i_2 L_2 + i_1 L_m$ winding flux linkages, respectively, where $L_1 = L_{\sigma 1} + L_m$, $L_2 = L_{\sigma 2} + L_m$ are the total stator and rotor winding inductance values.

$$i_1 = \frac{\psi_1 - \psi_2 L_m / L_2}{L_1 - L_m^2 / L_2} \quad \text{or} \quad i_1 = \frac{\psi_1 - \psi_2 k_2}{L'_1} = \frac{1}{L'_1}\psi_1 - \frac{k_2}{L'_1}\psi_2 ; \quad (6)$$

$$i_2 = \frac{\psi_2 - \psi_1 L_m / L_1}{L_2 - L_m^2 / L_1} \quad \text{or} \quad i_2 = \frac{\psi_2 - \psi_1 k_1}{L_2'} = \frac{1}{L_2'}\psi_2 - \frac{k_1}{L_2'}\psi_1, \tag{7}$$

where $k_1 = \frac{L_m}{L_1}$ is the coupling factor for stator; $k_2 = \frac{L_m}{L_2}$ – coupling factor for rotor;

$L_1' = L_1 - \frac{L_m^2}{L_2}$ – transient inductance of the stator winding in case of a short-circuit rotor winding;

$L_2' = L_2 - \frac{L_m^2}{L_1}$ – transient inductance of the rotor winding in case of a short-circuit starter winding.

Let us apply the stator (6) and rotor (7) current values, expressed in terms of flux linkages, to the stator current and induction machine voltage equations (3-5), then we get:

$$\frac{\psi_1 - \psi_2 k_2}{L_1'} = i_{02} - C_1 \frac{du_{c1}}{dt}, \quad \text{then} \quad i_{02} = \frac{1}{L_1'}\psi_1 - \frac{k_2}{L_1'}\psi_2 + C\frac{du_c}{dt}; \tag{8}$$

$$u_{c1} = \frac{\psi_1 - \psi_2 k_2}{L_1'} \cdot R_1 + \frac{d\psi_1}{dt}; \quad u_{c1} = \frac{R_1}{L_1'}\psi_1 - k_2 \frac{R_1}{L_1'}\psi_2 + \frac{d\psi_1}{dt}; \tag{9}$$

$$0 = \frac{\psi_2 - \psi_1 k_1}{L_2'} \cdot R_2 + \frac{d\psi_2}{dt}; \quad 0 = \frac{R_2}{L_2'}\psi_2 - k_1 \frac{R_2}{L_2'}\psi_1 + \frac{d\psi_2}{dt}. \tag{10}$$

Putting formulas (8 – 10) into the system of algebraic and differential equations (1 – 5), we get:

$$u_{22}(t) = R_{l_3} i_{02} + L_{l_3} \frac{di_{02}}{dt} + u_{c1}; \tag{11}$$

$$i_{02} = i_1 + i_{c1}; \tag{12}$$

$$i_{02} = \frac{1}{L_1'}\psi_1 - \frac{k_2}{L_1'}\psi_2 + C_1 \frac{du_{c1}}{dt}; \tag{13}$$

$$u_{c1} = \frac{R_1}{L_1'}\psi_1 - k_2 \frac{R_1}{L_1'}\psi_2 + \frac{d\psi_1}{dt}; \tag{14}$$

$$0 = \frac{R_2}{L_2'}\psi_2 - k_1 \frac{R_2}{L_2'}\psi_1 + \frac{d\psi_2}{dt}. \tag{15}$$

Let us differentiate instantaneous current and voltage values:

$u(t)_{22} = U_m \sin(\omega_0 t + \psi_u)$ – instantaneous voltage at the well entry;

$i(t)_{02} = I_m \sin(\omega_0 t + \psi_i)$ – instantaneous current of the pump locking device power cable;

$u(t)_{c1} = U_m \sin(\omega_0 t + \psi_u)$ – instantaneous voltage at the UPEC1 terminals.

Next, the instantaneous values of currents and voltages are differentiated:

$$\frac{d}{dt}u(t)_{22} = \frac{d}{dt}\left(U_m \sin\left(\omega_0 t + \psi_u\right)\right) = \frac{d}{dt}\vec{U}_{22} + j\omega_0 \vec{U}_{22}; \tag{16}$$

$$\frac{d}{dt}i(t)_{02} = \frac{d}{dt}\left(I_m \sin\left(\omega_0 t + \psi_i\right)\right) = \frac{d}{dt}\vec{I}_{02} + j\omega_0 \vec{I}_{02}; \tag{17}$$

$$\frac{d}{dt}u(t)_{c1} = \frac{d}{dt}\left(U_m \sin\left(\omega_0 t + \psi_u\right)\right) = \frac{d}{dt}\vec{U}_{c1} + j\omega_0\vec{U}_{c1}; \quad (18)$$

$$\frac{d\psi_1}{dt} = \frac{d}{dt}\vec{\psi}_1 + j\omega_0\vec{\psi}_1; \quad (19)$$

$$\frac{d\psi_2}{dt} = \frac{d}{dt}\vec{\psi}_2 + j\left(\omega_0 - \omega_2\right)\vec{\psi}_2. \quad (20)$$

The obtained values in the complex form are then applied to the original system of equations:

$$\vec{U}_{22} = R_{l_3}\vec{I}_{02} + L_{l_3}\left(\frac{d}{dt}\vec{I}_{02} + j\omega_0\vec{I}_{02}\right) + \vec{U}_{c1}; \quad (21)$$

$$\vec{I}_{02} = \frac{1}{L_1'}\vec{\psi}_1 - \frac{k_2}{L_1'}\vec{\psi}_2 + C_1\left(\frac{d}{dt}\vec{U}_{c1} + j\omega_0\vec{U}_{c1}\right); \quad (22)$$

$$\vec{U}_{c1} = \frac{R_1}{L_1'}\vec{\psi}_1 - k_2\frac{R_1}{L_1'}\vec{\psi}_2 + \frac{d}{dt}\vec{\psi}_1 + j\omega_0\vec{\psi}_1; \quad (23)$$

$$0 = \frac{R_2}{L_2'}\vec{\psi}_2 - k_1\frac{R_2}{L_2'}\vec{\psi}_1 + \frac{d}{dt}\vec{\psi}_2 + j\left(\omega_0 - \omega_2\right)\vec{\psi}_2, \quad (24)$$

where ω_2 is the rotation frequency of the pump locking device rotor; let us transform $\left(\omega_0 - \omega_2\right)\frac{\omega_0}{\omega_0} = s\omega_0$ – the product of the slip and the electric network cyclic frequency and denote $\frac{d}{dt} = p$ - the symbol of differentiation.

To simplify the recording format, the vector sign in the system of equations (21-24) can be omitted. Using the transformation and notation, let us put down this system in the rotating coordinates system with synchronous rotation velocity ω_0 :

$$U_{22} = R_{l_3}I_{02} + L_{l_3}\left(pI_{02} + j\omega_0 I_{02}\right) + U_{c1}; \quad (25)$$

$$I_{02} = \frac{1}{L_1'}\psi_1 - \frac{k_2}{L_1'}\psi_2 + C_1\left(pU_{c1} + j\omega_0 U_{c1}\right); \quad (26)$$

$$U_{c1} = \frac{R_1}{L_1'}\psi_1 - k_2\frac{R_1}{L_1'}\psi_2 + p\psi_1 + j\omega_0\psi_1; \quad (27)$$

$$0 = \frac{R_2}{L_2'}\psi_2 - k_1\frac{R_2}{L_2'}\psi_1 + p\psi_2 + js\omega_0\psi_2, \quad (28)$$

Projections of the initial parameter vectors on the x and y axes are obtained by asserting:

$U_{22} = U_{22x}$; $I_{02} = I_{02.x} + jI_{02.y}$; $\psi_1 = \psi_{1x} + j\psi_{1y}$;
$\psi_2 = \psi_{2x} + j\psi_{2y}$; $U_{c1} = U_{c1.x} + jU_{c1.y}$.

Let us put their projections into the system of equations:

$$U_{22.x} = R_{l_3} I_{02.x} + R_{l_3} jI_{02.y} + L_{l_3} \frac{\omega_0}{\omega_0} pI_{02.x} + L_{l_3} \frac{\omega_0}{\omega_0} pjI_{02.y} + j\omega_0 L_{l_3} I_{02.x} +$$
$$+ jj\omega_0 L_{l_3} I_{02.y} + U_{c1.x} + jU_{c1.y}; \tag{29}$$

$$I_{02.x} + jI_{02.y} = \frac{1}{L_1'}\psi_{1.x} + \frac{1}{L_1'} j\psi_{1.y} - \frac{k_2}{L_1'}\psi_{2.x} - \frac{k_2}{L_1'} j\psi_{2.y} +$$
$$+ C_1 \frac{\omega_0}{\omega_0} pU_{c1.x} + C_1 \frac{\omega_0}{\omega_0} pjU_{c1.y} + jC_1\omega_0 U_{c1.x} + jjC_1\omega_0 U_{c1.y}; \tag{30}$$

$$U_{c1.x} + jU_{c1.y} = \frac{R_1}{L_1'}\psi_{1.x} + \frac{R_1}{L_1'} j\psi_{1.y} - k_2 \frac{R_1}{L_1'}\psi_{2.x} - k_2 \frac{R_1}{L_1'} j\psi_{2.y} +$$
$$+ p\psi_{1.x} + pj\psi_{1.y} + j\omega_0\psi_1 + jj\omega_0\psi_1; \tag{31}$$

$$0 = \frac{R_2}{L_2'}\psi_{2.x} + \frac{R_2}{L_2'} j\psi_{2.y} - k_1 \frac{R_2}{L_2'}\psi_{1.x} - k_1 \frac{R_2}{L_2'} j\psi_{1.y} +$$
$$+ p\psi_{2.x} + pj\psi_{2.y} + js\omega_0\psi_2 + jjs\omega_0\psi_2, \tag{32}$$

Now, let us represent the system of differential equations by complex inductive and capacitive resistance values:

$$U_{22.x} = R_{l_3} I_{02.x} + R_{l_3} jI_{02.y} + \frac{Z_{l_3}}{\omega_0} pI_{02.x} + \frac{Z_{l_3}}{\omega_0} jpI_{02.y} +$$
$$+ jZ_{l_3} I_{02.x} - Z_{l_3} I_{02.y} + U_{c1.x} + jU_{c1.y}; \tag{33}$$

$$I_{02.x} + jI_{02.y} = \frac{\omega_0}{X_{L_1'}}\psi_{1.x} + \frac{\omega_0}{X_{L_1'}} j\psi_{1.y} - \frac{\omega_0 k_2}{X_{L_1'}}\psi_{2.x} - \frac{\omega_0 k_2}{X_{L_1'}} j\psi_{2.y} +$$
$$+ \frac{\omega_0}{Z_{C_1}\omega_0} pU_{c1.x} + \frac{\omega_0}{Z_{C_1}\omega_0} pjU_{c1.y} + j\frac{1}{Z_{C_1}} U_{c1.x} - \frac{1}{Z_{C_1}} U_{c1.y}; \tag{34}$$

$$U_{c1.x} + jU_{c1.y} = \frac{\omega_0 R_1}{X_{L_1'}}\psi_{1.x} + \frac{\omega_0 R_1}{X_{L_1'}} j\psi_{1.y} - k_2 \frac{\omega_0 R_1}{X_{L_1'}}\psi_{2.x} -$$
$$- k_2 \frac{\omega_0 R_1}{X_{L_1'}} j\psi_{2.y} + p\psi_{1.x} + jp\psi_{1.y} + j\omega_0\psi_1 - \omega_0\psi_1; \tag{35}$$

$$0 = \frac{\omega_0 R_2}{X_{L_2'}}\psi_{2.x} + \frac{\omega_0 R_2}{X_{L_2'}} j\psi_{2.y} - k_1 \frac{\omega_0 R_2}{X_{L_2'}}\psi_{1.x} - k_1 \frac{\omega_0 R_2}{X_{L_2'}} j\psi_{1.y} +$$
$$+ p\psi_{2.x} + jp\psi_{2.y} + js\omega_0\psi_2 - s\omega_0\psi_2, \tag{36}$$

For simplicity, the sign of the imaginary component in the system of equations can be omitted:

$$U_{22.x} = R_{l_3} I_{02.x} + \frac{Z_{l_3}}{\omega_0} pI_{02.x} - Z_{l_3} I_{02.y} + U_{c1.x}; \tag{37}$$

$$0 = R_{l_3} I_{02.y} + \frac{Z_{l_3}}{\omega_0} pI_{02.y} + Z_{l_3} I_{02.x} + U_{c1.y}; \tag{38}$$

$$I_{02.x} = \frac{\omega_0}{X_{L_1'}}\psi_{1.x} - \frac{\omega_0 k_2}{X_{L_1'}}\psi_{2.x} + \frac{1}{\omega_0 Z_{c_1}} pU_{c1.x} - \frac{1}{Z_{c_1}}U_{c1.y}; \tag{39}$$

$$I_{02.y} = \frac{\omega_0}{X_{L_1'}}\psi_{1.y} - \frac{\omega_0 k_2}{X_{L_1'}}\psi_{2.y} + \frac{1}{\omega_0 Z_{c_1}} pU_{c1.y} + \frac{1}{Z_{c_1}}U_{c1.x}; \tag{40}$$

$$U_{c1.x} = \frac{\omega_0 R_1}{X_{L_1'1}}\psi_{1.x} - k_2\frac{\omega_0 R_1}{X_{L_1'}}\psi_{2.x} + p\psi_{1.x} - \omega_0\psi_1; \tag{41}$$

$$U_{c1.y} = +\frac{\omega_0 R_1}{X_{L_1'}}\psi_{1.y} - k_2\frac{\omega_0 R_1}{X_{L_1'}}\psi_{2.y} + p\psi_{1.y} + \omega_0\psi_1; \tag{42}$$

$$0 = \frac{\omega_0 R_2}{X_{L_2'}}\psi_{2.x} - k_1\frac{\omega_0 R_2}{X_{L_2'}}\psi_{1.x} + p\psi_{2.x} - s\omega_0\psi_2, \tag{43}$$

$$0 = \frac{\omega_0 R_2}{X_{L_2'}}\psi_{2.y} - k_1\frac{\omega_0 R_2}{X_{L_2'}}\psi_{1.y} + p\psi_{2.y} + s\omega_0\psi_2. \tag{44}$$

Let us express the derived parameters of the equivalent circuit and divide them by coefficients:

$$pI_{02.x} = \frac{\omega_0}{Z_{l_3}}(U_{22.x} - R_{l_3}I_{02.x} + Z_{l_3}I_{02.y} - U_{c1.x}); \tag{45}$$

$$pI_{02.y} = \frac{\omega_0}{Z_{l_3}}(-R_{l_3}I_{02.y} - Z_{l_3}I_{02.x} - U_{c1.y}); \tag{46}$$

$$pU_{c1.x} = (I_{02.x} - \frac{\omega_0}{X_{L_1'}}\psi_{1.x} + \frac{\omega_0 k_2}{X_{L_1'}}\psi_{2.x} + \frac{1}{Z_{C_1}}U_{c1.y}); \tag{47}$$

$$pU_{c1.y} = \omega_0 Z_{C_1}(I_{02.y} - \frac{\omega_0}{X_{L_1'}}\psi_{1.y} + \frac{\omega_0 k_2}{X_{L_1'}}\psi_{2.y} - \frac{1}{Z_{C_1}}U_{c1.x}); \tag{48}$$

$$p\psi_{1.x} = U_{c1.x} - \frac{\omega_0 R_1}{X_{L_1'1}}\psi_{1.x} + k_2\frac{\omega_0 R_1}{X_{L_1'}}\psi_{2.x} + \omega_0\psi_1; \tag{49}$$

$$p\psi_{1.y} = U_{c1.y} - \frac{\omega_0 R_1}{X_{L_1'}}\psi_{1.y} + k_2\frac{\omega_0 R_1}{X_{L_1'}}\psi_{2.y} - \omega_0\psi_1; \tag{50}$$

$$p\psi_{2.x} = -\frac{\omega_0 R_2}{X_{L_2'}}\psi_{2.x} + k_1\frac{\omega_0 R_2}{X_{L_2'}}\psi_{1.x} + s\omega_0\psi_2, \tag{51}$$

$$0 = -\frac{\omega_0 R_2}{X_{L_2'}}\psi_{2.y} - k_1\frac{\omega_0 R_2}{X_{L_2'}}\psi_{1.y} + s\omega_0\psi_2. \tag{52}$$

The resulting system of equations (45-52) can be supplemented with the motion equation:

$$ps = \frac{1}{T_m}m_c - \frac{k_1}{T_m X_2'}\left(\psi_{1y}\psi_{2x} - \psi_{1x}\psi_{2y}\right), \tag{53}$$

where s is slip; T_m – rotational inertia constant; k_1 – stator coupling coefficient; X_2' – transient inductive resistance of the rotor winding; m_c – resistance torque in the slip function.

The input voltage is expressed using the function [3]:

$$U(t) = 1(t) - k \cdot 1(t - t_1) + k \cdot 1(t - t_2), \tag{54}$$

where t_1 – the moment when the voltage begins to decrease; t_2 - the moment when the voltage is restored; k – the voltage reduction degree coefficient.

Let us represent the resulting system of equations in the per unit format (53-52). Then we supplement this system of equations with the motion equation (53) and the stress function (54).

The basic values are taken as:

$$M_b = \frac{S_b}{\Omega_0} = \frac{\sqrt{3} \cdot U_{nom} \cdot I_{nom}}{2 \cdot \pi \cdot n_{nom}} = \frac{3 \cdot U_b \cdot I_b \cdot p}{2 \cdot \omega_0} = 8238.8,$$

where $\omega_r = \Omega_0 = 2 \cdot \pi \cdot f / p = \omega_0 / p$, $(\omega_r = \Omega_0 = 2 \cdot \pi \cdot n_{nom} / 60 = 101.3)$, rad/s – rotational speed,

$$U_b = \frac{\sqrt{2} \cdot U_{nom}}{\sqrt{3}} = 4899; \quad I_b = \sqrt{2} \cdot I_{nom} = 117.35; \quad Z_b = \frac{U_b}{I_b} = 41.73;$$

$$\Psi_b = \frac{U_b}{\omega_0}; \quad \Delta M_{dop*} = \frac{\Delta M_{dop}}{M_b};$$

$$T_m = \frac{J \cdot \omega_m^2}{M_{nom}}; \quad T_{mb} = \frac{J}{4J_b}; \; T_{m*} = \frac{T_m}{T_{mb}},$$

where $J = J_d + J_p$ is the total inertia moment of the mud pump unit.

4. Experimental results

The modeling has provided the following results: electromechanical, electromagnetic and resistance torques, as well as start-up current, voltage, true and reactive power.

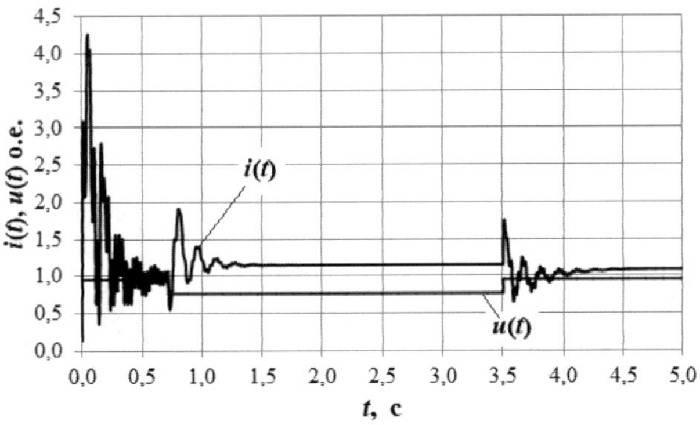

Figure 4. Graphs of current and input voltage dependence with the mud pump individual compensation unit enabled at input voltage decreased to 0.95 p.u. and under 20% voltage dip during 2.75 seconds

5. Discussion

The simulation results have shown that, in case of all possible disturbances of power distribution network dependent and independent variables, i.e. changes in the input voltage level and drop in the mains voltage level and changes in the resistance torque by 20% due to the drilling mud viscosity, the mud pump electric drive start is completed successfully.

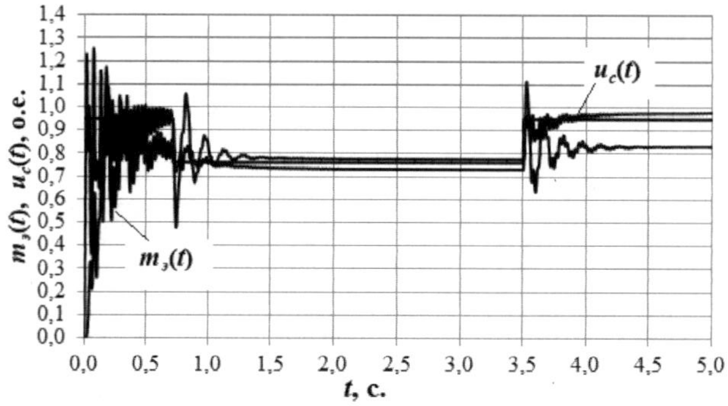

Figure 5. Graphs of electromagnetic torque, compensation device terminals voltage and input voltage dependence with the mud pump individual compensation unit enabled at input voltage decreased to 0.95 p.u. and under 20% voltage dip during 2.75 seconds

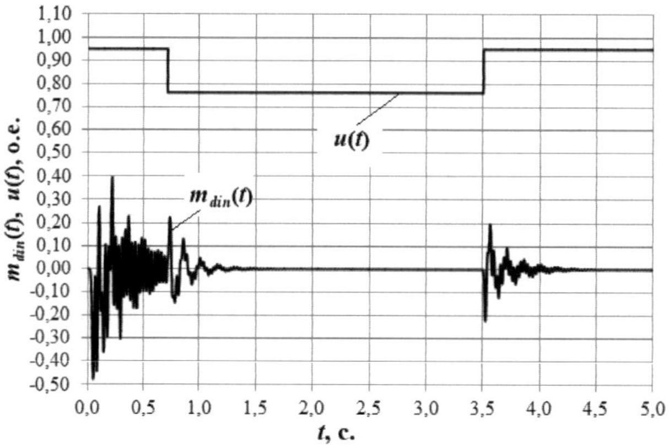

Figure 6. Graphs of electromechanical torque dependence with the mud pump individual compensation unit enabled at input voltage decreased to 0.95 p.u. and under 20% voltage dip during 2.75 seconds

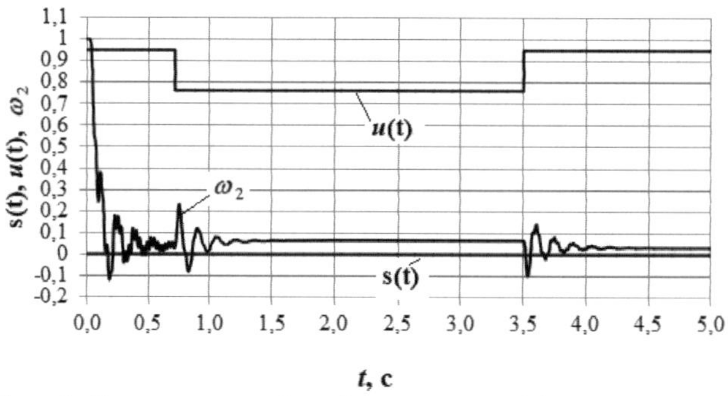

Figure 7. Graphs of slip, rotor rotation frequency and input voltage dependence with the mud pump individual compensation unit enabled at input voltage decreased to 0.95 p.u. and under 20% voltage dip during 2.75 seconds

6. Summary and conclusions

The introduction of a bypass is reasonable in order to improve the operation process reliability and continuity. It is also recommended to use an individual compensation unit to reduce losses in the power supply system. The results of mathematical modeling of the mud pump electric drive operating modes in the proposed power supply configuration prove guaranteed start taking into account all possible disturbances of power distribution network dependent and independent variables.

References

[1] Shvetskova L V 2016 *Improving the energy efficiency of the production well electrical complex with high viscosity oil* thesis of Ph.D. in Engineering Science 05.09.03, Samara

[2] Nurbosynov D N, Tabachnikova T V and Shvetskova L V 2015 *Optimising the electromagnetic torque of the production well electric drive start and self-start in case of viscous and high-viscosity oil recovery* Monthly production and engineering magazine Industrial Energy **10** 25–29

[3] Nurbosynov D N 1999 *Methods of calculation and mathematical modeling of voltage and power consumption in steady-state and transient processes: Monograph* (SPb.: Energoatomizdat)

[4] Konovalov Y V 2012 *Mathematical modeling of the AC motor start process* (Bulletin of Saratov State Technical University) **1 (68)** 146–149

[5] Kozyaruk A E, Belov M P and Zementov O I 2006 *Electric drive and automation system engineering: a study guide for higher educational institution students* (Moscow: Publishing house ACADEMIYA)

Statistical analysis of dynamics of decrease in oil production after well interventions

L R Zagitova

Almetyevsk State Oil Institute, 2A Lenin Street, Almetyevsk, 423452, Russia

E-mail: Liliya_zagitova@mail.ru

Abstract. The paper is based on the theoretical aspects of studying the success of oil production projects. To conduct clustering based on the empirical data from the producing well stock, a classifier has been compiled describing the types of interventions in 2013-2018. The selection of indicators was carried out on the basis of theoretical qualitative analysis. The following statistical indicators were included in the study information base: well profile, well diameter, field, horizon, and oil flow rate. The dynamics of the oil flow rate is translated into a comparable form using time synchronization. Since the initial attribute space is represented in various units of measurement, in order to ensure comparability, at the stage of formation of information array, standardization of indicators was carried out to the maximum. The interventions revealed the following combinations of factors which showed the greatest effect in terms of growth and total oil production: well profile, well diameter, field, as well as critical factors that can be interpreted as factors of ineffective interventions. The obtained factors were verified by various methods of multivariate statistical data analysis. The author suggests conclusions and recommendations for the most efficient interventions in oil wells.

1. Introduction

At oil fields, work is carried out on wells during their operation to regulate its development and maintain target levels of oil production. This set of works is called well interventions (WI) [1].

When selecting the type of WI, the issue of their effectiveness is brought to the fore and it is one of the grounds for the feasibility of the entire oil production project. For this purpose, the list of WI is planned and refined during the implementation process and is subject to systematic correction and refinement as current information on production volumes from the field becomes available. It is this process that largely influences the success of the implementation of projects of oil production.

2. Theory

For the purpose of evaluating the effectiveness of WI, wells are divided by the degree of reaction to the work performed. The well that received a positive effect or additional oil production is considered to have reacted, and the well that received a negative effect is considered not to have reacted to the WI.

The following principle is used to evaluate the effectiveness of WI: if separate production well of the exposure area has a positive effect, the duration is calculated up until the actual production of oil will not decrease below the baseline level; if the well has a negative effect, either immediately after the WI, or after a short-term positive effect, the calculation of the effect stops, that is calculated only positive component effect [2].

For statistical analysis, the initial data is the current production obtained as a result of the WI for the period under review for all activities.

The analysis is based on the assumption of a typical effect of WI. We assume that all wells work with some accuracy with the same dimensionless decrease rate of the flow rate $f(t)$, and $f(0) = 0$, but each i-th well has its own initial flow rate q_{0i}. Therefore, the additional oil rate received as a result of the WI depends on the accumulated operating time t according to the law $q_i(t) = q_{0i} f(t)$, or in logarithmic form:

$$ln\ q_i(t) = ln\ q_{0i} + ln\ f(t) + e_i(t). \quad (1)$$

where $e_i(t)$ is the remainder that is not explained within the specified model.

The quality criterion of the constructed model is to minimize the sum of squares of errors. Unknown values are the initial well flow rates $q_{0i}$and the values of the empirical function of the monthly flow rate drop $f(t_j)$. The resulting family of empirical decrease functions $f(t_j)$ is suitable for the determination of significant events and application of the results in the future WI planning process.

3. Results

In order to determine significant WIs, a cluster method of data analysis was used, the purpose of this method is to divide the initial set of WIs for wells with different characteristics into two clusters. The first cluster includes the best WIs for the total oil production and its growth, and the second cluster includes the worst of the interventions. Table 1 and Table 2 present the results of clustering of WI by total oil production rate and its increment for salt-acid treatment of wells and gas-dynamic fracturing (Figure 1). The following symbols are used in the table: ID – well number, PROF-well profile (HW-horizontal well, DS-Directional well, VW-vertical well), D-well diameter, A-area, HOR-horizon, ΔNT-increase in oil production, S_NT-total oil production.

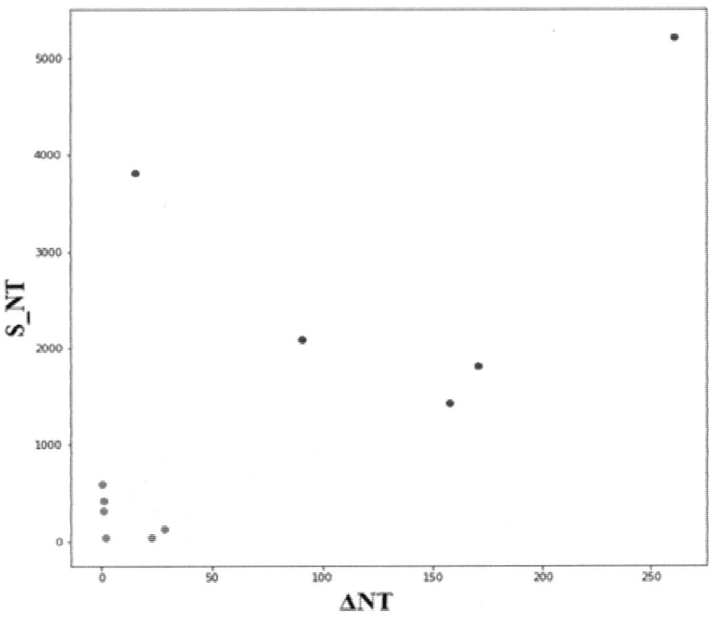

Figure 1. Result of the clustering of gas-dynamic fracturing

Table 1. Summary table of clustering of WI for oil production rate (the best ones).

Type of WI	ID	PROF	D	A	HOR	ΔNT	S_
Gas-dynamic fracturing	27649	DW	146	903	529	0.5	594
	28015	DW	168	903	529	1	417
	32546	DW	168	911	529	1	308
	32071	DW	168	911	529	261	5213
	32071	DW	168	911	529	171	1809
	28015	DW	168	903	529	158	1422
	27046	DW	146	1	476	17.9	1262.9
Salt-acid treatment	27194	DW	146	303	461	217	733
	27736	DW	146	201	481	139	329
	61693	HW	168	302	450	128	887.3
	27248	DW	146	201	481	1	64
	27151	VW	168	302	450	4	80
	27046	DW	146	1	476	20	1309
	27046	DW	146	1	476	17.9	1262.9
	32425	DW	168	246	481	30	1215
Traditional hydraulic fracturing	27261	DW	168	903	529	2	34
	32582	VW	168	911	529	22.5	42.5
	27404	DW	146	1	476	28.3	127
	32071	DW	168	911	529	261	5213
	33854	DW	146	903	529	15.5	3813.1
	59990	DW	146	902	529	91	2085

Tabel 2. Summary table of clustering of WI for the production of oil (the worst ones).

Type of WI	ID	PROF	D	A	HOR	ΔNT	S_
Gas-dynamic fracturing	27649	DW	146	903	529	0.5	594
	28015	DW	168	903	529	1	417
	32546	DW	168	911	529	1	308
Salt-acid treatment	63084	HW	168	302	450	0.7	110.3
	27248	DW	146	201	481	1	64
	59071	DW	114	303	461	1	22
Traditional hydraulic fracturing	27261	DW	168	903	529	2	34
	32582	VW	168	911	529	22.5	42.5
	27404	DW	146	1	476	28.3	127

In order to analyze the clustered data of the correct sample, an algorithm for constructing approximating functions by types of WI was used. Stages of the algorithm for constructing a family of approximating functions are:

- checking the correct data for uniformity and, if necessary, correcting it;
- selecting the type of approximating functions and calculating its coefficients;
- verification of the coefficients significance: if the corresponding coefficients are considered significant, they remain in the approximating function;

- if the coefficients are not significant, another type of approximating function is constructed for this sample, in which the coefficients are recalculated and checked for significance;
- calculation of the Fisher criterion for evaluating the adequacy of the obtained approximating functions and making a decision on its use.

Based on the results of testing the algorithm for constructing a family of approximating functions [4] for a control sample by type of WI for the reporting period from 2009 to 2017, a table of 3 approximating functions was obtained.

Table 3. Summary table of approximate functions of WI after clustering on the oil flow rate

Type of WI	Profile/diameter/ oilfield/ horizon	Approximating function	Model quality (r /R^2)
Gas-dynamic fracturing	(DW/146/294 /476)	y = 114.26ln(t) - 7.8832	0.89/0.94
Salt-acid treatment	(DW/146/293/475)	y = 25.621ln(t) + 0.4049	0.65/0.81
Traditional hydraulic fracturing	(DW/146/250/475)	y = -41.1ln(t) + 251.41	0.65/0.81

4. Conclusion

As a result of this work, the most significant geological and technical interventions were identified on wells with different technical characteristics using statistical methods. The proposed method allows evaluating the effectiveness of WI and implementing their planning.

References

[1] Tolstonogov A A 2014 Evaluation of the effectiveness of geological and technical interventions in the field of oil production *Fundamental research* **11-1** 150-154

[2] Kazakov A A 2003 Development of unified methodological approaches for evaluating the effectiveness of geological and technical interventions to improve oil recovery and intensify oil production (Moscow: Oil industry) **4** 26-29

[3] Kendall M G and Stuart A 1976 *Multivariate statistical analysis and time series* (Moscow: Nauka)

[4] Zagitova L R 2018 Mathematical model of statistical analysis of malfunctions and failures during equipment operation. Achievements, problems, and prospects of development of the oil and gas industry *International research and practical conference* (Almetyevsk) 328-332.

[5] Wilcox R R 2003 *Applying contemporary statistical techniques* (San Diego; London; Burlington - Academic Press)

[6] Freund R J, Wilson W J and Sa P 2006 *Regression analysis: statistical modeling of a response variable* (2nd ed.) (Burlington, MA - Elsevier Academic Press)

IOP Conf. Series: Materials Science and Engineering **860** (2020) 012017 doi:10.1088/1757-899X/860/1/012017

Bypassing annular gas by means of jet device in oil wells with electric submersible pumping units

D Saracheva, R Vakhitova and A Ivanov

Almetyevsk State Oil Institute (ASOI), 2, st. Lenina, Almetyevsk, 423450, Russia

E-mail: sarachevadiana85@mail.ru

Abstract. Nowadays the main equipment for oil production is the installation of electric centrifugal pumps. When operating oil wells with such installations, various complicating factors arise, in particular, the accumulation of free gas in the space between the outer surface of the tubing and the casing string. The experience of the ESP unit operation shows the negative impact of high gas pressure in the annular space on the efficiency of the unit. The technical solutions available today for the wells characterized by high GOR are the different types of valves, which have a number of operational shortcomings. To improve the efficiency of oil well operation, the authors suggest a jet device to bypass free gas above the dynamic level, thus reducing the gas pressure in annular space. The solution of the problem of reducing the negative effect of excess free gas in annular space is based on the study of experimental data from oilfields by using up-to-date information technology. The technology using a jet device in combination with a submersible electric centrifugal pump is proposed. The use of the jet apparatus enables reducing the gas pressure in the annular space of oil-producing wells even regardless of the discharge pressure and external ambient temperature, which increases the operating efficiency of the electric submersible pumping unit.

1. Introduction

Currently, many oil-producing companies use electric submersible pumps (ESPUs) as the main type of equipment for formation fluid production. These installations experience serious problems when pumping reservoir fluid containing an increased volume of free gas.

The degree of influence of the free gas determines the quality of operation of electric submersible pumps (ESPUs). This indicator is determined by the level of gas content at the ESPU reception.

The increase of free gas pressure in annular space facilitates pushing down the dynamic level of formation fluid to the bottomhole (to the level of submersible centrifugal pump's inlet level in extreme cases). This can trigger a failure of the submersible unit (pump starvation). That is why the development of devices and technologies for pressure reduction in the annular space of wells is relevant at this stage. The reduction of free gas pressure in annular space does not entail the reduction of bottomhole pressure, as this pressure is balanced by the increase of formation fluid's dynamic level. To prevent the process of formation fluid level fall to the level of ESP inlet and critical increase of gas pressure in annular space special valves can be installed at the wellhead. Experience analysis shows that stabilization of free gas pressure in annular space is not always possible by using wellhead valves (especially in case of freezing at low temperatures) and is not always effective.

The presence of free gas in the annular space between the pipes entails a reduction in the head capacity of a centrifugal pump. Presence of gas in the annulus increases the volume of formation fluid delivered to the first impellers and diffusers of the centrifugal pump and withdraws some portion of the energy that is supplied to the shaft of the ESP and is wasted on the compression of gas bubbles and their complete dissolution in the formation fluid. Then a certain amount of energy is returned to the fluid in the tubing. The released gas bubbles create a so-called "gas lift effect". They facilitate the lifting of formation fluid to the ground surface and reduce the required pressure head for the operation of an oil well. As the gas content increases, cavities are formed in the channels of ESPs' working stages (impellers and diffusers), which are not involved in the movement of gas-liquid mixture (GLM) through the channels in the general flow. The emergence of gas-filled cavities in a submersible centrifugal pump provides a rapid degradation of flow across impeller blades, reduces the flow capacity of the working channels, disrupts the energy exchange process of the pump with the pumped medium. The presence of gas bubbles on the operating surfaces of ESPs facilitates cavitation processes. The operation of the pump in forced cavitation modes is characterized by delivery failure, especially when the gas content increases.

The large-scale introduction of the pressure-assisted oil production system contributed to the increase in pressure at the wellheads of producing wells, which eventually increased free gas pressure in annular space. The increase in gas pressure at the wellhead is influenced by many different factors, such as unevenness of terrain lines, high pressure in the discharge lines due to remote location of automatic group metering station, increased viscosity of the produced fluid, etc. (figure 1).

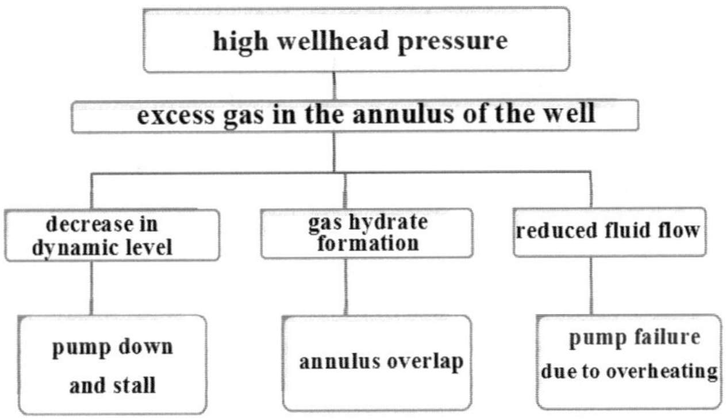

Figure 1. Consequences of high GOR influence on pumping equipment's operation in oil-producing wells

The excess amount of free gas in the space between the casing string and tubing facilitates heating of the submersible pump's housing components, the occurrence of gas-hydrate plugs, blocking of formation fluid flow, reduction of the effective volume of pumped fluid in ESP stages, lowering of dynamic level in a producing well. In case the dynamic level value reaches a critical point when the maximum gas content value at ESP inlet is exceeded, pump starvation and complete shutdown of the formation fluids production is expected.

The emergence of gas hydrates is facilitated by such processes during the operation of ESP units as the shutdown of production wells and the presence of a free gas phase at those sections of wells that are located above the level corresponding to the oil's bubble point pressure. When the well is shut down, the formation fluid starts cooling down and moisture is generated therein. Statistical analysis of field data on gas hydrate deposits indicates that such incidents are most frequently encountered in areas at the top of the tubing and annular space above the dynamic level. The formation of gas hydrates may lead to blocking of a wellbore's annular space between the tubes and to the reduction of formation fluid's flow characteristics, down to complete outage of the ESP unit.

Due to potential drop in the dynamic level in a producing well, it might be required to increase the location depth of a submersible unit. This entails additional material costs (submersible electric cable, tubing) and increased load on the tubing string.

Jet devices can be used to remove gas from the annulus of a producing well.

Jet devices came into use in various industries, as they can be used as standalone units or in combination with other pumping units [1, 2].

Gas injection is performed in jet devices. This process is featured by the transfer of kinetic energy from one stream to another during direct contact (through mixing method). Flows, in this case, may be in the same or different phases. The phase of the flowing streams may behave differently when they are mixed: either change or remain unchanged. A stream that starts the mixing process at a higher speed is called a working stream and another one at a lower speed is called an injected stream.

The following advantages are the reason for the wide popularity of the jet devices: small overall dimensions, no wear surfaces and moving parts, possibility of using even in hard-to-reach areas, high degree of reliability, no need for regular maintenance, possibility of pumping the fluids, which may have different phases, simple design, and high self-priming capacity.

Owing to their distinctive design features and operating principle, the jet devices have been adopted in many petroleum industry sectors, such as drilling and operation of oil and gas wells, formation fluid recovery, enhanced oil recovery from pay zones, bottom-hole zone treatment of production and injection wells [3 - 5].

2. Problem statement

Nowadays the technologies that involve the joint use of jet devices and submersible centrifugal units are effective. The joint use of jet devices and ESP units is reasonable in cases where there are complicated conditions: high content of free gas in formation fluid, aggressive pumped media, high temperature changes. Currently, the technology of jet devices implementation is widely used on oil-producing fields of Western Siberia in Russia, the USA, in Ukraine [6 - 9].

Certain requirements are imposed on the submersible unit as a whole when operating jet devices in oil wells. These requirements include the ability to provide three streams when the jet device and submersible unit are operated in conjunction: incoming injected and main flows, outgoing total flow. Then the following requirements should be considered: overall dimensions of individual submersible unit parts, located directly in the oil well, should match the internal diameter of the production string. The use of jet devices for recovery of downhole fluids contributed to the implementation of such principal types of equipment as a downhole power drive and surface power drive.

In the case of the surface power drive application, two independent flows are required – incoming main flow and the output total flow. If a submersible power drive is used, the axial lines of all three flows will coincide [10, 11].

To ensure the performance of oil wells, submersible power drive is the most suitable option, given the overall dimensions of the unit.

There is an application experience [6] of the submersible jet pump as a part of an ESP installation ("ESP installation - jet device" tandem technology).

Combined application of a jet device and a submersible centrifugal pump constitutes the tandem technology.

Application of tandem installations makes it possible [6], to effectively apply free gas at the inlet of a submersible centrifugal pump by bypassing it through a jet device into the tubing string from the annular space providing the creation of the optimal structure of the gas-liquid stream when recovering the formation fluid to the ground surface.

The jet device used in combination with an ESP makes it easier to maintain the required pressure in the annular space of an oil-producing well, which is close to the pressure created in the reservoir. This prevents the process of dynamic level lowering, reduces the consumption of tubing and submersible electric cable [12].

The method used to reduce the free gas pressure in the annular space is determined depending on various conditions of the formation fluid recovery process, such as oil well's flow rate profile, value of water cut in the formation fluid, gas-oil ratio and the amount of gas at the inlet of a submersible pump, the method of operating the producing well.

3. Research results

To implement the tandem technology, the authors propose a layout providing for the combined application of an electric submersible pump and a jet pump. The jet device is positioned above the dynamic level. The jet pump removes free gas from the annular space of a producing well. The jet pump shall have an automatic valve assembly as the main unit to reduce the pressure of annular gas in wells equipped with ESP installations. The automatic valve assembly bypasses the free gas into the inner space of the tubing string regardless of the value of free gas pressure in annular space and thermal conditions [13 - 15].

The jet pump is installed in the tubing string during set-up and running and pulling operations. The layout of the jet pump in the production well is shown in Figure 2.

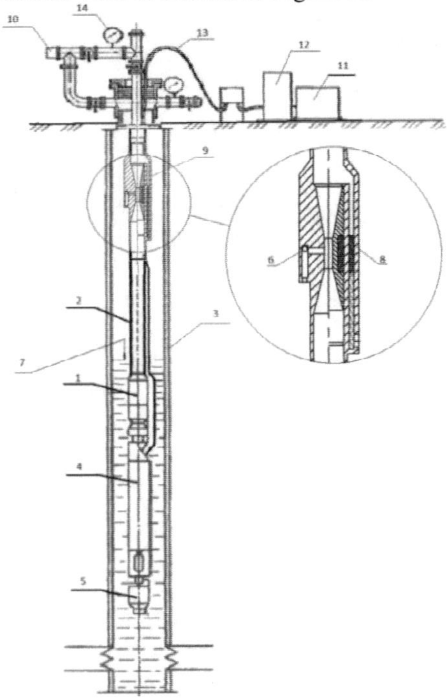

Figure 2. "Electric Submersible Pump – Jet Pump» tandem installation: 1 - submersible centrifugal pump; 2 - tubing string pipes; 3 - production string; 4 - submersible electric motor; 5 - seal section; 6 - check valve; 7 - fluid level; 8 - piston with magnetic components; 9 - jet pump; 10 - discharge pipe line; 11 - electric transformer; 12 - control station; 13 - surface part of the cable line; 14 - pressure control gauge.

The analysis of the theoretical and experimental research results [16] proves that in case the requirements to the technological process are fulfilled, there is an actual possibility to remove free gas from the wellbore annulus. To remove free gas from the annular space of the producing well and to further transfer this gas into the internal space of the tubing in accordance with the regulatory documentation, a jet pump is proposed (Russian Federation invention patent No. 2517287) [13, 16].

The proposed design provides for the jet pump which can be used to bypass free gas into the inner space of the tubing string. The jet pump is connected via a check valve to the inner space of the tubing string and the annular space. The jet pump is structurally designed as two symmetrical halves. The longitudinal section of the jet pump shows the following: one half is placed stationary in conjunction with the check valve, while the other half can move longitudinally inside the tubing string. The movable part of the device is connected by permanent magnets to a piston, which is spring-loaded from below. The piston is located in the cylinder parallel to the axis of the tubing string. In the lower part the cylinder is communicating with the annular space of the producing well, and in the upper part - with the inner space of the tubing [3, 13, 16].

IOP Conf. Series: Materials Science and Engineering **860** (2020) 012017 doi:10.1088/1757-899X/860/1/012017

The layout of the jet pump designed to remove free gas from the annular space is shown in Figure 3.

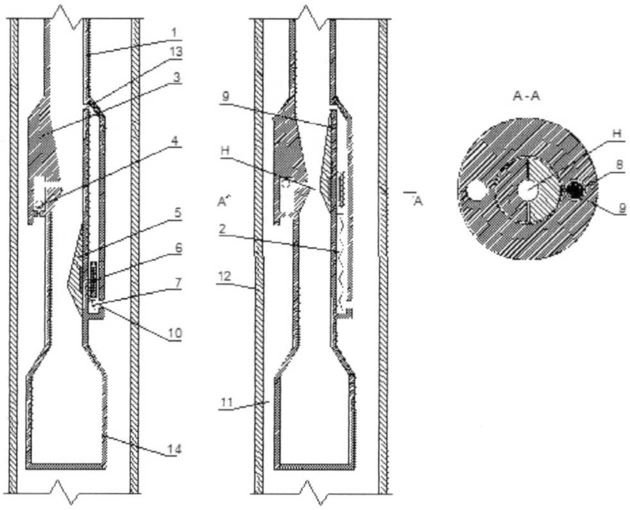

a) b)

Figure 3. Jet pump designed to remove free gas from the annular space 1 -tubing string; 2 - cavity under the piston; 3 - stationary unit; 4 - check valve; 5 - movable unit; 6 - magnets; 7 - spring; 8 - piston; 9 - steel cylinder; 10 - hole; 11 - annular space; 12 - casing string; 14 - submersible centrifugal pump; H - tapering hole: (a) Inactive position; (b) Active position

The principle of a jet pump's operation is as follows. When operating the submersible installation, the degassing of the formation fluid is activated at the ESP's inlet level. A certain portion of the gas mixture enters the electric submersible pump and is further pumped through the tubing to the primary oil treatment station. The remaining portion of the gas mixture is collected in annular space, accumulates therein and may rise above the dynamic level over time, thereby increasing the pressure at the wellhead of the producing well. If gas pressure rises in the annular space of the well, it will apply pressure to the bottom of the piston through the hole The piston moves upwards subject to the force of the spring and the pressure of free gas, which is higher than the pressure of formation fluid. The piston then displaces the movable symmetrical part of the jet pump designed to remove free gas, using permanent magnets. When the movable symmetrical part of the jet pump reaches the uppermost level, the pressure drop in the tapering section H begins (see Fig. 3, b). When pressure begins to drop, the check valve is opened and the gas mixture from the annulus flows into the inner space of the tubing string. Thus, the process of free gas pressure reduction is initiated in the annular space of the producing well. Further, the movable part of the jet pump begins moving downward under its weight. This increases the cross-sectional area between the stationary and movable parts, and also reduces hydraulic resistance of the formation fluid that moves along the tubing string.

4. Conclusion

1. Application of jet pump in combination with electric submersible pumping unit makes it possible to reduce the pressure of gas mixture in the annular space of producing wells. This makes it possible to increase the level of formation fluid in the well, which improves the operating conditions of the electric submersible centrifugal pump, namely, it increases the daily average output characteristics of the well. By reducing the gas mixture's pressure creation of gas hydrate plugs in the annular space of the well can be avoided.

2. Application of a jet pump designed to remove free gas from the annular space shall improve the performance of the submersible unit, namely: increase the efficiency of the submersible pump, reduce the submersible unit's installation depth by raising the formation fluid level in the annular space, i.e.

decrease the consumption of tubing pipes and increase the time between maintenance of electric submersible centrifugal units.

References

[1] Eames I, Milazzo A, Paganini D and Livi M 2013 The design, manufacture, and testing of a jet-pump chiller for air conditioning and industrial application *Applied Thermal Engineering* **58** 234-240

[2] Taubert D, Schinner G, Tomaras C, Ludwig S, Tranitz H and Wegscheider W 2011 An electron jet pump: the venturi effect of a fermi liquid *Journal of Applied Physics* **109 (10)** 102-112

[3] Saracheva D, Vakhitova R, Davletov M 2013 On reducing the gas pressure in the annulus of wells equipped with submersible centrifugal pumps *Materials of the scientific-practical conference with international participation "Earth Sciences: current status and development priorities* **9** 38 (in Russian)

[4] Carvalho P, Podio A, Sepehrnoorik H 1999 An Electrical Submersible jet Pump for Gassy Oil Well *J. of Petroleum Technology* 34-36

[5] Shtoff A 1999 Prediction of the oil production rate of jet pumping oil wells from the sample well data *SPE Production and Facilities* **14 1** 77-80

[6] Mishchenko I 2003 *Downhole Oil Production: A Textbook for High Schools* (Moscow: Publishing House "Oil and Gas" of Russian State University of Oil and Gas named after I.M. Gubkin) (in Russian)

[7] Panevnik A 2002 Classification of designs of borehole ejection systems *Chemical and Oil and Gas Engineering* **2** 19-20 (in Russian)

[8] El-Sallak M and Hefny M 2010 Experimental investigation of the performance of liquid gas jet pumps with inlet swirling *Proceedings of the Institution of Mechanical Engineers Part A: Journal of Power and Energy* **224 3** 363-372

[9] Xiang Q, Wu Y, Li H and Wang C 2014 Experimental study on flow in the diffuser of liquid-jet gas pump *Paiguan Jixie Gongcheng Xuebao* **32** 7 558-562

[10] Ansari A, Sylvester N, Sarica C, Shoham O and Brill J 1994 A Comprehensive Mechanistic Model for Two-Phase Flow in Wellbores *SPEPF 143 Trans. AIME* 143-152

[11] Hasan A and Kabir C 1992 Two-phase flow in vertical and inclined annuli *International Journal of Multiphase Flow* **18** 279-293

[12] Vakhitova R, Saracheva D, Abramova E 2012 Gas removal from the annulus of wells equipped with ESP, jet pumps. *Materials of the scientific session of scientists of Almetyevsk State Oil Institute* **1** 139 – 141 (in Russian)

[13] Urazakov K, Vakhitova R, Saracheva D and Abramova E 2014, RU, Patent No. 2517287

[14] Topolnikov A, Urazakov K, Vakhitova R, Saracheva D 2011 The method of calculating the parameters of the jet pump during a joint operation with ESP *Electronic scientific journal "Oil and gas business"* **3** 134-146 (in Russian)

[15] Kuisheng W and Zhongqiang T 2001 Revised characteristics curve optimizes jet pumping. *Oil & Gas Journal* **99** 52

[16] Saracheva D, Molchanova V and Vakhitova R 2014 Jet apparatus for annular gas bypass in wells equipped with ESP *Innovative oil and gas equipment: problems and solutions: proceedings of III All-Russian scientific and technical conference* 129-132 (in Russian)

IOP Conf. Series: Materials Science and Engineering **860** (2020) 012018 doi:10.1088/1757-899X/860/1/012018

Compensator-vibration isolator

R I Vakhitova, D A Saracheva and Z F Zaripova

Almetyevsk State Oil Institute, 2A, Lenin Str., Almetyevsk, 423452, Russia

Abstract. Nowadays electrical submersible pump (ESP) units remain the main type of formation fluid production equipment. There may be complicating factors during the operation of oil production wells equipped with such units. Complicating factors include the fall of the submersible unit downhole to the bottom of the production well. The damage from such falls is great. Despite the accumulated experience in this area, effective solutions need to be developed. In the work, the sources of vibration in the ESP are noted, and the places of breaks in the unit are indicated. The reasons for the occurrence of vibration processes in a submersible unit are summarized. The authors propose a design of a compensator intended for quenching vibration energy in a submersible unit. To evaluate the effectiveness of the compensator as part of an ESP unit, tests on industrial sites were conducted. At Open Joint-Stock Company Alnas, the test stand was a vertical well. Tests in Almetyevsk Central Base of Electric Submersible Production Service were carried out on a horizontal test stand section by section. Analysis of the results of the tests showed that the use of a compensator can reduce vibration levels.

1. Introduction

Quite often emergencies in electric submersible pump units (ESP) occur in oil production: SD-failures (spontaneous dismemberment) or "falls". This problem of the last decades has not yet been solved radically. The financial damage from the "falls" is great, so any possibility of solving this problem is justified. More than a thousand accidents occur in oil production regions. Even if the average amount of damage per accident is more than 500,000 rubles, the total losses per year will be more than 500 million rubles [1, 2]. At the present stage, some experience has been accumulated in dealing with emergencies, but discussions are still ongoing about the causes and measures to eliminate "falls".

2. Problem statement

The causes of SD-failures are not fully understood at the moment, but the available analysis of operational data suggests that they are mainly associated with vibration processes [3, 4].

Sources of vibration include [4, 5]:

1) the main sources of vibration processes that occur during the operation of equipment that contribute to the development of a critical level of vibration – the wrong choice, wear of working surfaces, the curvature of the wellbore, etc.;

2) additional vibration levels in the ESP unit that occur after repair and assembling works;

3) the permissible level of initial vibration, which is laid down in the unit design of 6.5 mm/s.

Note the places where most often there are breaks in the ESP units (Figure 1):

- connection between the module head and the upper section of the electrical submersible pump,
- connection between the compensator and the submersible motor,
- on the body part of the electric submersible pump,
- on the body part of the protector,

IOP Conf. Series: Materials Science and Engineering **860** (2020) 012018 doi:10.1088/1757-899X/860/1/012018

- by threaded connections of the check valve,
- for threaded connections of tubing,
- on flange connections.

The analysis of the operating history of submersible centrifugal units allowed establishing the relation between such indicators as the operating time between failures of submersible units and the distribution of vibration levels.

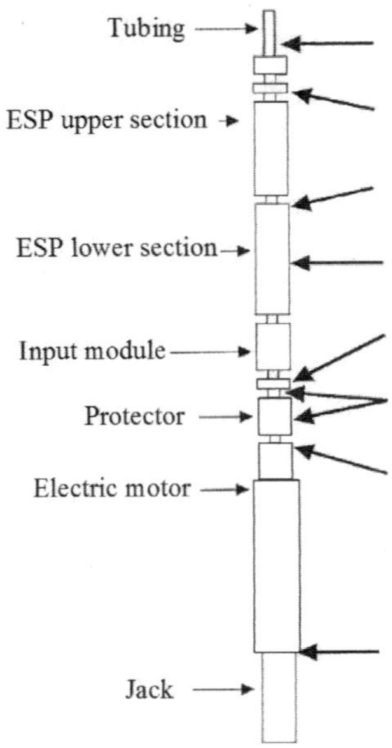

Figure 1. Places of possible breakages of the submersible unit

Many researchers have the same view that falls in the ESP occur as a result of processes such as pressure pulsation and increased vibration of a worn-out submersible pump. This leads to the destruction of the main fasteners and various weak points in the electric submersible pump unit [1]. Wear resistance and strength of the main components of the structure are the most important indicators of the resource and operation of the submersible unit. Structural x-ray analyses, fractographic and other studies of worn parts of details have shown the presence of almost all types of wear mechanisms in electric submersible pumps: abrasive, adhesive, erosive, fatigue and corrosive [6, 7].

To reduce the level of accidents in oil production wells, analysis of the operational history of submersible units indicates that it is necessary to develop devices that reduce vibration levels.

There are two ways to reduce harmful dynamic phenomena: vibration absorption and vibration isolation.

The use of vibration absorbers is one of the most common ways to reduce the level of vibration [8].

The friction element is the main part of any absorber. Friction contributes to the absorption of vibration energy and forced action oscillations are carried out with a smaller amplitude than in the absence of friction. This effect is most noticeable in the resonance area [9].

The units' vibration is common for the operation of oil production wells equipped with ESPs, namely the occurrence of oscillating waves of the longitudinal section in the lifting column of the tubing, which, having risen at the wellhead return to the submersible unit.

The overlap of oscillatory contours of forwarding and reverse waves is dangerous for the occurrence of resonant phenomena, which can contribute to the breakage of the main nodes of the ESP or the lifting column.

IOP Conf. Series: Materials Science and Engineering **860** (2020) 012018 doi:10.1088/1757-899X/860/1/012018

3. Research Results

The authors propose a specific design of the compensator installed at the outlet of the submersible unit to extinguish the vibration energy in the ESP and prevent dangerous vibrations in the lifting column.

The design of the compensator (Figure 2) includes housing, friction elements, fasteners. The compensator housing is connected to the tubing using a coupling and an adapter. In the inner part of the housing, there is an eccentric pipe with friction elements. The compensator is connected to the submersible pump unit by a branch pipe. The pipe is connected to the eccentric pipe using a thread. Friction elements are made in the form of cylindrical bushings. The material of these bushings is an oil-resistant rubber. The key locks the bushings from turning the bushings inside the housing. The contact surfaces of the eccentric pipe and the bush are coated with the appropriate lubricant. The presence of radial and longitudinal play of the eccentric pipe is not allowed. Axial play is eliminated by tightening the adapter with the ring until it stops. The presence of oil-resistant rubber rings helps to maintain tightness between the adapter and the eccentric pipe. It is possible to ensure the tightness of the compensator due to the own weight of the submersible unit and the presence of pressure in the ESP. The pressure drop is expected to be maintained at least 200 kgf/cm^2.

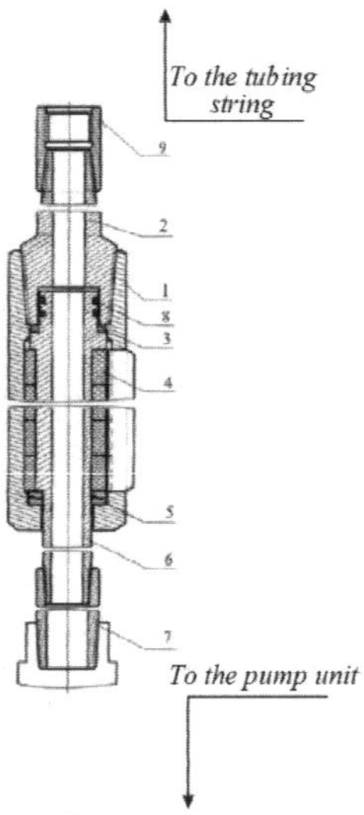

Figure 2. Radial compensator design: 1 - housing, 2 - adapter, 3 - rings, 4 - rubber bushings, 5 - washers, 6 – eccentric pipe, 7 - connecting pipe, 8 - rings, 9 - coupling

It is recommended to use the radial compensator with the ESP unit. As a drive, it is better to use submersible electric motors with a power of up to 40 kW. The surface area intended for compensation is 308 cm^2. The compensation torque is 14 kg·m. The axial load is allowed at a value of 2830 kg.

The compensator's principle of operation, designed to reduce the vibration level of the submersible unit, is as follows. During the start of the submersible pump unit's electric motor, a torque occurs. The starting torque is perceived by the friction elements and the eccentric pipe. This prevents twisting of the electric cable of the submersible unit, the possibility of limiting concentration stresses in the pipe column, their breakage, and the permissible limits of starting currents are not exceeded.

If the check valve is omitted due to contamination or de-energization of the electric pump unit, the friction elements and the eccentric pipe perceive the reverse torque that occurs from the rotating masses of the submersible pump unit and the inertial forces of the reservoir fluid column in the tubing column, which in turn prevents the tubing from turning.

To assess the effectiveness of the use (reduction of the vibration level of the submersible unit) of the radial compensator, tests were conducted in Open Joint-Stock Company Alnas and Almetyevsk Central Base of Electric Submersible Production Service in Almetyevsk, Republic of Tatarstan.

Tests at Open Joint-Stock Company Alnas of the compensator intended for lowering the vibration level of the submersible unit were carried out on a vertical automated stand of the VIG-98 series, which externally simulates a well. During the tests, measurements of vibration, pressure, and energy parameters were made.

The results of the tests carried out on the test station of the Open Joint-Stock Company Alnas of the compensator for lowering the vibration levels of the submersible unit on the V1G series vertical well stand are shown in figures 3 and 4. A submersible sectional centrifugal pump Electric Centrifugal Pump A5-45-2050 type was selected for the experiments.

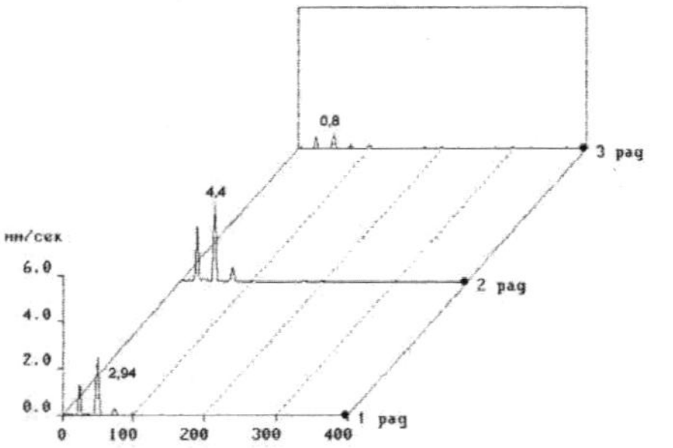

Figure 3. Results of Electric Centrifugal Pump A5-45-2050 tests without using a compensator and measuring vibration levels (4.4067 mm/sec)

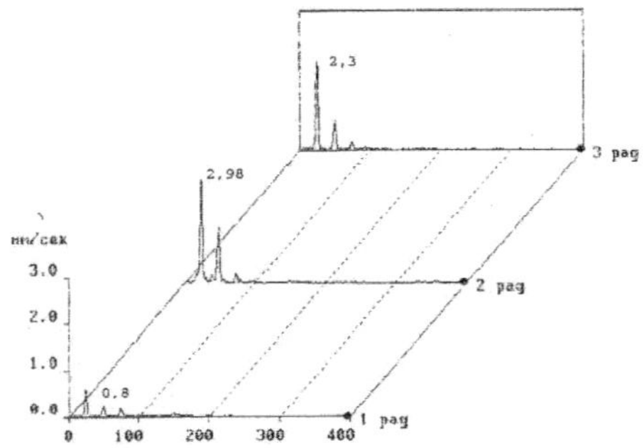

Figure 4. Results of Electric Centrifugal Pump tests A5-45-2050 using the compensator and measurement of vibration levels (2.9801 mm/sec)

The following results were obtained during the tests: the vibration level of the submersible unit without the use of a compensator was 4.4067 mm/s (Figure 3), with the use of compensator – 2.9801 mm/s (Figure 4).

The analysis of the results of measuring the vibration levels of the ESP unit shows that the use of a radial compensator makes it possible to reduce the value of vibration speed.

The deviation of the pressure characteristic of the submersible pump Electric Centrifugal Pump A5-45 during tests without the use of a compensator was mainly 2.73 m, with the use of a compensator - 1 m. A comparison of the results of deviations of the pump pressure characteristics showed that the use of the compensator contributed to the minimum head deviations in contrast to the passport data.

The power consumption of the submersible unit during tests without the use of a compensator was mainly 11.27 kW, with the use of a compensator-10.82 kW.

The efficiency of the submersible unit during tests without the use of a compensator was mainly 15.45 %, with the use of a compensator-15.64 %.

The analysis of the obtained results of energy characteristics showed that the use of the compensator together with the ESP increases the efficiency and reduces the power consumption of the unit.

Tests at the Almetyevsk Central Base of Electric Submersible Production Service of compensator units for lowering the vibration level were carried out on the horizontal test stand of the REDA company. Tests of electric submersible pumps were carried out in sections on a horizontal stand after washing (Figure 5).

During the tests, the pump section and the compensator for reducing the vibration of the submersible unit are installed horizontally on the stand, and the clamping chain devices are tightened to hold the nodes on the stand during the test period. The discharge adapter and the corresponding head are connected to the centrifugal pump. The personal computer is turned on, and the HPTS program is automatically loaded and started. This computer program makes it possible to test the entire range of commercially available electric submersible pumps. The colour printer displays test results with a graphic and tabular image of the power consumed by the unit, the Q-H characteristics, and the efficiency. A horizontal test bench records the vibration level and checks the tightness of the pump section. During steady-state operation of the pump section with a load when measuring vibration levels at three points: REDA company vibration analyzer is used on the base, in the middle of the section and on the head as shown in Figure 5.

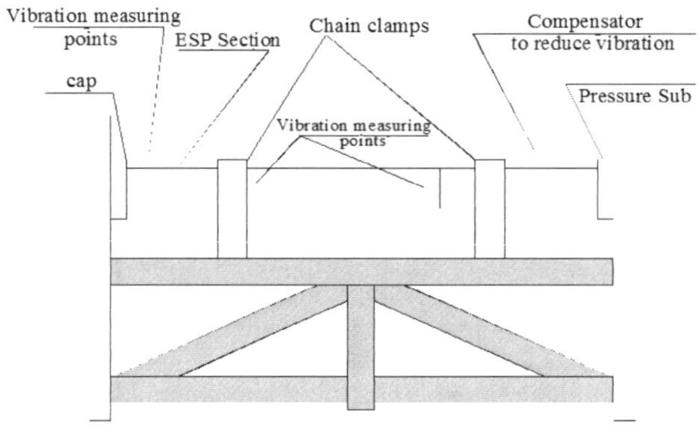

Figure 5. Horizontal test stand of REDA company

The Electric Centrifugal Pump A5-125-1300 pump section, restored after repair work, was selected for testing. A number of stages in the pump's section 131.

The results of the tests on Almetyevsk Central Base of Electric Submersible Production Service of the compensator designed to reduce the vibration level of the submersible unit on the horizontal test stand of submersible pumps of REDA company are presented in tables 1 and 2.

Table 1. Test results of the Electric Centrifugal Pump A5-125-1300 pump without the use of a compensator

Places for measuring vibration levels	Rotation speed of the shaft, amp.
The head of the section	1.524
The middle of the section	2.102
The base of the section	1.452

Table 2. Test results of the Electric Centrifugal Pump A5-125-1300 pump using a compensator

Places for measuring vibration levels	Rotation speed of the shaft, amp.
The head of the section	1.519
The middle of the section	2.098
The base of the section	1.441

During the tests, the rotation frequency was 50 Hz. The vibration level of the pump section during the tests without the use of a compensator in general was 1.69 mm/s, with the use of a compensator – 1.68 mm/s. A comparative analysis of measuring the vibration levels results of the tested sections of the submersible pump allowed concluding that the use of a radial compensator can reduce the amount of vibration speed.

4. Conclusion

A radial compensator is proposed for the reduction of the vibration levels of the submersible unit and to absorb the vibration processes of the tubing lifting column. The tests of this compensator on the stand - on the well belonged to Open Joint-Stock Company Alnas has shown that the use of the compensator reduced the magnitude of the unit' vibration from 4.41 mm/s to 2.98 mm/s. A comparison of the energy characteristics' measurement results showed that the use of the compensator in conjunction with submersible units allowed reducing the consumed unit power capacity by 4%. The conducted tests of the compensator on the Almetyevsk Central Base of Electric Submersible Production Service horizontal test stand also showed that the use of the compensator reduced the values of the vibration speed of the pump section.

References

[1] Volkov V 2000 New studies of RS - failures of the ESP *Materials of the IX All-Russian technical conference "Production and operation of the ESP"* (in Russian)

[2] Tsykin I 2002 ESP operation in the fields of the Tyumen Oil Company. Experience, problems, prospects *Materials of the Xl All-Russian Technical Conference "Production and Operation of ESP"* (in Russian)

[3] Muzipov Kh, Savinykh Yu, Dunaev S 2005 Acoustic technology to reduce vibration of tubing equipped with centrifugal electric pump installations *Oil industry* **11** 82 – 83 (in Russian)

[4] Shelkovnikov E 2000 Analysis of the causes of accidents at wells equipped with ESPs at OJSC "Surgutneftegas" *Materials of the IX All-Russian Technical Conference "Production and Operation of ESPs"* (in Russian)

[5] Zalyatov M 2000 Problems of accident rate at wells with ESP in TATNEFT *Materials of the IX All-Russian Technical Conference "Production and Operation of ESP"* (in Russian)

[6] Panko Ya 1990 Fundamentals of the applied theory of oscillations and shock *(Leningrad: Polytechnic)* (in Russian)

[7] Vakhitova R, Zotov A, Urazakov K 2005 Justification of the possibility of using a mechanical compensator to reduce the level of vibration in submersible electric centrifugal pump installations *Problems of geology, geophysics, drilling and oil production* pp 147 – 154 (in Russian)

[8] Vakhitova R, Zotov A., Urazakov K 2005 Mechanical compensator to reduce vibration in submersible electric centrifugal pump installations *Oilfield business* **10** 34 – 37 (in Russian)

[9] Urazakov K, Topolnikov A, Doomler E, Vakhitova R 2017 Diagnosing the technical condition of electric centrifugal pumping units by their vibration level *Oil and gas business: scientific. journal USTU* **15 (1)** 103 – 107 (in Russian)

IOP Conf. Series: Materials Science and Engineering **860** (2020) 012019 doi:10.1088/1757-899X/860/1/012019

Structural and intelligent scheme of navigation system of a ground-based mobile robot for forming a traffic route

A A Sharafutdinov and A Y Timasheva

[1] FSBEI HE «Ufa State Petroleum Technological University» 4, M. Pinskii str., Ufa, 450064, Russia

E-mail: azat_sharaf@mail.ru

Abstract. At the beginning of Russian studies on the task of creating mobile ground fire robots for use in extreme situations in late 1980, it was justified by the fact that firefighters need protection when dealing with the consequences of accidents in areas with high levels of radioactive pollution. The use of fire robotics was increase because of a large number of technogenic accidents. There was a need for use of fire ground robots, where emergency situations, fire, and technogenic accidents arise because it's dangerous for the life and health of firefighters and people. Fire mobile ground robot is an automated self-propelled technical device which does given commands of human and other actions without human. People use this robot at the accident on area, where is dangerous for the health of firefighters things are situated like oil, nuclear, chemical objects. Now we are considering a structural and intelligent navigation system of fire mobile ground robot, in which a structure of fire robot must create a good way to the center of ignition, control parameters of motions, track own location for safe navigation and improvement tactical of opportunities of the firefighters. The introduction of a fire mobile ground robotics device is needed for an increase in the tactical capabilities of fire detachments during the extinguishing of the fire. The goal is to the definition of tasks and development structural and intelligent scheme of the navigation system of fire mobile robot for the shortest path to fire and extinguish a fire.

1. Introduction

A robotics device is an automatic device which is designed to perform motor and highly intelligent human functions.

Robotics devices became an important part of our industry in the twenty-first century.

Now we are going to consider a mobile robot for use in emergency situations which is name EL-4.

The robot's task is to do emergency rescue work on the free part in the territory, on the warehouse and on the production facilities. Also, it does transport fire extinguishers things and does fire extinguishing work in the conditions of technological accidents. Besides, it disassembles constructions for passage to fire.

The mobile robot is regulated by a routing computer or by radio with support for the TV radio module or by a cable that connects with the sensor system, control system, and communication system. Program product and the structure of regulation of the robotic device have a component scheme, do improvement in part of revisions, implement noise immunity, to test of reliability improvement, self-diagnosis. Besides, they carry out auxiliary goals and increase other tactical and technical parameters [1].

The Mobile robot has sensors and with them, it estimates an environment that surrounds it, also it has a chain of precise effects for the sake of influencing the environment and the concept of control, which provides to do directed and practical operations for robotics device.

A robotics device uses remote detectors, temperature sensors, chemical element sensors, radiation sensors in emergency situations, and it uses motor devices for the perception of its environment.

In a circular structure, elements make the control mechanism work, in relation to changes in the environment. Consequently, it is feedback. If the control mechanism is exposed at the right time, the sensors indicate this change by sending a message about the environment to the control system.

2. Materials and Methods

The use of such a robotics device in emergency situations will be able to eliminate the threat to the health and life of the human operator. Also, it is important to form a navigation system structure that will make a map of the environment where mobile robotics reproduces the route leading to the landmark and can bypassing the obstacles encountered on the way.

The main method of controlling the navigation of a mobile robotics device is performed in the return of tasks, where from values with the highest degree of abstraction to levels with the lowest degree of abstraction, and data about the state of the robot pass in the opposite direction. At any given degree, navigation saves a map of the robot's own working area and contains special video sensors with visual probability.

Imagine that the outer environment isn't considered in the source period and the mobile robotics device is in the middle of an arbitrary area (in picture 1 it's numeral 0), where it is the formation of the coordinates of the constitute map, as shown in picture 1 [2].

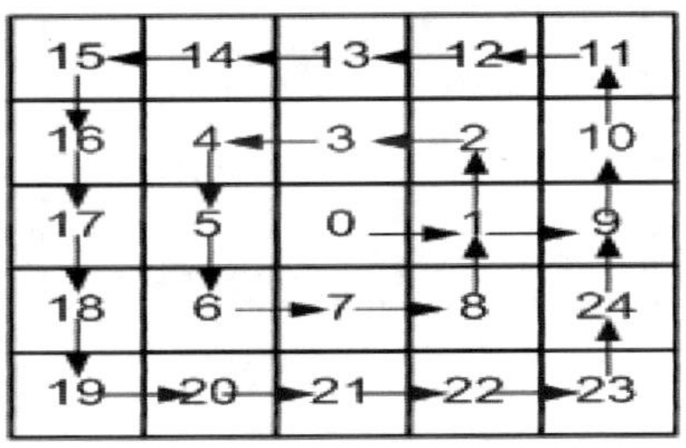

Figure 1. Map of the robot's trajectory

Areas 1,2,3,4,5,6,7,8 are accepted as potentially passable. The inspection begins with the first section. Movement is performed in the direction of the middle of the initial section only if this section is free.

Based on this, we can make a map of the territory, as shown in figure 2 [3].

IOP Conf. Series: Materials Science and Engineering **860** (2020) 012019 doi:10.1088/1757-899X/860/1/012019

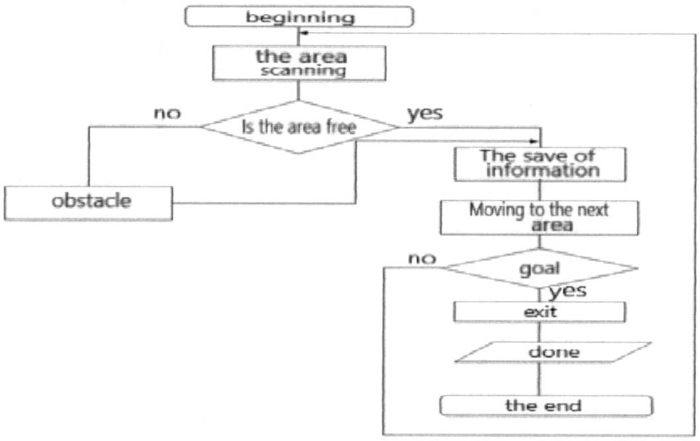

Figure 2. Sequence of drawing up a map of the area

Ultimately, we extracted the result, as shown in picture 3 [3].

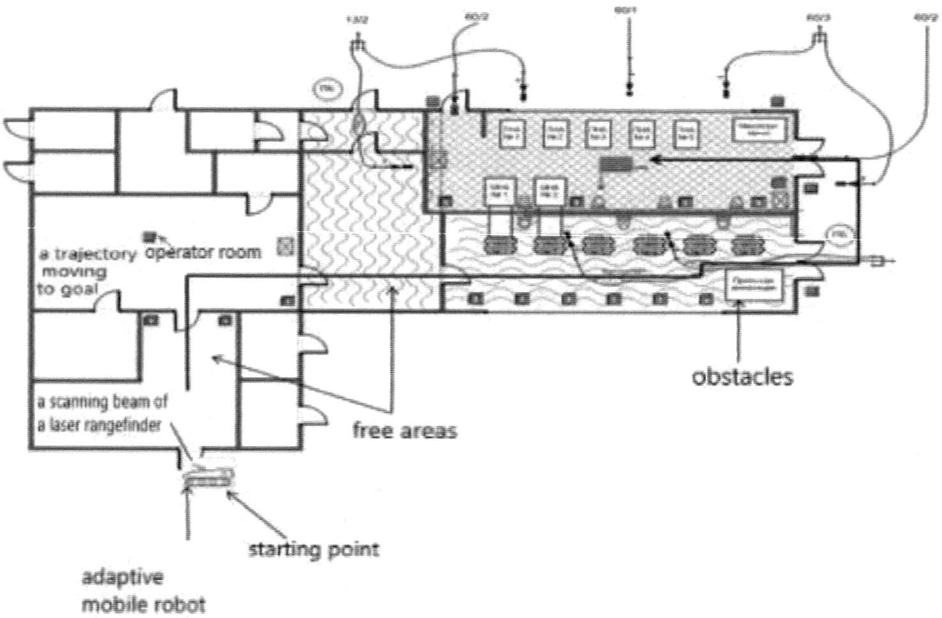

Figure 3. Passage of the optimal path of the linear production and dispatch station "Cherkasy».

After creating a map of the environment, the robot uses it in the process of moving to the goal.

The structure of regulation by a mobile robotics device solves given tasks:

- data processing to collect information about the work and the environment around it;
- processing data from the interface to the operator;
- event planning for the understanding of the main task and the sequence of subtasks for further execution of the main task;
- creating a software path for moving the mobile robot;
- drawing up actions at the robot's executive mechanisms [4].

3. Discussion

One of the peculiarities of constructing concepts for managing a highly intelligent mobile robotic device is that it is based on a hierarchical multi-level principle and with an increase, the hierarchical rank of subsystems its level of intelligence development also increases. The action management system is considered the highest component of this hierarchy, next is the movement management system, and the punctual mechanism management system is considered the lowest component of this hierarchy [5]. In addition to these subsystems, there is an information and measurement system in the structure, which is still required to possess some mental probabilities and an interface with the operator.

The action control system is pre-defined to create a reasonable maneuver of the robotic vehicle to complete the task set before it.

The motion control system is designed to plan the software trajectory of the mobile robot.

The control system of the precise device centralizes the plan of the actuators of the mobile robotic device [6].

The information and measurement system is designed to collect, process, and transform sensory data into signals.

We can conclude that the creation of mobile robots for production purposes will be relevant. Thus, the robot's task is to study the environment in order to make a map and find a goal, form a map of the environment, and find the optimal path. Based on this, the application of the described method allows quick building of a map of the external environment and effectively plans the trajectory of movement to the goal. This application of the mobile robot will improve the tactical capabilities of firefighters, strengthen the level of protection against fire hazards [7].

4. Conclusion

The use of a mobile robot will remove the health and life threats of people. In addition, a system of navigation can make up a map, in which a mobile robot is functional, and robot plans a route, which leads to the goals and can avoid obstacles, which encountered on the way.

References

[1] Hafizov I F and Sharafutdinov A A 2016 Designing technical training tools for oil and gas specialists based on an optimal set of trainings *in the collection: Modern technologies in oil and gas business-2016* 366-369

[2] Sharafutdinov A A and Khafizov F I 2015 Development of training for the simulator complex for training fire control personnel *News of higher educational institutions. Oil and gas* **5** 136-141

[3] Sharafutdinov A A and Hafizov F Sh 2015 Application of an automated communication system and operational management of fire departments of the state fire service in extinguishing large fires *Electronic scientific journal of oil and Gas business* **1** 345-363

[4] Sharafutdinov A A and Hafizov I F 2017 Development and creation of a web application for modeling emergency situations at hazardous production facilities of the oil and gas complex *Problems of collection, preparation and transport of oil and petroleum products* **1 (107)** 210-218

[5] Sharafutdinov A A and Hasanova A F 2015 Use of a training complex to optimize personnel actions in the event of a fire *Problems of safety in liquidation of consequences of emergency situations* **2 1 (4)** 319-323

[6] Timasheva A Y 2019 Structural-intellectual diagram of the navigation system of the ground mobile robot for forming the route of motion *Proceedings of twenty eighth international scientific-technical conference « Safety systems - 2019»* N G Topolsky (Moscow November 28 2019)

IOP Conf. Series: Materials Science and Engineering **860** (2020) 012019 doi:10.1088/1757-899X/860/1/012019

[7] Timasheva A Y 2019 Block diagram of a mobile robot navigation system *Modern technologies in the oil and gas business – 2019* (Electronic Materials vol 2) Mukhametshin V Sh (Ufa USPTU)

IOP Conf. Series: Materials Science and Engineering **860** (2020) 012020 doi:10.1088/1757-899X/860/1/012020

Underground storage facilities for helium concentrate in porous medium

A O Golovachev and I V Starokon

Gubkin Russian State University of Oil and Gas, 65, Leninsky Ave., Moscow, 119991, Russian Federation

andreigolovachev@mail.ru

Abstract. The gas of many fields in Eastern Siberia (Chayandinskoye, Kovyktinskoye, Sobinskoye, Chikanskoye, etc.) has a complex component composition with a high helium content. In order to realize the advantage of a unique raw material base, it is necessary to put into practice solutions for the industrial implementation of technologies for the extraction of helium, storage and transportation. One of the directions for the solution of these problems is the creation of temporary underground storage of helium concentrate in a porous medium next to gas chemical plants. This will ensure supply flexibility and reduce helium losses. The global demand for helium is continuously increasing from year to year and in the long term, a favorable situation will emerge on the global helium market for Russia, a positive effect from which can be achieved, provided the benefits of the raw material base are realized. A positive effect means not only the monetary effect, but also takes into account the formation of prerequisites for the development of new spheres of influence: access to new trading platforms and sales markets, the development of high-tech industries, basic and applied sciences, etc. In order to realize the advantage of the raw material base, Russia needs in a fairly short period of time to put into practice the decisions on the industrial introduction of gas separation technologies and underground helium storage.

1. Introduction

In accordance with the Eastern gas program of PJSC Gazprom and the energy strategy of Russia in the East, new large gas production centers and a unified gas transportation system are being formed. This centers in the long term will allow organizing a new channel for exporting Russian gas to the countries of the Asia-Pacific region. Currently, several regional gas production centers have been formed in accordance with the location of large fields:

- Yakut on the basis of the Chayandinskoe field with the prospect of developing the neighboring fields - Sobolokh-Nedzhelinskoe, Verkhnevilyuchanskoe, Tas-Yuryakhskoe and Srednetyungskoe;
- Irkutsk based on the Kovyktinskoe field with the development of the Chikansky field and fields in the north of the Irkutsk Region;
- Krasnoyarsk based on the Sobinsko-Paiginsky and Yurubcheno-Tokhomsky fields with the prospect of developing the Omorinsky, Kuyumbinsky, Agaleevsky and other fields. (These fields do not belong to Gazprom).

2. Reserves of Russian helium

The gas of most of these fields has a complex component composition, but their main distinguishing feature is the high helium content [1].

Figure 1. Large fields in eastern Russia

The total reserves of Russian helium in the main prospective helium fields are estimated at *14.5 billion* m^3. Of these, helium reserves in the A + B + C_1 categories amounted to about *7.6 billion* m^3, and in the C_2 category - *6.9 billion* m^3 [2]. The largest helium fields of Russia in categories A + B + C_1 include [3]:

- Kovyktinskoe field - volume of explored reserves of helium 3.9 billion m^3 (Irkutsk region, helium content - 0.26 - 0.28%)
- Chayadinskoe field - 1.85 billion m^3 (Republic of Sakha, helium content - 0.43 - 0.65%)
- Sobinskoe field - 0.8 billion m^3 (Evenki Autonomous Area, helium content - 0.50 - 0.70%)
- Srednebobuinskoye field - 0.6 billion m^3 (Republic of Sakha, helium content - 0.2-0.67%)
- Tas-Yuryakhskoye field - 0.4 billion m^3 (Republic of Sakha, helium content - 0.38%)

Russian reserves system is based solely on the analysis of geological attributes. Explored reserves are represented by categories A, B, and C1; preliminary estimated reserves are represented by category C2; potential resources are represented by category C3; and forecasted resources are represented by categories D1 and D2. Natural gas reserves in categories A, B and C1 are considered to be fully extractable [4].

Characteristics of the most promising helium-containing fields are presented below.

Table 1. Information about Chayandinskoe field

Field name	**Chayandinskoe**
Country	*Russia*
Region	*Yakutia (Sakha Republic)*
Company owner (license)	*PJSC Gazprom, 2008*
Field content	*Gas, oil, gas condensate*
Rank	*Large (unique)*
Discovery date	*1989 г.*
Beginning of field development	*2015 г.*
The field's annual design output	*25 billion m^3 of gas, 1.9 million tons of oil and 0.4 million tons of gas condensate*
Depth of the field	*1450-1850 м*
Reserves	*B_1-B_2, 1,4 trillion m^3 of gas and about 76,7 million tons of oil and gas condensate*
Helium content	*0,43-0,65%*

Table 2. Information about Kovyktinskoe field

Field name	**Kovyktinskoe**
Country	*Russia*
Region	*Irkutsk region*
Company owner (license)	*JSC RUSIA Petroleum, 1992* *PJSC Gazprom, 2011*
Field content	*Gas, gas condensate*
Rank	*Large (unique)*
Discovery date	*1987 г.*
Beginning of field development	*2001 г.*
The field's annual design output	*25 billion m^3 of gas*
Depth of the field	*2838-3388 м*
Reserves	*C_1-C_2, 2.7 trillion m^3 of gas and 90.6 million tons of gas condensate*
Helium content	*0,26-0,28%*

In accordance with the classification, these fields are rich and very rich helium-containing gases

Table 3. Classification of natural gas depending on the helium content

Group name	Helium concentration, %
Poor	0,02-0,05
Rich	0,05-0,30
Very rich	0,30-1,0
Uniquely rich	>1,0

IOP Conf. Series: Materials Science and Engineering **860** (2020) 012020 doi:10.1088/1757-899X/860/1/012020

These fields are key elements in the formation of the resource base for the Power of Siberia gas pipeline and are included in the contract for the supply of natural gas to China.

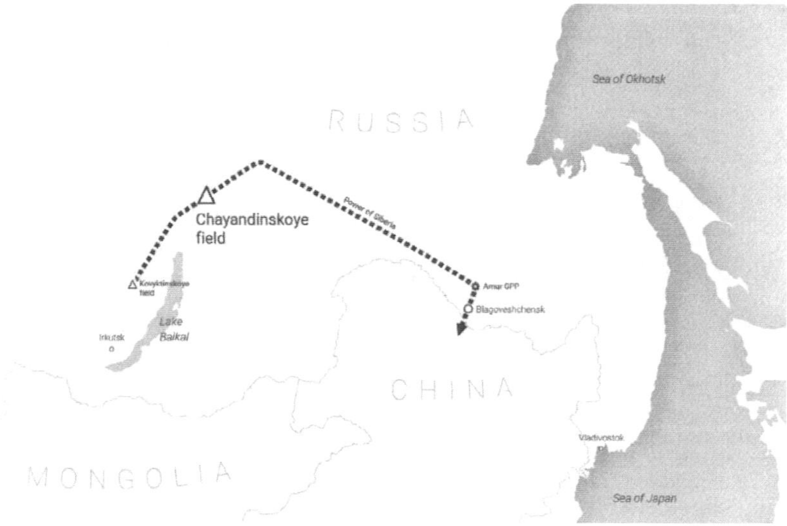

Figure 2. Gas pipeline Power of Siberia

In October 2015, the beginning of the construction of the Amur GPP for processing and extraction of valuable components from natural gas before delivery to China. The stage-by-stage commissioning of its production trains will be synchronized with the development of Gazprom's production capacities in Yakutia and the Irkutsk Region [5].

Table 4. Characteristics of the Amur GPP

Design processing capacity	*42 billion m^3 of natural gas per year.*
Helium production	***60 million m^3 per year.***
Ethane production	*about 2.5 million tons per year.*
Propane production	*about 1 million tons per year.*
Butane production	*about 500 thousand tons per year.*
Plant's area	*800 hectare*

The supplier of the main technological equipment for cryogenic gas separation with the production of helium and other components for petrochemistry is the ***German company Linde AG***. The design capacity for the production of helium will be about *60 million m^3/year* or *10 thousand tons/year*, the degree of extraction - ***99.2%.***

Based on the assumptions of specialists of Gazprom VNIIGAZ LLC, the maximum annual production of helium concentrate will be in 2025. This year the maximum is *80 million m^3/year*, the average value is *70 million m^3/year*, and a minimum of *59.6 million m^3/year* [5]. Thus, when dealing with helium, it is necessary to be guided by the principle of rational distribution and accumulation of the extracted helium reserves of Eastern Siberia and the Far East. The demand for helium in the domestic market is quite low, and the sale of large volumes of helium on the world market can cause problems with the frequency and volume of purchases of Russian helium. Therefore, helium accumulation is required.

One of the ways to resolve this problem is the creation of underground storage facilities for helium concentrate (UHCSF) in a porous medium in close proximity to gas chemical plants. This will ensure supply flexibility and reduce helium losses [6].

Currently, the technical and technological aspects of the creation of UHCSF are not well researched.

3. Promising objects of helium underground storage

Analyzing domestic and foreign experience in underground storage of helium and helium concentrate in porous media, it was found that smaller fields, which were always more than large ones, were well suited for storing helium concentrate. And therefore it is possible to choose an object closer to the GPP, freeing up space for helium concentrate, either by developing the field, or by transferring the required volume of gas to the deposit of a large field [7].

When selecting objects for storage of helium concentrate, it is necessary to implement the requirements for the construction of underground gas storage facilities in porous media, but there are also specific requirements [1]. The geological and technical group of requirements imposed on helium storage facilities is primarily determined by the fact that helium is a gas with high penetrating power. A favorable factor in the creation of helium concentrate storage facilities is the existence of an overlying reservoir containing natural gas. When creating a storage facility in the lower deposit, even if helium leaks from the storage facility, it will be entrapped into the overlying gas reservoir and can be produced. Also, quite high requirements are placed on the tightness of wells in the storage and, therefore, it is necessary to strive to reduce the number of wells in the field [8]. Thus, on the basis of the literature analysis, the criteria for the initial selection of prospective deposits were formulated:

1. most suitable for long-term storage of gas field with a high initial content of helium. This is a criterion for the tightness of helium caprocks;
2. existence of an overlying reservoir containing natural gas.
3. fields of small sizes are suitable as a storage facility.
4. a little number of wells
5. objects for storage of helium concentrate should be located in the immediate vicinity of the gas processing plant.

Several promising fields were selected according to these criteria:

Table 5. Promising objects of underground storage of helium concentrate in porous media

Field	Productive formation	Storage volume of helium concentrate billion m^3
Tas-Yuryahskoe (unallocated fund Yakutnedra)	Talakh sandstones horizon	7
Hoto-Murbayskoe (Gazprom neft-Angara LLC)	Sandstones Venda	4
Chayandinskoe (PJSC Gazprom)	Sandstones of the Khamakinsky Vendian of the Southern II or Samanchakitsky blocks	9
Adnikanskoe (unallocated fund)	Sandstones, aleuroliths with interlayers of mudstone of the Kindal Formation (Cenomanalb)	2

4. Conclusion

Considering the fact of a significant volume of exports through the city of Blagoveshchensk, it is necessary to consider the option of developing UHCSF in the area of this city.

The question of creating UHCSF in the Khabarovsk region is most actual. This UHCSF will be able to regulate the irregularity of helium consumption and ensure the reliability of export helium supply to China, the APR countries and the USA.

References

[1] Han S A, Igoshin A I, Kazaryan V A, Scriabin A S and Sohranskiy V B 2015 *Underground storage of helium* (Izhevsk: Institute for Computer Research)

[2] Scriabin A S 2016 The experience of underground storage of helium *International research journal* **10 (52) 2** 155-159

[3] U.S. Geological Survey 2016 *Mineral commodity summaries*, available at: http://minerals.usgs.gov/minerals/pubs/mcs/2016/mcs2016.pdf

[4] Gluyas J G, Macpherson C G, Abraham-James T H, Bluett J J, Barry P H and Ballentine C J 2016 New High-Grade Helium Discoveries in Tanzania *Goldschmidt Conference Abstracts*, available at: http://goldschmidt.info/2016/uploads/abstracts/finalPDFs/A-Z.pdf

[5] Milovanov S V, Kislenko N N and Troynikov A D 2016 Development and implementation of innovative technologies for the extraction of helium from natural gas *Scientific journal of the Russian Gas Society* **2** 10-17

[6] Bondarev V L 2015 Storage prospects helium enriched natural gas in Eastern Siberia and the Far East *News of Gas Science: Problems development and operation of gas gas condensate and oil and gas condensate deposits Gazprom VNIIGAZ* **3 (23)** 63–67

[7] SemenovaK M, Churikova I V and Lopatin A Y 2018 Geological and gas-dynamical provisions for underground storing of helium in the terrigenous Vendian sediments at the north-east of Nepa-Botuoba anteclise *Scientific and technical collection NEWS OF GAS SCIENCE* **3 (35)** 68

[8] Starokon I V 2017 The influence of the processes of transient vortices formation and the resulting alternating loads on the stress state in the offshore oil and gas structures *Construction of oil and gas wells on land and at sea* **11** 49-53

Identification of transition processes parameters in flotation machine by computer vision

A V Zatonskiy

Perm National Research Polytechnic University, 29, Komsomolsky Ave., Perm, 614000, Russia

E-mail: zxenon@narod.ru

Abstract. The efficiency of potassium fertilizers production needs to improve industrial control systems because the fertilizers are very important for agriculture and economy all over the world. Now a control of flotation quality is made by a visual operator's observation, and a human factor could influence on the results. This paper is about of researching of foam recognize algorithms based on determination of flares from bubble surfaces. Statistics of flares are used for determination of parameters of transition processes in the flotation machine. A method based on bubble flares detection instead of bubble borders recognition. The method allows recognizing bubbles even on a surface of specific potassium foam, non-contrast, solid and color. This research solves the tasks of an optimal binarization threshold approximation, foam driver imagination exclusion, transition process detection. The optimal threshold is developed from medium frame luminosity quite linear. The foam driver can be excluded from the frame by specific character of luminosity trend. The transition process leads to change of bubbles count easy detected by software. A time of processing is about 15-30 ms per frame. Thus, the method can be used as in alarm systems, so in control systems on decision support systems.

1. Introduction

Potassium fertilizers are very important to agriculture. They increase productivity, resistance against diseases and a growth velocity of plants. The production of potassium fertilizers is a valuable export deal for producing countries such as Canada or Russia. So increasing of production processes is an important and actual task. One of most valuable production stage is a flotation. A sludge flotation is used to split useful and insoluble ore components. A next stage is sylvine flotation when KCl and NaCl are split. Both processes are flowing with a foam layer creation. The foam must be smooth, must consist bubbles of equal determined size for maximum quality of KCl extraction. A productivity of the machine and a quality of the flotation is closely linked with foam parameters. But there are no devices for foam testing useful in potassium flotation conditions. A control of the foam parameters are made by human – an operator of a flotator. It is leads to human factor expressions such as mistakes, or errors, of delay in an operator's reaction. The flotator is observing the foam visually time by time when he is going near the machine. A flotation plant can have many machines (for example, 20-30), so the operator can not observe all of them immediately. A view of the foam (color, organoleptic density, disturbances, smooth etc) is the information for decision support [1, 2, 3]. That is why a task of decreasing of human factor influence is very relevant as a part of production efficiency increasing.

One of possible ways is a computer vision systems using. Non-breakable observation could avoid of delays in identification of transition processes. Computer vision systems are widely used in polymetal or coal flotation machines. The coal foam has a high contrast surface, but even here many computation problems are occurs. The foam in potassium flotation machines is non contrast, solid, with a wide range of bubbles size. It leads to difficulties in a bubbles borders detection, which is applied by most authors.

Some authors (for example [4]) suggested to design of three dimensional foam models for a high-quality definition of processes and reasonable decision support, but they met with computation complexity usually. Authors [5] have led to result about irresistible complexity of a video stream processing. One of conclusions in the paper is possibility of reconstruction algorithms can be used only for static pictures with a successful speed of processing. Many algorithms useable for coal flotation nearly in real time cannot be used for the potassium flotation because the coal foam is black and white, but the color of potassium foam and RGB relation are important to decision support.

The authors [6] have suggested a method of bubble clusters detection and small and big bubbles separately. But many methods are mostly ideas then real using in enterprises. The methods cannot be applied in control systems directly.

A paper [8] is about of compensation disturbances in pictures by intermediate matrixes of color and grayscale pixels. A propos, examples of coal flotation pictures listed in the paper are high quality, and they are obtained in a laboratory machine with a very good lighting. Authors [9] also created a laboratory coal flotation machine with multisource lighting to increase a quality of polyhedral bubbles detection. But they reached only 5-second period of captured video processing. This period is very suitable but it is impossible to create such lighting conditions in an enterprise.

Authors [7] have suggested using of flares from bubbles instead of bubbles borders identification. This method is developed to an experimental software realization in [10], where additional problems linked to real lighting conditions and computation complexity are described. Some of them are solved in [11] using a video shot on a real flotation machine in JSC "Uralkali" (Russian Federation, Perm region). So the main aim of the paper is to improve the algorithms and to apply them to transition processes parameters identification.

2. Materials and methods

The experiment is made in 1st and 3nd chambers of enterprise flotation machine with a compact lighting source. Flotation reagents are injected in these chambers. Precision values of flows cannot be posted here, but flow in the 1st chamber is quite bigger then in 3nd. We used a wide-angle video recorder with a 1980*1200 resolution and 50 fps. We cut a rectangle region of interest between an axe of a foam driver and a chamber wall (Figure 1) for flare identification.

Figure 1. Source frame captured from video.of the Al_2O_3 powder.

A reagent flow to the 1st chamber was decreased by 1/3 when recording. A transition process begins immediately. We can observe some changes in the foam visually. After the process ends we restored the flow and recorded the video until changes are end. We can not see some transitional processes in 3nd chamber when the reagent flow was stopped at all, so only results was obtained on 1st chamber are discussed below.

We solved the next tasks when processing the video:

1. The foam driver exclusion from the frames by identifying flares features on points "A" and "B" (Figure 2).
2. An optimal binarization threshold calculation when maximum flares are determined in the frame.
3. Transition process detection by change of flares number with the optimal threshold.

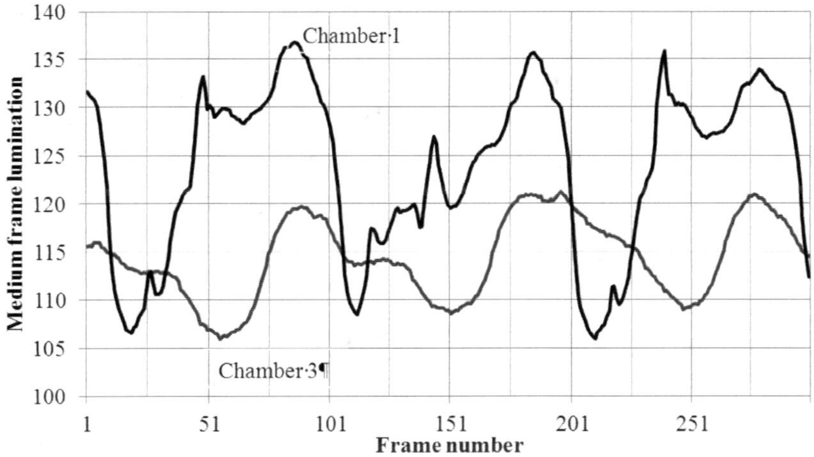

Figure 2. Specific character of the medium luminosity caused by the foam driver.

The optimal threshold of binarization could be determined for every frame but it is impossible in real time by a long time of computations (70-300 sec for one frame depending of luminosity and bubbles count). That is why we approximated it from the medium luminosity of the frame. The approximation must be made one time for every chamber and every lighting condition (Figure 3)

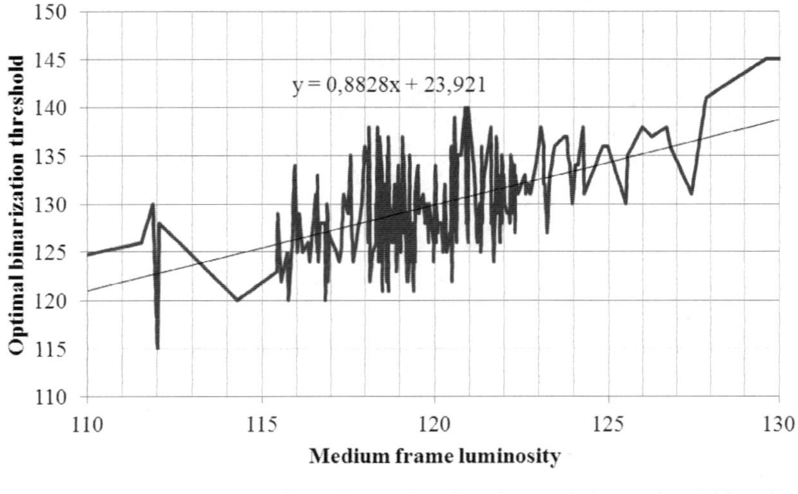

Figure 3. Dependence between the medium frame luminosity and the optimal binarization threshold.

Using of approximation threshold leads to error about 7-10% in bubbles count. It is not so small, but is it possible to identify a transition process, is a question.

3. Results and discussion

We combined a trend of estimated transition process with a trend of bubbles count in 1st and 3nd chambers after the reagents flow to the 1st chamber was changed (Figure 4).

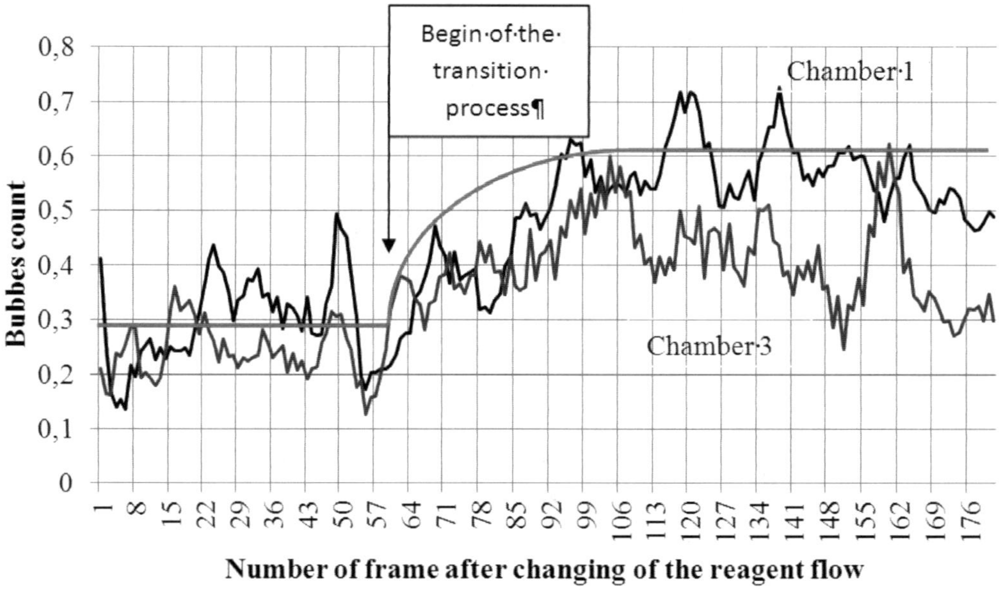

Figure 4. Ideal and real trends of transition processes

4. Conclusion

As a result we obtained obvious dependence between the real transition process and the count of bubbles flares in the frame. If we apologize the chamber is a first stage (ideal mixing) object with transition function $W(s) = \frac{K}{T \cdot s + 1}$, we obtained a value T ≈ 8.3 sec for the 1st chamber and T ≈ 8.6 sec for the 3nd chamber. It is very close to T = 7.6…10 sec from other researchers ([for example, 12]). So there are no contradictions with a previously known data at least. Meanwhile the bubbles count is near constant when there are no changes in reagent flows. New processing algorithms are quite quickly as for 15-30 ms per frame, it is quite enough for the real using.

So the aim is reached. We modified algorithms of video processing so the transition processes can be detected automatically. As a result we can use the recognizing as for alarm systems so for decision support too.

References

[1] Aldrich C, Smith L K, Verrelli D I, Bruckard W J and Kistner M 2017 *Mineral Processing and Extractive Metallurgy* **127(3)** 146-156

[2] *Minerals Processing Solutions Automation,* available at: http://www.metso.com/miningandconstruction/MaTobox7.nsf/DocsByID/0AF54610F6C67DDBC2257B030037F030/$File/Automation%20ENGLISH%20Low%20Res.pdf

[3] *Outotec FrothSenseTM,* available at: http://outotec.com/ImageVaultFiles/id_735/d_1/cf_2/OTE_Outotec_FrothSense_eng_web .PDF

[4] Romanenko S A and Olennikov A S 2014 *Obogashchenie Rud* **2** 23-28

[5] Sadr-Kazemi N and Cilliers J J 1997 *Minerals Engineering* **10** 1075-1083
[6] Tan J, Liang L, Xia W and Xie G 2018 *Separation Science and Technology* **53 (9)** 1434-1444
[7] Xu D, Chen Y, Chen X, Xie Y, Yang C and Gui W 2016 *Chemometrics and Intelligent Laboratory Systems* **154** 112-121
[8] Zatonskiy A V and Malysheva A V 2017 *Obogashchenie Rud* **2** 35-41
[9] Zatonskiy AV and Varlamova S A 2015 *Modeling and Information Technologies: Selected Papers of the International Scientific School "Paradigma"* 52-56
[10] Zhao L, Peng T, Zhao Y, Xia P and Xie Y 2016 *IFAC-PapersOnLine* **49** 90-95
[11] Becker V F 2012 *The solution of technological problems of existing production by means of automation. T.1: Production of potash fertilizers* (Perm BF PNIPU)
[12] Malysheva A V 2018 *Modern Science: Actual Problems of Theory and Practice* **5** 110-116

IOP Conf. Series: Materials Science and Engineering **860** (2020) 012022 doi:10.1088/1757-899X/860/1/012022

Problems of evaluating the stressed state of welded joints of offshore fixed platforms of type "jacket" through the example of the Black sea region

I V Starokon

Russian State University of Oil and Gas named after I.M. Gubkin, 65/1, Leninsky Prospekt, Moscow, 119991, Russia

Abstract. Offshore stationary platforms are actively used for oil and gas production on offshore fields in the world. Thus only in the Gulf of Mexico, approximately 4,500 such structures are used. Actively offshore stationary platforms are used by Norway, Azerbaijan, Iran and other Persian Gulf countries, China, Brazil, the USA and many other countries. The offshore platforms considered in the article are structures welded from pipes of various diameters. These platforms are affected by wind, wave and other loads, which cause them to vary in direction and magnitude of voltage. The article analyzes methods for calculating stress values for welded joints of platforms. Based on these methods, the author carried out numerical and analytical modeling of the stress state of the welded joints of the offshore stationary platform, obtained practical results on the example of a "T" type welded joint

1. Introduction

The world has significant marine resources of oil and gas located on offshore and oceanic shelves that are produced using offshore fixed platforms. Since the platforms operate in adverse marine conditions with significant loads, significant stresses arise in welded joints, which subsequently lead to fatigue failure. The purpose of this article is to study methods for assessing the stress state of welded joints of an offshore platform and to select the most effective of them [1-14].

The world has significant marine resources of oil and gas located on the sea and ocean shelves. In recent years, offshore deposits of the Black Sea have been actively developed, the production of which is carried out using offshore oil and gas facilities. These structures are offshore fixed platforms of the "JACKET" type for various purposes. Offshore stationary platforms are actively used for oil and gas production on offshore fields in the world. So, for example, in the Gulf of Mexico alone, approximately 4500 such structures are used. Actively offshore stationary platforms are used by Norway, Azerbaijan, Iran, some of Gulf countries, China, Brazil, the USA and many other countries. Those we can confidently say that these structures are widespread and play a key role in the extraction of oil and gas resources on the shelf [15-20].

Since the platforms operate in adverse marine conditions, characterized by significant loads from the effects of waves, currents and other climatic factors, significant stresses arise in welded joints, which subsequently lead to fatigue failure. The purpose of this article is to study the stress state of the welded joints of the offshore platform, which will allow further assessment of their resource. This article studies the stress state of two types of compounds: T-joint or in foreign literature the so-called "T"-connection, pipe connection at a certain angle, called a brace or in a foreign "K"-connection.

2. Methods

2.1. Strength Stress Analysis Method

Nowadays, various assessments of the stress state of structures made of round and square pipes have been developed. Based on the theory of the strength of welded joints and methods for calculating and designing structures made of round pipes, the following methodology can be proposed to study the stress state of welded joints of OFP:

1. First of all, it is necessary to numerically calculate the forces and stresses arising from the action of the gravity of the structural elements of the OFP, the installed technological equipment and the buoyancy force of the aquatic environment.
2. Determine the parameters of wind and waves in the studied field.
3. Calculate the loads arising from the wave action.
4. Determine the load moments from the wave action in the welded joint of the OFP.
5. Calculate the total stress from the wind-wave load in the welded section.
6. Determine the sum of the stresses from the wave load and the total weight of the structure with the technological equipment installed on it, taking into account the buoyancy force and the weight of the sea fouling.

We illustrate the developed methodology. To do this, we choose a welded joint at the OFP installed at the Subbotinskoye field. This welded joint is a T-joint, it is a horizontal element made of a cylindrical pipe with a diameter of 530 mm, cut at a right angle into a vertical column (column) - a pipe with a diameter of 720 mm (figure 1). The schematic diagram of the connection is shown

Figure 1. Connection "T" -type offshore platform.

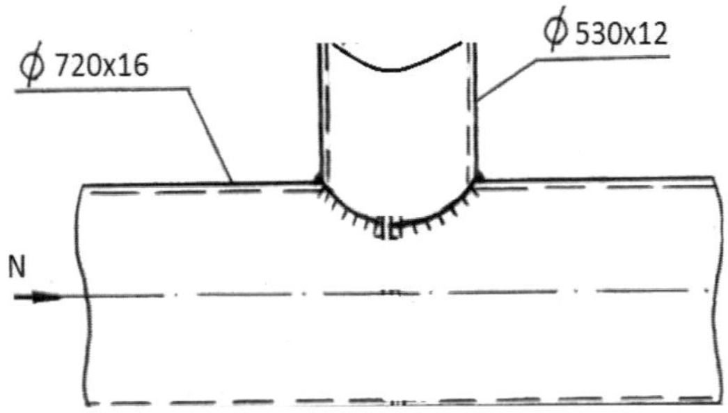

Figure 2. Connection of the horizontal element to the column.

According to the project of the OFP, welded joints are isosceles triangles with legs k_{fz} = 1.56 cm. When calculating the strength of corner joints, including T-joints, it is assumed that, regardless of the current load, they work for a conditional cut. Destruction occurs along the smallest slip area of the joint under the influence of both bending and torque moments, as well as longitudinal or transverse forces. The calculated cross section of the fillet is taken in the form of an isosceles triangle inscribed in the cross section of the weld. The side of the triangle will be considered equal to the height of the seam, and the influxes are not taken into account. Consider a simplified scheme of the action of the following forces on a welded joint: 1. The force P and the bending moment from the wave action on a vertical column (column) with a diameter of 720 mm, perpendicular to the surface of the column; 2) The distributed force q and its moment caused by the wave action on a horizontal element with a diameter of 426 mm. This force is directed perpendicular to the surface of the horizontal element and has its own calculation features that are different from determining the wave load on the vertical element. The welded joint, as well as all the structural elements of the OFP support block, is affected by the gravity of the elements of the support block and equipment, the weight of the sea fouling, as well as the Archimedean force on the elements immersed in water. The resultant of all forces is denoted by N. The total moment arising in the welded joint is M_N. The stress in the welded joint, in the case of the perpendicularity of the force and moment action vectors, is calculated by the formula:

$$\tau = \frac{N}{\beta_{zf} k_f L} + \frac{6M_N}{\beta_{zf} k_f L^2}, \tag{1}$$

where: M_N -moment of force from the action of the total load; N - is the resultant force; L-seam length, in this example, equal to 1.74 m; k_{fz} -a cathetus of the seam, which should be at least 6 mm, and in accordance with GOST 16037-80, is assumed to be 1.3 times the thickness of the thinner part. Considering that the thickness of the pipe with a diameter of 426 mm is 12 mm, we get 12 mm x 1.3 = 15.6 mm = 0.0156 m; β_{zf} -coefficient depending on the type of welding, position of the welded joint, weld leg and the choice of calculation method (i.e., for the metal of the fusion boundary or the metal of the welded joint), taken equal to $\beta_{zf} = 0.7$.

The described method for determining the stress state follows from the theory of calculating the strength of a welded joint according to the given values of the maximum forces and moments arising from the offshore platform. However, this does not take into account the heterogeneity of the stress state of this compound, the presence of points with stress concentration, and many other factors, which are analyzed in detail below. Therefore, when calculating the stress state of a welded joint, this method should be used with caution.

2.2. Determination of the stress state of welded joints using StructureCAD software

The determination of the stress state in a welded joint is a difficult task, for the solution of which it is necessary to take into account the mutual influence of many structural elements of the support block. According to the author, it is advisable to use the displacement method and the automated design complex StructureCAD, which allows automatic calculation of the values of displacements and angular deformations. If we build a model of the support block in it and subject it to the appropriate loads, it becomes possible to determine the magnitude of the linear and angular deformations that occur in the nodes. However, to obtain these deformation values does not mean solving the problem of the stress state in the welded joint of the offshore platform. For this, according to the author, a transition is necessary from the values of linear and angular strains to the values of "equivalent strains", and already from them to the values of equivalent stresses.

Let us consider this question in more detail. As it is known know, at any point it is possible to distinguish three orthogonal directions (with indices 1, 2, 3), called the principal axes of deformation, for which the angular deformations are equal to zero, while the linear ones get their extreme values (ε_1 - is the maximum, ε_3 –is the minimum, ε_2 - minimax). To find the values of the main strains, it is necessary to solve the following equation:

$$\varepsilon^3 - J_1\varepsilon^2 - J_2\varepsilon - J_3 = 0 \tag{2}$$

The invariants of the strain tensor are defined as follows:

$$J_1 = \varepsilon_x + \varepsilon_y + \varepsilon_z \tag{3}$$

$$J_2 = -\varepsilon_x\varepsilon_y - \varepsilon_y\varepsilon_z - \varepsilon_x\varepsilon_z + \frac{\gamma_{xy}^2+\gamma_{yz}^2+\gamma_{xz}^2}{4} \tag{4}$$

$$J_3 = \varepsilon_x\varepsilon_y\varepsilon_z - \frac{1}{4}\gamma_{xy}\gamma_{yz}\gamma_{zx} - \frac{1}{4}\left(\varepsilon_x\gamma_{yz}^2 + \varepsilon_y\gamma_{zx}^2 + \varepsilon_z\gamma_{xy}^2\right) \tag{5}$$

where: $\varepsilon_x, \varepsilon_y, \varepsilon_z$ - linear deformations along the corresponding axes, $\gamma_{xz}, \gamma_{yz}, \gamma_{xy}$- values of angular deformations.

IOP Conf. Series: Materials Science and Engineering **860** (2020) 012022 doi:10.1088/1757-899X/860/1/012022

When the values of the principal strains are calculated, it becomes possible to go on to determine the equivalent stress through the dependences

$$\sigma = \frac{2G}{\sqrt{3}}\sqrt{(\varepsilon_1 - \varepsilon_2)^2 + (\varepsilon_2 - \varepsilon_3)^2 + (\varepsilon_1 - \varepsilon_3)^2} \tag{6}$$

$$G = \frac{E}{2(1+\mu)} \tag{7}$$

where: G- shear modulus, E- elastic modulus, $\varepsilon_1, \varepsilon_2, \varepsilon_3$- major deformations, σ - equivalent stress, μ - Poisson's ratio.

2.3. Analysis of the stress state of the welded joints of the platform in the SolidWorks software package

SolidWorks – it is a computer-aided design and engineering analysis system that uses the so-called finite element method as a calculation method. This software package has a long history of application and has proven itself in mechanical engineering and solving other problems of engineering analysis.

3. Result

Let us consider the stress state of the support block of the offshore fixed platform using the example of an OFP installed at the Subbotinskoye field. The design experience and the practice of operating such platforms indicate that taking into account the mutual influence of various platform elements is an extremely difficult task. To analyze the stress state of the welded joints of the platform, a computer model was built in the StructureCAD software package (figure 3).

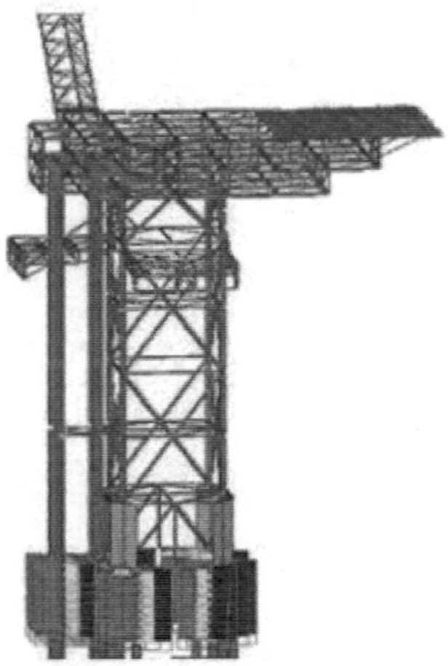

Figure 3. Model of the offshore platform built in the software package StructureCAD.

In order to assess the stresses arising in the platform elements, the model was subjected to loads from its own weight, the severity of the equipment and marine fouling, as well as the buoyancy of the aquatic environment and from wind-wave exposure. Moreover, to take into account the mutual influence of the elements, the direction of development of the wind-wave load was set both along the X axis (combination of loads K_1) and in the direction at an angle of 45^0 to the axis X (combination of loads K_2). Using the StructureCAD software package, we simulate the effect of a wind-wave load in three loading combinations at a wind speed of 25 m/s and a wave height of 6.8 m, the values of linear and angular deformations in the nodes, which are the connection of the column with horizontal elements and braces. Table 1 shows the maximum results obtained for the "K" - connection.

Table 1. Stress state of the welded joint with various load combinations calculated by the deformation method

Combinations downloads	ε_x	ε_y	ε_z	γ_{yz}	γ_{xz}	γ_{xy}	J_1	J_2	J_3	ε_1	ε_2	ε_3	σ, MPa
K_1	$6\cdot10^{-4}$	$4.2\cdot10^{-6}$	$6.12\cdot10^{-6}$	$1.5\cdot10^{-5}$	$-3.69\cdot10^{-5}$	$-3.3\cdot10^{-5}$	$6\cdot10^{-4}$	$-5.22\cdot10^{-9}$	$-2.6\cdot10^{-14}$	$5.9\cdot10^{-4}$	$1.2\cdot10^{-5}$	$3.5\cdot10^{-6}$	82
K_2	$8\cdot10^{-4}$	$6.2\cdot10^{-6}$	$8.98\cdot10^{-6}$	$2.2\cdot10^{-5}$	$-5.41\cdot10^{-5}$	$-4.8\cdot10^{-5}$	$8\cdot10^{-4}$	$-1.12\cdot10^{-8}$	$8.1\cdot10^{-14}$	$8\cdot10^{-4}$	$1.9\cdot10^{-5}$	$5\cdot10^{-6}$	115

The author also analyzed the values of the interconnection of linear and angular displacements of the "T" - connection and the "K" - connection with a brace, which made it possible to compare the nature of the stress-strain state in both.

Table 2. Values of the relations of linear and angular displacements of the T-joint and the connection with the brace

Combinations downloads	*The values of the relations of linear and angular displacements of the T-joint and the connection with the brace*					
	X	Y	Z	Ux	Uy	Uz
K_1	2.69	1.52	1.05	-0.75	-1.90	2.13
K_2	1.54	1.34	0.93	-5.23	0.65	1.44
K_3	1.28	1.43	7.06	0.53	0.23	2.43

In addition, the author built a platform model in the SolidWorks software package. We studied both the general stress state of the platform and the stresses in individual welded joints of "T" and "K" types (figure 4).

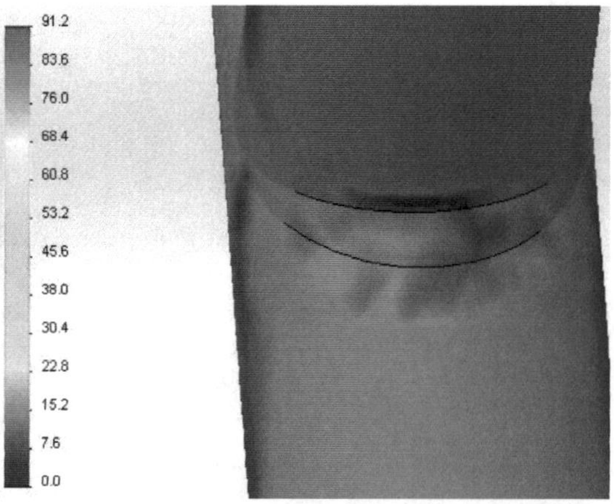

Figure 4. Weld model in SolidWorks software package.

4. Conclusion

After modeling using various methods, the author comes to the conclusion that the solution based on the strength method does not meet modern requirements, because this method does not take into account local overstress in the weld, which distorts the real picture of the stress state of the welded joint, presenting it as uniformly distributed stresses. SolidWorks software package showed good results, which allows simulating the geometry of each individual welded joint, which allows obtaining fairly accurate results. However, to obtain accurate results, it is necessary to specify a high-density finite element grid, which requires very serious computing resources, which not all design organizations possess. In addition, it is worth noting the complexity of calculating large structures in SolidWorks, such as a platform more than 70 meters high with a large number of large diameter elements. Very accurate results were shown by the StructureCAD software package, which allows working with large-sized structures, however, the calculation of the stress state of welded joints required an additional labor-intensive data processing procedure. In recent years, the development of new CAE-programs, specializing in the analysis of the stress state of building structures. The results of their experimental application in relation to the assessment of the stress state of welded joints of offshore platforms will be published by the author later.

References

[1] Efimenko L, Kapustin O, Gusev V, Merkulova A and Nikiforov N 2014 Evaluation of structural and phase composition and corrosion resistance properties of coatings made by laser surfacing *Chemical and Petroleum Engineering* **50** 402-411

[2] Efimenko L 2015 Methodological fundamentals of the evaluation of the weldability of structural materials using oil ahd gas pipes as an example *Welding Interhational* **9 (28)** 727-731

[3] Efimenko L, Elagina O, Ramus A, Volkov I, Ladyzhanskii A, Rakhmanov D, Konovalov A and Kurkin A 2015 Repair features of mechanical and corrosion damage in a main gas pipeline under pressure *Chemical and Petroleum Engineering* **50** 749-757

[4] Elagina O, Efimenko L and Volkov I 2015 Study of the strain ageing on corrosion failure and submicrostructure of low-carbon steels for gas-distriution pipelines *Chemical and Petroleum Engineering* **51** 433-437

[5] Efimenko L, Kapustin O, Merkulova A, Subbotin R, Puyko O and Shchkalova A 2016 Features of transformation kinetics of austenite in heat treatment of stamped and welded components made of 10G2FBYU steel *Welding International* **30** 472-478

[6] Efimenko L and Ramus A 2016 Effect of the morphology of structure on the resistance of welded joints ofhigh-strength pipe steels to brittle fracture *Metal Science and Heat Treatment* **57** 559-563

[7] Efimenko L, Kapustin O, Ramus A and Ramus R 2016 Control of softening processes in the heat-affected zone during welding of high-strength steels *Metal Science and Heat Treatment* **5** 435-441

[8] Efimenko L, Kapustin O, Merkulova A, Subbotin R, Puyko O and Shchkalova A 2016 Features of transformation kinetics of austenite in heat treatment of stamped and welded components made of 10G2FBYU steel *Welding International* **30** 472-478

[9] Makarov G 2014 Modernization of systems of pipeline transport of gas and oil *Welding International* **28 (9)** 744-747

[10] Starokon I 2019 Varieties and kinetics of fatigue cracking of fixed offshore platforms for oil and gas production *Materials Science and Engineering* **734** 012055

[11] Antonov A 2014 Investigation of fields of residual stresses in welded structures *Welding International* **28** 966–969

[12] Starokon I 2020 Features of fatigue processes occurring in restored welded joints of offshore fixed platforms *Materials Science and Engineering* **734** 012171

[13] Starokon I 2020 Resource of repaired welded compounds of marine oil and gas structures *Materials Science and Engineering* **734** 012193

[14] Guseynov Ch 2020 In Arctic - new technical means and technologies for the development of oil and gas fields in long-term freezing deepwater areas *Materials Science and Engineering* **734** 012174
[15] Starokon I 2019 On the results of studying the influence of vibration-fluctuations processes on the resource of welded joints of marine stationary platforms for oil, gas and other useful minerals on the shelf *J Akustika* **285** 148-151
[16] Bezkorovayniy V, Bayazitov V and Bobov D 2018 Management of the Design and Construction of Offshore Oil and Gas Facilities with Bim Base *Materials Science and Engineering* **463 (4)** 042056
[17] Reitenbach V, Ganzer L and Albrecht D 2014 Influence of Hydrogen on Underground Gas Storage *Research Report DGMK-752* **73** 6627-6937
[18] Reitenbach V, Ganzer L, Albrecht D and Hagemann B 2015 Influence of added hydrogen on underground gas storage: a review of key issues *Conf. Ser.: Environ. Earth Sci.* **73 (10)** 6927–37
[19] Ganzer L, Reitenbach V, Pudlo D, Albrecht D, Singhe AT, Awemo K N, Wienand J and Gaupp R 2014 Experimental and numerical investigations on CO2 injection and enhanced gas recovery effects in Altmark gas field (Central Germany) *Acta Geotechnica* **9 (1)** 39-47
[20] Guseynov C 2010, RU, *Patent* No 10-281 Methodological recommendations for developing a strategy for the development of offshore oil and gas fields (Moscow: Rospatent)

Estimation of efficiency of the thermal diagnostic method for detecting corrosion defects of fixed offshore platforms

I V Starokon

Russian State University of Oil and Gas named after I.M. Gubkin, 65/1, Leninsky Prospekt, Moscow, 119991, Russia

Abstract. Currently, the world is actively developing offshore fields. This production is carried out using fixed offshore platforms. During operation, offshore platforms are exposed to various influences, including corrosion. As a result of this effect, caverns are formed on the surface of the structural elements of the platform, which are stress concentrators and capable of causing critical overvoltage of the platform elements. To prevent this situation, it is necessary to identify hazardous corrosion defects in a timely manner. The article describes a comparative analysis of modern non-destructive testing methods, as well as the results of a method developed by the author on the basis of experimental studies for assessing the depth of corrosion defects.

1. Introduction

In the world there are significant reserves of oil and gas located offshore. Oil and gas in these conditions is conducted with the use of offshore platforms of various types, the most common of which is the fixed offshore platform of gravitational type. Nowadays, most of these platforms have been in operation for more than 30 years, which cannot but raise the question about the safety and reliability of operation of the facilities. The solution to this question, according to the author, lies in taking timely measures to identify and eliminate constructive elements of FOPs, to develop their resources and adversely affecting the possibility of further safe operation of offshore platform. During the entire period of operation of FOPs, its constructive elements (CE) are subjected to the influence of various loads and impacts, which leads to the formation and development of various defects, contributing to the destruction of TBE FOPs. Defects of structural elements of offshore stationary platforms have different origins and are also categorized according to its location (weld or base metal) and excellent in the degree of danger. These can be defects industrial production of the base metal, such as sinks, hairline cracks, thermal cracking; the weld defects like pores, lack of fusion, undercuts; the defects obtained in the process of operation such as dings, stress cracks, nicks etc. Of course, all of these defects under the influence of variable loads and impacts, can be further developed and to turn TBE FOPs in a potentially dangerous element. However, the biggest action since the specificity of the marine deposits, have corrosion defects. In figure 1- figure2 presents photographs of various FOPs exposed to the corrosive effect of illustrating their actual condition.

Figure 1. Support block, production module

Figure 2. Support block, accommodation module

The analysis of operating offshore platforms have shown that the highest corrosion of the structural elements of SMEs, equal to approx 12 % of the initial wall thickness is achieved in the zone of variable wetting. In the underwater area, corrosion of the lower and varies in the range of 5-8,8 % (depending on the underwater currents). In the atmospheric zone corrosion average is 7%. Assessing the degree of damage of the steel structures supporting the SME units, it should be noted that all elements of the platforms corroded, however for SMEs it can be noted a very important trend: the actual state of the metal surface parts of the production blocks in the zone of alternating wetting is characterized by the highest corrosive wear. The average wear of the elements in this zone is 25 to 40 %. The maximum wear of the individual elements reaches 75-85 %. These data are confirmed normative and technical documentation in which the rate of corrosion in the zone of alternating wetting is estimated at 0.16 mm/year, and in submarine and surface areas of 0.12 mm/yr and 0.10 mm/year, respectively. Numerous data indicate that the maximum wear is in the zone of periodic wetting. Based on this statement we can conclude that the greatest amount of potentially faulty elements is in the zone of periodic wetting. The identification of such potentially defective items in the opinion of the author, possible only through a comprehensive diagnostic survey of the whole platform of modern methods of nondestructive testing.

2. Methods

Currently there are developed and effectively applied various methods of nondestructive control, but the specific conditions of the marine deposits, the scale of construction FOPs and many other factors create certain requirements for various NDC methods. Let us formulate the requirements for NDC methods with respect to FOPs: 1) High speed of operation of the control; 2) Accuracy of detection of dangerous defects; 3) Corrosion layer SMEs should not impede the inspection; 4) Full control should be 100% of the surface and welded joints of structural elements of FOPs rather than their individual plots; 5) the Possibility of conducting monitoring in remote areas; 6) Low cost of control operations; 7) this method of NDC, it is possible to diagnose large-sized object, like FOPs; 8) A successful application of this method in various industries. It is necessary to note the fact that due to the nature of the environment CE underwater part, almost there is no substitute for a visual diving survey. None of the following NDC techniques cannot be used in submerged conditions, with the exception of maybe ultrasonic thickness measurements. The author is aware of such successful control operations; however, the ultrasonic underwater thickness measurements were accompanied by frequent breakage of the expensive ULTRASONIC thickness gauges, due to the ingress of moisture into the electrical circuit of the device. In addition, it is difficult to find a well-prepared ultrasonic testing inspector and diver at the same time, it is technically difficult to provide the controlled surface sweep from the corrosion layer, etc.

Let us consider the modern methods of NDC. First in popularity among the modern techniques is ultrasonic inspection. This method has shown good results in the detection of defects in aircraft, quality control of welded joints of gas pipelines and other industries, but in reality the whole surface of the FOPs covered with a layer of corrosion thickness 3-4mm, which prevents the spread of ultrasonic vibrations and reliable detection of defects. In addition, the presence of a significant number of corrosion cavities will give a lot of false defects. Remove the same corrosion layer on hard-to-reach and large size elements FOPs are technologically feasible. The second most popular among the NDC methods is the x-ray inspection method (XK), which is not as sensitive to corrosion damage. However, the complexity of equipment location XK control on difficult structural elements of SMEs, a very long time to prepare to conduct operations control and removal of platform personnel to a safe distance in accordance with the safety requirements in the conduct of Kazakhstan makes the method is practically unusable. The next popular application followed by eddy-current (ECC), the magnetic (MK) and capillary (CC) control methods. Just note that these methods can only detect certain defects. For example, the ECC method only detects surface defects located at a depth of no more than 3mm from the surface of the structural element, the CC method may be used only provided that the output of the defect to the surface and the ratio of defect depth to its width not less than 10, etc. In General, all these methods have strict requirements to the quality of Stripping the structural elements of FOPs, roughness of surface, etc., that does not allow effective use of them for complex diagnostics. In addition, the cost of such control operations and supplies will greatly exceed the cost of the FOPs. In recent years, considerable popularity so-called thermal control (TC). This control method is based on the fact that the functioning of many objects connected with the effect of wind, hydrodynamic, etc. loads, so with the action variable and fixed heat (temperature) fields. However, their internal structure has a significant impact on the nature of the emitted thermal field, which allows the analysis of anomalies to judge the change of metal structure as a whole or its individual zones. Thermal field of the surface of the object is a source of information about the features of the process of heat transfer, which in turn depends on the design and shape of the object and the presence of the defect. To date, various industries have successful experience of identifying with the heat of cracks, voids, delaminations, pores, lack of fusion, cavities and other defects are also typical for FOPs. The experience of using the thermal control method in the aircraft industry, shipbuilding, energy and other industries helped to create the methodology and consistency of its implementation. However, for the diagnosis of FOP methodology thermal control is currently lacking. In this regard, the author seeks to develop such a methodology that would require an examination of processes of heating and cooling for SMEs under the action of solar radiation, study of the influence of corrosion layers on the processes of heat transfer, and most importantly, studies of the distortion of the thermal field at different parameters of the defect. From the point of view of thermal control, all defects can be divided into passive ones, i.e. not generating heat, and active, which are a source of heat. The detection of passive defects characteristic of SMEs is best possible with the active method. The active control method involves heating the controlled object. When a controlled object is heated, the defect resists the heat flux, because its thermal conductivity differs from the thermal conductivity of the metal of the structural element of the FOP. Spreading deeper into the spacecraft core, the heat flux flows around the defect along the surrounding layers of the base metal. In this case, there is an accumulation of heat in the layer before the defect and its lack in the layer behind it, which is manifested in a local change in temperature on the surfaces of the controlled object. The described phenomenon is also true for the cooling process. The author proposes to investigate not a one-time thermogram of an FOP, but several thermograms sequentially executed at equal intervals of time, which will allow characterizing the rate of change of the temperature state, which will allow better interpretation of the size of the defect. In addition to this, it is also necessary to establish the effect of the corrosion layer on the heat transfer processes in the CE of the FOP. Currently, such techniques are absent and the author sets the task to develop them. Here is a comparative table of NDC methods as applied to performing diagnostic work on FOPs (table 1).

Table 1. Comparative table of non-destructive testing methods for the implementation of diagnostic work on FOPs

Parameter	Name of the control method					
	USC	**XC**	**ECC**	**MC**	**CC**	**TC**
Speed of control operations	Low	Low	Average	Low	Low	High
Defect detection accuracy	Average	High	Average	Low	Average	Average
Sensitivity of the NDC method to the corrosion layer	High	High	High	High	High	Low
Possibility of 100% control of base metal and welds	Flimsy	Flimsy	Flimsy	Flimsy	Flimsy	High
The ability to control in inaccessible places	Average	Low	Average	Average	Average	High
Cost of control	Average	High	Average	Average	High	Low
The ability to control large-scale objects	Flimsy	Flimsy	Flimsy	Flimsy	Flimsy	High
Experience in applying the NDC method	Long-life	Long-life	Long-life	Long-life	Long-life	Average

Thus, according to the author, the identification of dangerous corrosion defects is possible by conducting comprehensive diagnostics using modern non-destructive testing methods, the best of which is thermal monitoring.

3. Result

The author proposes to determine the depth and their equivalent volume based on data on temperature contrasts arising during heating of OOGS elements with defects. In order to implement this proposal, an experiment was performed (Figure 3 and Figure 4).

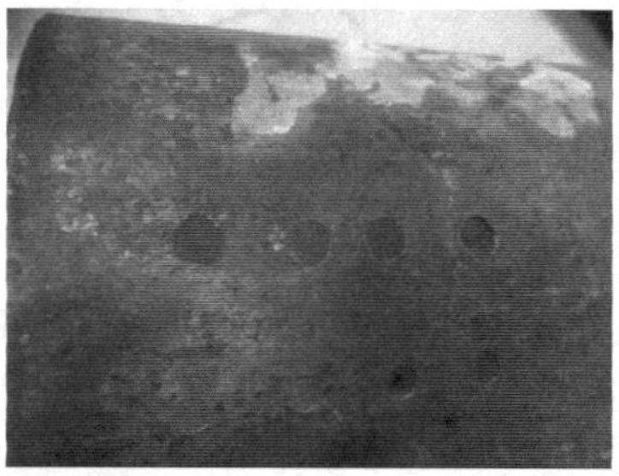

Figure 3. Preparation of the experimental sample

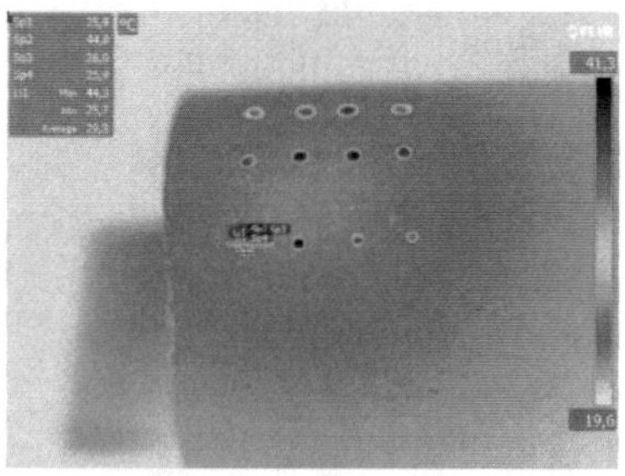

Figure 4. View of the experimental sample through the thermal imager

In the wall of pipes with different diameters and wall thicknesses (in Figure 3 and Figure 4, a pipe with a diameter of 133 mm and a wall thickness of 8 mm is shown) with a thickness of a uniform corrosion layer of 1 mm, holes of different diameters were drilled, simulating corrosion damage in depth that were filled with Fe2O3. The relationship between the following parameters was studied:

the ratio of wall thickness to the depth of corrosion damage and the effectiveness of the heat flux; the ratio of the temperatures of defect-free and corrosion-modeling areas.

As a result of the experiment, the following was established: 1) The magnitude of the temperature contrast is greater, the deeper the defect and the greater its geometric dimensions in the projection; 2) The temperature contrast is affected not only by the depth of the defect, but also by its size in the projection onto the area. Most often, corrosion cavities have a projection shape that is close to rounded (circle, ellipse, etc.). Because of the projection area of a corrosion defect can be visually estimated using a thermal imager, it becomes possible to calculate its depth. According to the author, the most dangerous from the point of view of stress concentration will be a defect of "V" -shaped form. The volume of such a defect can be described by the formula:

In case of corrosion defects with the shape of its projection onto a plane close to a circle:

$$l = \frac{3(k_1[D]^2+k_2[D]+k_3[C]^2+k_4[C]+a)}{\pi[r]^2}, \quad (1)$$

where: l is the depth of the defect [mm]; π = 3.14, r is the radius of the defect, [mm]; D is the dynamics of the heat flux, $[\frac{W}{s \cdot m^2}]$; C is the temperature contrast, constants [scalar number]; k_1=0,00463937; k_2=-8,70561; k_3=132,525; k_4=1432,5; a=3748,46.

In of corrosion defects with the shape of its projection onto a plane close to an ellipse:

$$l = \frac{3(k_1[D]^2+k_2[D]+k_3[C]^2+k_4[C]+a)}{\pi hb}, \quad (2)$$

where: l-depth of the defect, [mm]; h and b-axis of the elliptical base of the defect, [mm] and the remaining values are the same as in the previous formula.

4. Conclusion

1) It was shown that the thermal method has higher efficiency compared to other modern methods of non-destructive testing in the detection of dangerous corrosion defects.
2) According to the conducted experimental studies, the author developed new formulas that allow determining the depth of a corrosion defect during an active heat control procedure.

References

[1] Efimenko L, Kapustin O, Gusev V, Merkulova A and Nikiforov N 2014 Evaluation of structural and phase composition and corrosion resistance properties of coatings made by laser surfacing *Chemical and Petroleum Engineering* **50** 402-411
[2] Efimenko L 2015 Methodological fundamentals of the evaluation of the weldability of structural materials using oil ahd gas pipes as an example *Welding Interhational* **9 (28)** 727-731
[3] Efimenko L, Elagina O, Ramus A, Volkov I, Ladyzhanskii A, Rakhmanov D, Konovalov A and Kurkin A 2015 Repair features of mechanical and corrosion damage in a main gas pipeline under pressure *Chemical and Petroleum Engineering* **50** 749-757
[4] Elagina O, Efimenko L and Volkov I 2015 Study of the strain ageing on corrosion failure and submicrostructure of low-carbon steels for gas-distriution pipelines *Chemical and Petroleum Engineering* **51** 433-437
[5] Efimenko L, Kapustin O, Merkulova A, Subbotin R, Puyko O and Shchkalova A 2016 Features of transformation kinetics of austenite in heat treatment of stamped and welded components made of 10G2FBYU steel *Welding International* **30** 472-478
[6] Efimenko L and Ramus A 2016 Effect of the morphology of structure on the resistance of welded joints ofhigh-strength pipe steels to brittle fracture *Metal Science and Heat Treatment* **57** 559-563
[7] Efimenko L, Kapustin O, Ramus A and Ramus R 2016 Control of softening processes in the heat-affected zone during welding of high-strength steels *Metal Science and Heat Treatment* **58** 435-441

IOP Conf. Series: Materials Science and Engineering **860** (2020) 012023 doi:10.1088/1757-899X/860/1/012023

[8] Efimenko L, Kapustin O, Merkulova A, Subbotin R, Puyko O and Shchkalova A 2016 Features of transformation kinetics of austenite in heat treatment of stamped and welded components made of 10G2FBYU steel *Welding International* **30** 472-478
[9] Makarov G 2014 Modernization of systems of pipeline transport of gas and oil *Welding International* **28** **(9)** 744-747
[10] Starokon I 2019 Varieties and kinetics of fatigue cracking of fixed offshore platforms for oil and gas production *Materials Science and Engineering* **734** 012055
[11] Antonov A 2014 Investigation of fields of residual stresses in welded structures *Welding International* **28** 966–969
[12] Starokon I 2020 Features of fatigue processes occurring in restored welded joints of offshore fixed platforms *Materials Science and Engineering* **734** 012171
[13] Starokon I 2020 Resource of repaired welded compounds of marine oil and gas structures *Materials Science and Engineering* **734** 012193
[14] Guseynov Ch 2020 In Arctic - new technical means and technologies for the development of oil and gas fields in long-term freezing deepwater areas *Materials Science and Engineering* **734** 012174
[15] Starokon I 2019 On the results of studying the influence of vibration-fluctuations processes on the resource of welded joints of marine stationary platforms for oil, gas and other useful minerals on the shelf *J Akustika* **285** 148-151
[16] Bezkorovayniy V, Bayazitov V and Bobov D 2018 Management of the Design and Construction of Offshore Oil and Gas Facilities with Bim Base *Materials Science and Engineering* **463** **(4)** 042056
[17] Reitenbach V, Ganzer L and Albrecht D 2014 Influence of Hydrogen on Underground Gas Storage *Research Report DGMK-752* **73** 6627-6937
[18] Reitenbach V, Ganzer L, Albrecht D and Hagemann B 2015 Influence of added hydrogen on underground gas storage: a review of key issues *Conf. Ser.: Environ. Earth Sci.* **73** **(10)** 6927–37
[19] Ganzer L, Reitenbach V, Pudlo D, Albrecht D, Singhe AT, Awemo K N, Wienand J and Gaupp R 2014 Experimental and numerical investigations on CO2 injection and enhanced gas recovery effects in Altmark gas field (Central Germany) *Acta Geotechnica* **9** **(1)** 39-47
[20] Guseynov C 2010, RU, *Patent* No 10-281 *Methodological recommendations for developing a strategy for the development of offshore oil and gas fields* (Moscow: Rospatent)

IOP Conf. Series: Materials Science and Engineering **860** (2020) 012024 doi:10.1088/1757-899X/860/1/012024

Study of efficiency of methods for determining fatigue limits through the example of its determination for welded joints of marine stationary platforms

I V Starokon

Russian State University of Oil and Gas named after I.M. Gubkin, 65/1, Leninsky Ave., Moscow, 119991, Russia

Abstract. Currently, the world is actively engaged in the extraction of oil, gas and other useful resources at offshore fields using offshore stationary platforms. Due to the fact that the platforms are operated for a long time, fatigue damage accumulates in them. This issue becomes especially important when assessing the residual life of welded joints of the offshore platform, the repair of which is carried out using various technologies. The resource of such compounds is determined on the basis of the fatigue diagram, the key value of which is the fatigue limit. A large number of theories have been developed to determine the fatigue limit. However, calculations based on these theories do not always give a correct estimate. Therefore, the purpose of this article is to study the effectiveness of various methods for determining endurance limits using welded joints of offshore stationary platforms as an example and to select a method that allows obtaining the most accurate value of fatigue limit.

1. Introduction

Currently, the world is actively producing oil, gas and other useful resources in offshore fields [1-9, 16-20]. This production is carried out using offshore stationary platforms (figure 1). To date, such a situation has developed that these platforms have a long service life, which leads to the accumulation of fatigue damage in them. This problem is especially acute when assessing the resource of welded joints of the offshore platform, the repair of which is carried out using various technologies. The resource of such compounds is determined on the basis of the fatigue diagram, the key value of which is the fatigue limit (also called the endurance limit in the literature). Therefore, the purpose of this article is to study the effectiveness of methods for determining endurance limits in relation to its determination for offshore stationary platforms [1-9].

Typically, the endurance limit (or fatigue limit) is determined based on laboratory tests. An important practical point of view is the task of scaling the results obtained during testing of laboratory samples to real welded joints of fixed offshore platforms (FOPs). Modern research offers various theories for assessing the influence of the scale factor, depending on the surface roughness of the real structural element of the FOP and the welded joint as compared with a smooth sample (roughness within 0.16-0.32 microns) As well as the real geometric dimensions of the structural and welded joints of the offshore platform in comparison with laboratory samples, the concentration of stresses due to the type of weld and possible defects, residual stresses in the seam and the heat-affected zone, as well as the dependence on the asymmetry coefficient of the cycle, the probability of the presence of metal sections with different mechanical properties, cross-sectional dimensions, etc.

Figure 1. Stationary offshore platform.

The objective of this article is to choose a method for scaling the fatigue limit obtained as a result of experimental studies on real welded joints of the offshore platform. A fatigue diagram has been developed for tubular welded joints of offshore stationary platforms (figure 2). According to this diagram, the endurance limit is at the level of 20 MPa.

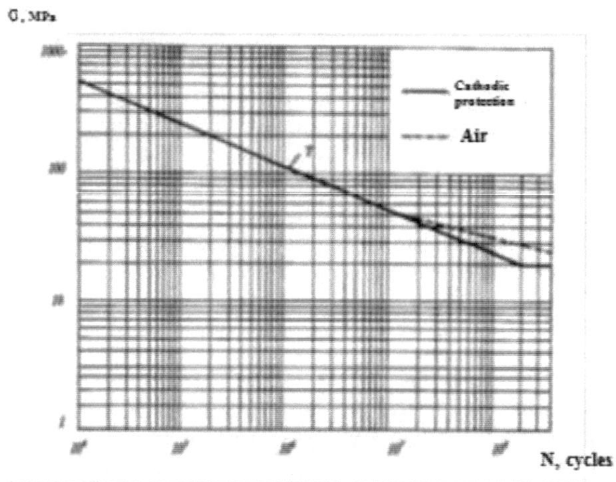

Figure 2. Fatigue diagram of welded joints of offshore platforms

Taking this diagram as a basis, we analyze how various methods of calculating the endurance limit allow obtaining a result close in value to 20 MPa.

2. Methods

To implement this technique, initially it will be necessary to determine the tensile strength of the welded joint. An analysis of the project materials showed that in the construction of offshore platforms for welding elements made of steel grade VSt3Sp5, electrodes of the E42A grade and the UONI-13/55 grade were used. The tensile strength of welded joints made in accordance with GOST 9467-75 is 420MPa. And for welded joints obtained from pipes of steel 09G2S (D) with electrodes E50A of the UONI-13/55 brand according to GOST 9467-75, the tensile strength of the weld should be 490MPa. To observe the similarity of the experiment, pipes made of material according to DIN 2393-1994 RSt37-2 standard were used as the column model, and 9MnSi5 steel was used for modeling the waist pipe.

Various methods are proposed to determine the tensile strength of an experimental weld.

IOP Conf. Series: Materials Science and Engineering **860** (2020) 012024 doi:10.1088/1757-899X/860/1/012024

2.1. Method of determining the fatigue limit depending on the value of the hardness of the metal
The following formula is well known:

$$\sigma_{-1} = 95.5(\sqrt{149 + H} - 12.2) \quad (1)$$

where$H = \frac{HV}{8.2}$ (HV- hardness, MPa); σ_{-1}-fatigue limit, MPa.

In turn, the hardness of welded joints can be determined by the equation of B. D. Lebedev $HV = 4160 \cdot C_e^{1,08} \cdot \vartheta_{cool}^{0.17}$ (where$C_э$ – carbon equivalent; ϑ_{cool} – cooling speed).

It is quite obvious that to better match the experimental sample to the real object, not only geometrically similar samples with similar loading conditions should be used, but also materials with similar mechanical characteristics. Therefore, during the experiment, welding materials were used that provide similar values of tensile strength, grades E4303-P for the "T" connection and E4903-P for the "K" connection according to ISO2560-2009, providing a tensile strength of the welded joint equal to 490 MPa. The results of this experiment are given in [12-15].

We calculate the coefficient of influence of the scale factor by the formula:

$$K_{D\sigma} = 0.5\{1 + [(1/88.3)(L/\bar{G}_1)]^{-0.2}\} \quad (2)$$

Where: $K_{D\sigma}$-scaling factor; $\bar{G}_1$- the relative gradient of the first principal stress; L-characteristic size of the working part of the seam perimeter. The relative gradient of the first principal stress is determined by the formula:

$$\bar{G}_1 = \left(\frac{1,6}{r}\right)\sqrt{\sin\varphi} + 2/\delta \quad (3)$$

where: r - the height of an isosceles triangle formed by an angular seam (according to project data); φ - an angle equal to 45; δ - thickness of the smallest of the welded SE.

Let us calculate the characteristic dimensions of the working part of the perimeters of the seams formed by connecting structural elements of various diameters. Then, according to the above formulas, we calculate the voltage gradient and scaling factor, and the results are summarized in table 1:

Table 1. The results of calculating the dimensions of the working part of the perimeters, the relative gradient of the first principal stress and the scaling factors of typical connections of the offshore platform

	The size of the working part of the perimeter of the seam in meters			***The relative gradient of the first principal stress***			***Scalingfactor***		
Diameters of connected elements	***325***	***426***	***530***	***325***	***426***	***530***	***325***	***426***	***530***
720	1.035	1.373	1.742	346.3	288.6	288.6	3.81	3.55	3.46
1080	1.027	1.352	1.693	346.3	288.6	288.6	3.78	3.51	3.4

It should be noted that this method of calculating the scaling factor has shown itself well in relation to machine parts. However, for welded joints, in the presence of residual stresses, the condition for the distribution of maximum stresses in the elastic region is not fulfilled and it should be used with caution for welded joints of FOPs.

2.2. Method of determining endurance limit depending on temporary tensile strength
It's well known, that the scale factor may be due to the influence on the destruction process of the total elastic energy accumulated in the loaded system. This statement is rather well confirmed in practice. As experimental data show, with an increase in the size of the sample, the propagation speed of the

crack increases, since larger samples also have a large reserve of elastic energy. However, certain characteristics of the time dependence of fatigue on the sample size and the actual weld size disappears, i.e. in other words, there is a certain boundary, after which the differences in the fatigue characteristics of the test specimens and the actual welded joints disappear. So, on the basis of experimental studies, it is known that when the cross-sectional area of the studied samples is more than 4000 mm^2, the scaling factor should be taken equal to 0.6. Modern energy theories make it possible to correct this value; a coefficient of 0.58 can be used in relation to the welded joints of the MSP support block.

Since the experiment was carried out on samples from pipes with an untreated surface, the results of this experiment are of practical interest, determining the endurance limit for samples with an untreated surface by the formula:

$$\sigma_{-1} = K_{D\sigma}(92 + 0{,}125\sigma_{\text{в}}) \tag{4}$$

Having obtained the values described above, we proceed to the process of transferring the obtained experimental results to a real welded joint. In [13], the following main parameters were identified that should be taken into account: the ratio of sample sizes to real objects, roughness, asymmetry of stress cycles, and residual stresses. In particular, formulas are given that make it possible to calculate the amplitudes of the first principal stresses, taking into account the above parameters, depending on the characteristics of the cycle. So, in the area of an alternating cycle, this limit amplitude will be determined by the formula:

$$\sigma_{1a}^{\text{п}} = \frac{\sqrt{6}\alpha\sigma_{-1}}{K+[M_a/(1-R_{\sigma_0})](\eta_1+R_{\sigma_0}^{\text{ост}}\eta_2)} \tag{5}$$

And in the field of permanent:

$$\sigma_{1a}^{\text{п}} = \frac{\sqrt{6}\alpha(1-c_1 R_{\sigma_0}^{res} M_{max}^{res})\sigma_{-1}}{K+\eta_1 M_a} \tag{6}$$

Where α is determined by the formula:

$$\alpha = \frac{0.249\sigma_{\text{в}}+2.5}{0.35\sigma_{\text{в}}+70} \tag{7}$$

It is important to pay attention to the fact that residual stresses play a large role in determining the endurance limit. And, depending on the size of the elements to be welded, they can create a linear, flat or volumetric stress state. As shown in [1-6] in elements of considerable length and with a thickness of less than 30 mm, they create a plane stress state. When external loads act on the welded element, plastic deformations occur and the fibers elongate (which received plastic compression deformations during welding). As a result of this, the initial residual stresses, if they coincide along the line of action and sign with the stresses from the external load, after unloading, in the case of a linear stress state they decrease by σ_{max}:

$$\sigma_{res} = \sigma_{\text{т}} - \sigma_{max} \tag{8}$$

Where σ_{max} – maximum stresses from external load at the site of the initial residual. With a flat stress state, the main residual stresses $\sigma_{1\,res}$, $\sigma_{2\,res}$coincide in sign and line of action with the main $\sigma_{1\,max}$, $\sigma_{2\,max}$from external loads. In this case, we can write that after unloading, the main residual stresses are equal to:

$$\sigma_{1\,res}' = \sigma_{\text{т}} - \sigma_{1\,max};\ \sigma_{2\,res}' = \sigma_{\text{т}} - \sigma_{2\,max} \tag{9}$$

We calculate the values of the asymmetry coefficients of the cycles, taking into account the residual stresses according to the formula:

$$R_{\sigma^0}^{res} = \frac{(\sigma_{min}^0)res}{(\sigma_{max}^0)res} = \frac{\sigma_{1\,min}+\sigma_{2\,min}+\sigma_{3\,min}+\sigma_{1\,res}+\sigma_{2\,res}+\sigma_{3\,res}}{\sigma_{1\,max}+\sigma_{2\,max}+\sigma_{3\,max}+\sigma_{1\,\text{ост}}+\sigma_{2\,res}+\sigma_{3\,res}} \tag{10}$$

Where$\sigma_{1\ res}$, $\sigma_{2\ res}$, $\sigma_{3\ res}$ – main residual sresses; $(\sigma_{min}^{0})_{res}$, $(\sigma_{max}^{0})_{res}$ – octahedral, equal to the sum of residual and stresses from external loads; $\sigma_{1\ max}, \sigma_{2\ max}, \sigma_{3\ max}$- maximum main cycle stresses; $\sigma_{1\ min}, \sigma_{2\ min}, \sigma_{3\ min}$-minimum principal stresses of the cycle. The remaining values are given in [12-15].

3. Result

We carry out numerical and analytical modeling of the endurance limit according to the method described in part 2.1. We calculate the values of the limiting amplitudes of the principal stresses taking into account the influence of the asymmetry of the cycle of principal stresses, characterized by the values of the coefficients $R_{\sigma 1}$, $R_{\sigma 2}$, $R_{\sigma 3}$.

We carry out numerical and analytical modeling for the first case. We will build a model of the experimental setup in the software packages StructureCAD (figure 3) and SolidWorks (figure 4). Both models show approximately the same stress distribution, with SolidWorks having a built-in algorithm for calculating tubular joints taking into account local stress concentration. We determine the nominal and ultimate stresses.

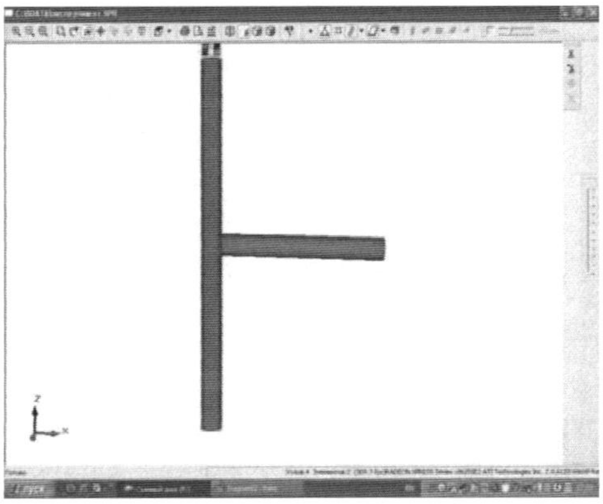

Figure 3. Analytical model of an experimental setup simulating tee joints made in a software package StructureCAD.

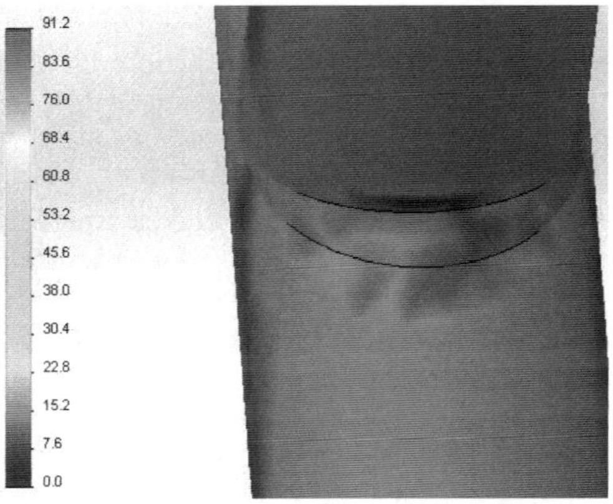

Figure 4. Stress distribution in the analytical model of an experimental setup simulating tee joints in a software package Solid Works.

As a result of the load simulation, it was found that the maximum stress amplitude in the conditions of the Black Sea deposits for the T-type connection is 91.2 MPa. We calculate the fatigue limit of a welded T-joint under the conditions of the first fracture. We determine the values of residual stresses. In accordance with the project materials for the welded spatial nodes of the support blocks in the area of non-shaped joints made of 09G2S steel, the yield strength of the weld material with a thickness of 10 to 80 mm should be at least 280 MPa. We obtain the values of the main residual stresses $\sigma_{1\ res}$= 188 MPa and σ_{2res} = 273 MPa (under the condition that the maximum load from the bending moment is 4 kNm, with other values of the bending moment the figures of the residual stresses will also change).

As it is shown by numerical and analytical modeling, the first major stresses make the main contribution to the determination of the limiting voltage amplitudes. The effects of the second and third major stresses are less. At the same time, it should be noted that, if we follow the rule of distribution of a greater voltage along an algebraic sign, then we will violate the physical meaning of the process under study, since in this situation, we get not a symmetric cycle, but zero-to-compassion. We determine the values of the limiting amplitudes of the main stresses. We will make all the necessary calculations and bring the results to the table

Table 2. Results of calculating a fatigue range of T-joints according to method 2.2.

σ_{1a}	R_a^{ocm}	$M_{\max}^{ocm}$	C_1^{ocm}	K	a	η_1	η_2	η_3	σ_{-1}	σ_{1a}^n
91.3	0.63	2	0.53	1.41	0.5	0.1	0.4	0.24	91.95	24.9

Designation:

σ_{1a} - a limiting amplitude for the first primary stress;

R_a^{ocm} - an asymmetry coefficient of residual stresses;

$M_{\max}^{ocm}$, C_1^{ocm}, K - auxiliary coefficients

η_1, η_2 and η_3 - an influence coefficient of tension, compressive and bending normal stresses, respectively

σ_{-1} – a fatigue range

σ_{1a}^n – a limiting amplitude for the first primary stress

4. Conclusion

If we compare the diagrams obtained in normative documents [239] and [459], we can see that the authors use such concepts as the voltage span equal to the difference between the maximum and minimum voltage of the cycle, as well as the concept of the amplitude of the cycle equal to half of this difference. In both cases, the results are similar and correspond to the values. Therefore, the fatigue range$\boldsymbol{\sigma_{-1}}$can be approximately taken equal to the value of the limiting amplitude of the first principal stress equal to 25 MPa. Thus, we can conclude that the described method for determining the fatigue limit depending on the temporary tensile strength well satisfies the requirements for calculating the fatigue limit of welded joints of offshore stationary platforms. If we compare the diagrams obtained in normative documents [239] and [459], then we can see that the authors use such concepts as the range of stresses equal to the difference between the maximum and minimum stresses of the cycle, as well as the concept of the amplitude of the cycle, equal to half of these differences. In both cases, the results are similar and correspond to the values. Therefore, the fatigue range$\boldsymbol{\sigma_{-1}}$ can be approximately taken equal to the value of the limiting amplitude of the first principal stress equal to 25 MPa. Thus, we can conclude that the described method for determining the fatigue limit depending on the temporary tensile strength well satisfies the requirements for calculating the fatigue limit of welded joints of stationary offshore platforms.

References

[1] Efimenko L, Kapustin O, Gusev V, Merkulova A and Nikiforov N 2014 Evaluation of structuraland phase composition and corrosion resistance properties of coatings made by laser surfacing *Chemical and Petroleum Engineering* **50** 402-411

[2] Efimenko L 2015 Methodological fundamentals of the evaluation of the weldability of structural materials using oil and gas pipes as an example *Welding Interhationa l***9 (28)** 727-731

[3] Efimenko L, Elagina O, Ramus A, Volkov I, Ladyzhanskii A, Rakhmanov D, Konovalov A and Kurkin A 2015 Repair features of mechanical and corrosion damage in a main gas pipeline under pressure *Chemical and Petroleum Engineering* **50** 749-757

[4] Elagina O, Efimenko L and Volkov I 2015 Study of the strain ageing on corrosion failure and submicrostructure of low-carbon steels for gas-distriution pipelines *Chemical and Petroleum Engineering* **51** 433-437

[5] Efimenko L, Kapustin O, Merkulova A, Subbotin R, Puyko O and Shchkalova A 2016 Features of transformation kinetics of austenite in heat treatment of stamped and welded components made of 10G2FBYU steel *Welding International* **30** 472-478

[6] Efimenko L and Ramus A 2016 Effect of the morphology of structure on the resistance of welded joints ofhigh-strength pipe steels to brittle fracture *Metal Science and Heat Treatment* **57** 559-563
[7] Efimenko L, Kapustin O, Ramus A and Ramus R 2016 Control of softening processes in the heat-affected zone during welding of high-strength steels *Metal Science and Heat Treatment* **58** 435-441
[8] Efimenko L, Kapustin O, Merkulova A, Subbotin R, Puyko O and Shchkalova A 2016 Features of transformation kinetics of austenite in heat treatment of stamped and welded components made of 10G2FBYU steel *Welding International* **30** 472-478
[9] Makarov G 2014 Modernization of systems of pipeline transport of gas and oil *Welding International* **28 (9)** 744-747
[10] Starokon I 2019 Varieties and kinetics of fatigue cracking of fixed offshore platforms for oil and gas production *Materials Science and Engineering* **734** 012055
[11] Antonov A 2014 Investigation of fields of residual stresses in welded structures *Welding International* **28** 966–969
[12] Starokon I 2020 Features of fatigue processes occurring in restored welded joints of offshore fixed platforms *Materials Science and Engineering* **734** 012171
[13] Starokon I 2020 Resource of repaired welded compounds of marine oil and gas structures *Materials Science and Engineering* **734** 012193
[14] Guseynov Ch 2020 In Arctic - new technical means and technologies for the development of oil and gas fields in long-term freezing deepwater areas *Materials Science and Engineering* **734** 012174
[15] Starokon I 2019 On the results of studying the influence of vibration-fluctuations processes on the resource of welded joints of marine stationary platforms for oil, gas and other useful minerals on the shelf *J Akustika* **285**148-151
[16] Bezkorovayniy V, Bayazitov V and Bobov D 2018 Management of the Design and Construction of Offshore Oil and Gas Facilities with Bim Base *Materials Science and Engineering* **463(4)** 042056
[17] Reitenbach V, Ganzer L and Albrecht D 2014 Influence of Hydrogen on Underground Gas Storage *Research Report DGMK-752* **73** 6627-6937
[18] Reitenbach V, Ganzer L, Albrecht D and Hagemann B 2015 Influence of added hydrogen on underground gas storage: a review of key issues *Conf. Ser.: Environ. Earth Sci.***73(10)** 692737
[19] Ganzer L, Reitenbach V, Pudlo D, Albrecht D, Singhe AT, Awemo K N, Wienand J and GauppR 2014 Experimental and numerical investigations on CO2 injection and enhanced gas recovery effects in Altmark gas field (Central Germany) *ActaGeotechnic a***9(1)** 39-47
[20] Guseynov C 2010бRU, Patent No 10-281 *Methodological recommendations for developing a strategy for the development of offshore oil and gas fields* (Moscow:Rospatent)

IOP Conf. Series: Materials Science and Engineering **860** (2020) 012025 doi:10.1088/1757-899X/860/1/012025

Evaluation of efficiency of the "K" welded repair technology of marine stationary platform with account ofaccumulated damages

I VStarokon

Russian State University of Oil and Gas named after I.M. Gubkin, 65/1, Leninsky ave., Moscow, 119991, Russia

Abstract. The article discusses the problems associated with calculating the residual life of welded joints of offshore stationary platforms using the example of "K" -type joints. It should be noted that the residual life of such repaired compounds is still not well understood. The author built computer models of offshore platforms, which made it possible to determine the values of the maximum amplitudes of the alternating stresses that arise during the operation of the platform. Based on this information, the author justified the conditions of the experiment. The article describes the method of repairing a crack by the method of its welding with the establishment of so-called "crack traps" along the edges of the crack. The experiment was carried out under conditions of primary and repeated fracture of the welded joint "K" – type. As a result of the studies for the described repair technology, the endurance limits of the repaired K-type welded joints were calculated, which is 13.3 MPa based on the cycle Ng = 3 * 10^6.

1. Introduction

The Russian Federation has rich oil and gas deposits at the bottom of marine areas [1, 6-10]. In the Black Sea, offshore oil and gas facilities (OOGF) are used to extract these valuable minerals, which are stationary platforms of various design purposes (figure 1), with various production cycles, etc. Most of these platforms have been in operation for more than 30 years - they were established back in Soviet times. It should be noted that the wear of welded joints of offshore fixed platforms (OFP) is very large, which is explained by their work in adverse conditions and the destructive effect of sea waves, currents and other adverse factors [1, 3, 6-10].

In connection with such difficult operating conditions of offshore platforms, it is of great practical interest to study the residual life of welded joints taking into account accumulated damage [2] and the effectiveness of various methods of their repair in terms of extending their life. Currently, there is no methodology for assessing the residual life of such repair joints as applied to the support block of the OFP. Such compounds, in addition to the negative factors arising as a result of the welding process itself (residual stresses, heat-affected zone, etc.) are also affected by fatigue processes (for example, accumulated fatigue damage) and other factors [1, 3, 6- 10].

Figure 1. Offshore fixed platforms (OFP)

Consider the main types of welded joints used in the construction of offshore stationary platforms, which are shown in figure 2. Generally speaking, two types of compounds are used in offshore oil and gas facilities (OOGF). The joint of two pipes at a right angle (T-joint or in foreign literature the so-called "T"-connection), shown in figure 3 and the joint of pipes at an angle (brace or in foreign literature - "K"- connection), shown in figure 4. Each of them has its own performance features. In a number of cases, the T-joint and brace do not have a common nodal joint, since, as proved by the author's calculations, the stresses in the knot increase slightly. Insomecases, the senodes will share in common.

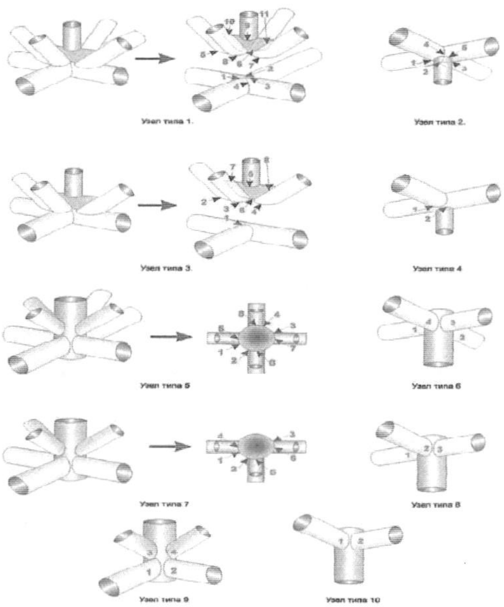

Figure 2. Various manufacturing schemes for welded joints of offshore fixed platforms.

It should be noted that in recent years the projects of offshore oil and gas facilities (OOGF) feature special nodal connections. However, in the offshore oil and gas facilities (OOGF), built more than 30 years ago, these technologies were not applied, and the designs shown in figure 3 and figure 4 were used.

Figure 3. Unpaired t-joint and brace.

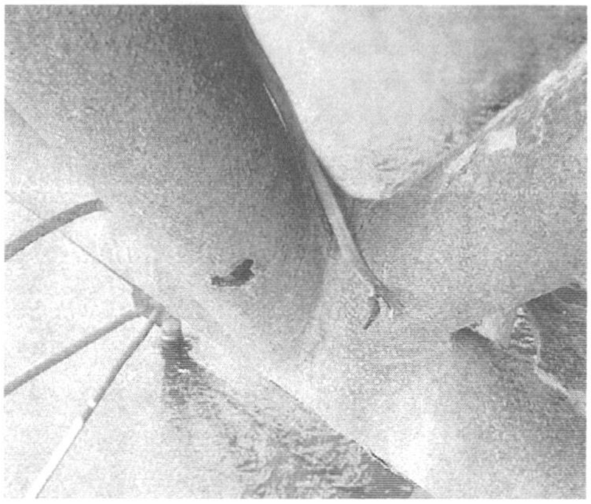

Figure 4. Paired “K”-Connection.

Earlier, the author published the results of a study of the resource of a repaired T-joint. This article will discuss the study of "K" -connection. As mentioned earlier, the resource of both a new and a repaired welded joint is determined by the accumulation of fatigue damage in it. Fatigue damage is caused by the action of alternating stresses that occur under the action of external forces and environmental influences. Therefore, to assess the residual life of welded joints, it is necessary to obtain a picture of their stress state. The solution to this problem is impossible without the use of modern computer technology. Therefore, the author built a model of an offshore stationary platform in the StructureCAD software package, on which the effect of loads on the welded joint was simulated based on the following conditions: loads from the platform’s own weight, from the weight of the equipment and sea fouling, the buoyancy force of the aquatic environment (combination of K_1 loads) and from wind and wave exposure (combination of loads K_2 and K_3). Moreover, to take into account the mutual influence of the elements, the direction of development of the wind-wave load was set both along the X axis (combination of loads K_2) and in the direction at an angle of 450 to the X axis (combination of loads K_3). The value of the wave load was chosen corresponding to the wave action at a wind speed of 25 m/s and a wave height of 6.8 m. The calculation was carried out by the method of determining the main stresses in order to go on to the values of equivalent stresses. As a result of calculating the maximum-stress “K”-connection (at a static voltage of 80 MPa), the following values of the main stresses were obtained:

Table 1. Values of the main stresses during the direct and reverse action of the wave load acting in the "K"-connection (σ_1, σ_2, σ_3- first, second and third major stresses).

Loadcombination	***Direct exposure to wave load***			***Inversewaveload***			***Cycleasymmetrycoefficient***
	σ_1	σ_2	σ_3	σ_1	σ_2	σ_3	***R***
K_1	160	12.4	2.1	1	-	-	-
K_2	194	17.8	3.6	-33	7.8	1.2	≈0.17

σ_1, σ_2, σ_3- major stresses, MPa

The obtained values of the principal stresses were reduced to equivalent values. The analysis of these values suggests that in the case under consideration, the maximum values of the amplitudes of the variable equivalent stresses, which cause fatigue failure, are equal for the combination K_1=81 MPa and for the case K_2=115 MPa. Having determined the values of the stresses acting in the welded joint, we turn to the question of assessing the resource of welded joints with accumulated damage.

2. Materials and methods

Due to the fact that the endurance limit of such a welded joint cannot be determined theoretically, the author developed an experimental setup and performed a series of six experiments.

The experiment was carried out as follows. From the high-pressure pipeline with compressed air, a tap was made to the control module for the experimental pressure values. Then the pressure was reduced to a certain specific value, which was set in the control module. After that, the pressure was transferred to the switching unit of the experimental load vector, where the pressure was transformed due to a certain value, which acted in different directions along the vertical axis, destroying the welded joint. The experiment consisted of two stages. Bringing the compound to primary destruction (figure 5), repairing it and then secondary destruction. It should be noted that the duration of the extension of the resource very much depends on the chosen method of repair. For research purposes, the following repair method was chosen. After receiving the primary crack, a hole with a diameter of 8 to 12 mm was drilled at a distance of 20 mm from its ends (figure 6). The metal along the crack was removed with abrasive wheels with at least 25 mm beyond each side of the crack visible. And at the crack site, the metal was chosen in such a way as to simulate a V-shaped groove of the edges (to ensure penetration of the crack mouth). It should be noted that the hole created during the repair process itself is a stress concentrator. When mating pipes 720x20 and 530x12 in the presence of a hole with a diameter of 8 mm, the voltage increases by 1.3 times, and if a hole with a diameter of 12 mm - 1.7 times. With the pairing of pipes 1020x25 and 530x12 and the presence of a hole with a diameter of 8 mm, the voltage increases 1.25 times, and with a hole with a diameter of 12 mm - 1.6 times. Therefore, the author sets the task to scale the applied load so that the voltage values remain unchanged. In view of the fact that the diameters of real structural elements differ from elements of a laboratory sample, and the length of an experimental welded joint is significantly less than the real one, the diameter of the crack collector is 8 mm. In order to preserve the values of the operating stresses, the experimental load was reduced from 3.9 kN and 4.8 kN to 2.3 kN and 3.25 kN respectively.

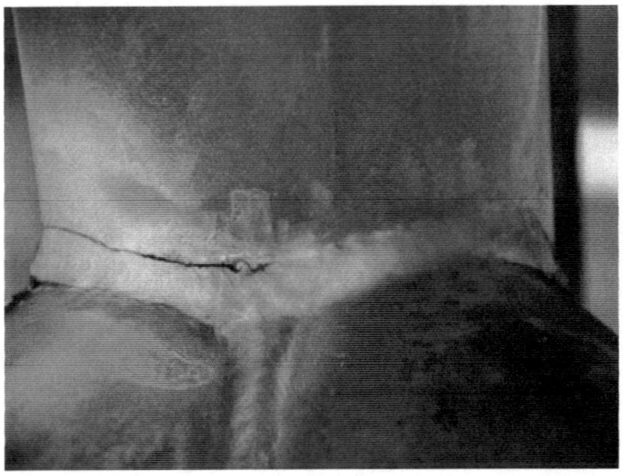

Figure 5. Primary crack in the experimental "K"-connection.

Figure 6. Experimental "K"-connection after repair.

Briefly describe the methodology for processing the results of the experiment. When conducting multiple measurements of the same type of a certain value, it is necessary:

1. Carry out these multiple measurements under the same conditions and write them in a table.
2. Calculate the average value according to the formula:

$$\overline{x} = \frac{1}{n}\sum_{i=1}^{n} x_i \quad (1)$$

3. Calculate the variance estimate

$$\hat{\sigma}^2 = s_n^2 = \frac{\sum (x_i - \overline{x})^2}{(n-1)} \quad (2)$$

4. Calculate the standard error of the mean

$$s_x = \frac{s_n}{\sqrt{n}} \quad (3)$$

5. At the required level of confidence ρ, the student's t test $t(\rho, n-1)$ and the module of the confidence interval. In industry, a probability level of 95% is acceptable, if the number of experiments is four, the student's t test is 3.18.

$$\Delta x = s_x \cdot t(\rho, \nu) \quad (4)$$

6. Rounding off the corresponding results, write down the answer in the form

$$X = \overline{x} + \Delta x \quad (5)$$

3. Results

Series of six samples of repaired welded joints of the "diagonal" type was tested at a level of stress amplitudes of 82 MPa and 115 MPa, and additional points at a level of 44 and 27 MPa. The following results were obtained (see table 2)

Table 2. Number of N cycles obtained during the experimental test of the repaired diagonal brace

Model №1	Model №2	Model №3	Model №4	Model.№5	Model №6	$\overline{x}$	s_n	s_x	Δx	X±Δx
					Voltageamplitudeσ_1 =115MPa					
4332	3975	4672	4348	4301	4359	4332	221	111	352	**4332** ±352
					Voltageamplitudeσ_2=82 MPa					
10253	9071	10798	10803	10164	10504	10253	643	321	1022	**10253** ±321
					Voltageamplitudeσ_1 =44 *MPa*					
699444	702333	700123	704553	692341	702324	700363	4251	2125	6759	**700363** ±6759
					Voltageamplitudeσ_2= 27 *MPa*					
1006330	1011202	1007330	1010920	1009486	1009497	1009128	1941	971	3086	**1009128** ±3086

The endurance limit was calculated by the formula:

IOP Conf. Series: Materials Science and Engineering **860** (2020) 012025 doi:10.1088/1757-899X/860/1/012025

$$\sigma_R = \sqrt[2]{(\sigma_1^2 N_1 - \sigma_2^2 N_2)/(N_1 - N_2)} \tag{6}$$

Where: σ_1- is the maximum voltage amplitude taken equal to 115 MPa, and σ_2- is the voltage amplitude taken equal to 82 MPa; N_1- is the experimentally set number of cycles to failure equal to 10253 and N_2 -is the experimentally set number of cycles to failure, equal to 1009128.

4. Discussion

As a result, the endurance limit equal to 43 MPa was calculated. Using the Klykov method [4], which takes into account the limiting amplitudes of the principal stresses, the coefficients of the influence of normal stresses, residual stresses, and other factors, it was found that the endurance limit of the compound under study is 13.3 MPa.

Let us construct a schematic diagram of the fatigue of repair welded joints of the "diagonal" type of the brace block.

The fatigue curve (figure 7) equation for a "K"-connection with accuracy sufficient for practice can be represented as follows:

$$\sigma_a = -14.5539 \cdot \ln(1.51603 \cdot 10^{-7} N) \tag{7}$$

where: σ_a- the amplitude of the alternating stresses acting in the welded joint; N - is the number of cycles. Having made the calculation based on the Baskvin equation, we obtain the value of the fracture point of the fatigue curve $N_G=3*10^6$.

We calculate the values of endurance limits for some cycle asymmetry coefficients. With a zero cycle, the endurance limit will be equal to σ_0=26MPa, and with asymmetry coefficients R=0.17, the endurance limit calculated on the basis of the Goodman equation will be $\sigma_{R=0,17}$=10.79MPa.

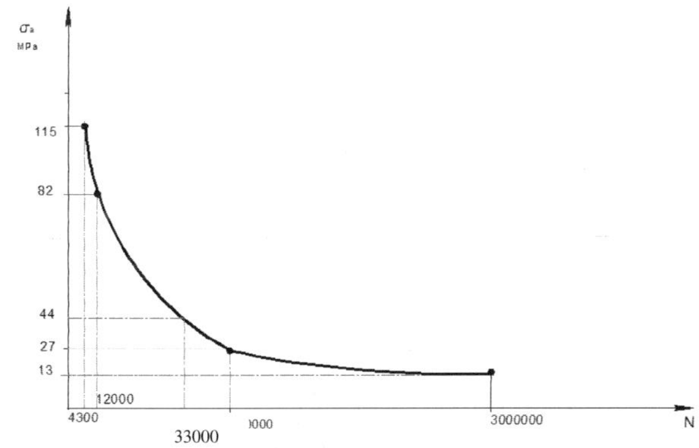

Figure 7. Schematic representation of the fatigue diagram of the welded joint "K" -connection of the OFP after repair.

5. Conclusion

The results of the study showed that the value of the endurance limit, as well as the number of cycles for both types of joints studied are almost the same, from which we can conclude that the σ_{-1} parameters of offshore oil and gas facilities welded joints are determined by the mechanical properties of the material and the type of repair.

References

[1] Efimenko L, Kapustin O, Gusev V, Merkulova A and Nikiforov N 2014 Evaluation of structural and phase composition and corrosion resistance properties of coatings made by laser surfacing *Chemical and Petroleum Engineering* **50**pp 402-411

[2] Efimenko L2015 Methodological fundamentals of the evaluation of the weldability of structural materials using oil ahd gas pipes as an example *Welding Interhational* **9 (28)** 727-731

[3] Efimenko L, Elagina O, Ramus A, Volkov I, Ladyzhanskii A, Rakhmanov D, Konovalov A and Kurkin A2015 Repair features of mechanical and corrosion damage in a main gas pipeline under pressure *Chemical and Petroleum Engineering* **50** 749-757

[4] Elagina O, Efimenko L and Volkov I 2015 Study of the strain ageing on corrosion failure ahd submicrostructure of low-carbon steels for gas-distriution pipelines*Chemical and Petroleum Engineering* **51** 433-437

[5] Efimenko L, Kapustin O, Merkulova A, Subbotin R, Puyko O and Shchkalova A2016 Features of transformation kinetics of austenite in heat treatment of stamped and welded components made of 10G2FBYU steel*Welding International* **30** 472-478

[6] Efimenko L and Ramus A2016 Effect of the morphology of structure on the resistance of welded joints ofhigh-strength pipe steels to brittle fracture *Metal Science and Heat Treatment* **57** 559-563

[7] Efimenko L, Kapustin O, Ramus Aand Ramus R 2016 Control of softening processes in the heat-affected zone during welding of high-strength steels*Metal Science and Heat Treatment* **58** 435-441

[8] Efimenko L, Kapustin O, Merkulova A, Subbotin R, Puyko O andShchkalova A2016 Features of transformation kinetics of austenite in heat treatment of stamped and welded components made of 10G2FBYU steel*Welding International* **30** 472-478

[9] Makarov G 2014 Modernization of systems of pipeline transport of gas and oil *Welding International* **28 (9)** 744-747

[10] Starokon I 2019 Varieties and kinetics of fatigue cracking of fixed offshore platforms for oil and gas production *Materials Science and Engineering* **734** 012055

[11] Antonov A2014 Investigation of fields of residual stresses in welded structures *Welding International* **28** pp 966–969

[12] Starokon I2020 Features of fatigue processes occurring in restored welded joints of offshore fixed platforms *Materials Science and Engineering* **734** 012171

[13] Starokon I 2020 Resource of repaired welded compounds of marine oil and gas structures *Materials Science and Engineering* **734** 012193

[14] Guseynov Ch 2020 In Arctic - new technical means and technologies for the development of oil and gas fields in long-term freezing deepwater areas *Materials Science and Engineering* **734** 012174

[15] Starokon I2019 On the results of studying the influence of vibration-fluctuations processes on the resource of welded joints of marine stationary platforms for oil, gas and other useful minerals on the shelf *J Akustika* **285** 148-151

[16] Bezkorovayniy V, Bayazitov V and Bobov D 2018 Management of the Design and Construction of Offshore Oil and Gas Facilities with Bim Base *Materials Science and Engineering* **463(4)** 042056

[17] Reitenbach V, Ganzer L and Albrecht D 2014 Influence of Hydrogen on Underground Gas Storage *Research Report DGMK-752* **73** 6627-6937

[18] Reitenbach V, Ganzer L, Albrecht D and Hagemann B 2015 Influence of added hydrogen on underground gas storage: a review of key issues *Conf. Ser.: Environ. Earth Sci.***73(10)** 6927-37

[19] Ganzer L, Reitenbach V, Pudlo D, Albrecht D, Singhe AT, Awemo K N, Wienand J and Gaupp R 2014 Experimental and numerical investigations on CO2 injection and enhanced gas recovery effects in Altmark gas field (Central Germany) *ActaGeotechnica* **9 (1)** 39-47

[20] Guseynov C 2010 RU, Patent No 10-281 *Methodological recommendations for developing a strategy for the development of offshore oil and gas fields* (Moscow: Rospatent)

IOP Conf. Series: Materials Science and Engineering **860** (2020) 012026 doi:10.1088/1757-899X/860/1/012026

Experimental research of electrical loads in residential and public buildings

Y I Soluyanov[1,2], A I Fedotov[1,3], D Y Soluyanov[1,2] and A R Akhmetshin[1,3]

[1] Association «Roselectromontazh», 13, Ak. Koroleva str., Moscow, 129515, Russia
[2] JSC «Tatelectromontazh», 50A, Adoratskogo str., Kazan, 420132, Russia
[3] Kazan State Power Engineering University, 51, Krasnoselskaya, str., Kazan, 420066, Russia

E-mail: ahmetshin.ar@mail.ru

Abstract. The Association "Roselectromontazh" conducted a study of electrical loads for residential and public buildings in the Republic of Tatarstan showed that actual electrical loads are 1.5 - 2.5 times lower than in regulatory documents. Recently constructed cable networks and transformer substations of 0.4 / 10 kV in fact turn out to be underloaded. Moreover, this underload is such that most of the transformers rarely operate at 30% of their capacity during hours of maximum power consumption. The calculation of electric load is the basis for designing the power supply system of any capital construction facility. According to the open joint-stock company "Network Company" and with the support of the President of the Republic of Tatarstan, the Association "Roselectromontazh" carried out an experimental study on updating the requirements for calculating electricity consumption in public and public buildings of the Republic of Tatarstan. The actualization of electrical loads will lead to a significant economic effect by reducing the declared capacity and accordingly the power and number of power transformers reducing the cross-sections of power cables and consequently, capital costs during construction and operation.

1. Introduction

The practice of building and operating urban power supply systems has shown that in most cases real loads are less than calculated, this leads to an overpricing of the cost of technological connection [2]. To calculate the electric power consumed by residential and public buildings, use the normative values from the section "Designed electrical loads" SP 256.1325800.2016 "Electrical equipment of residential and public buildings. Rules of design and erection" (SP), which in turn have not been reviewed for decades. Studies of the loading of transformer substations of 0.4/10kV showed that most of the transformers are underloaded.

2. Experimental researches of electrical loads

In most cases, real loads are less than calculated. As a result, the constructed cable networks and transformer substations of 0.4 / 10 kV are in fact underloaded. Moreover, this underload is such that about 80% of transformers operate with a maximum load of less than 30% during the year, Figure 1.

Consequently, transformers operate with large losses [4-6] in relation to the transmitted electric power, Figure 2.

Figure 2 shows that the less loaded the transformer, the more share the loss in transmitted power (ΔP).

To reduce losses in relation to the transmitted power, it is proposed to install transformers with lower power or to turn off part of the transformers to ensure a greater load of the remaining ones [3].

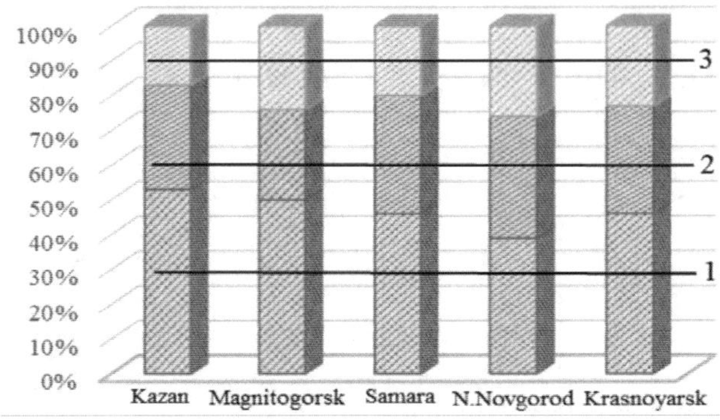

Figure 1. Loading transformer substations in cities.
(1 - up to 15%; 2 - from 16 to 30%; 3 - from 31 to 85%).

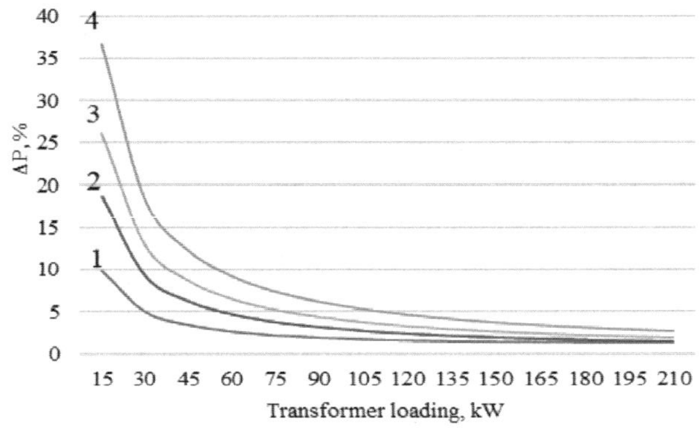

Figure 2. Losses in transformers (ΔP) regarding their loading
(1–TM250/10; 2–TM630/10; 3–TM1250/10; 4–TM2500/10).

The need for work on the adjustment of specific calculated electrical loads is confirmed by the following documents:

- The State Program of the Russian Federation "Energy Development" as amended on March 28, 2019 No. 335, one of the goals of which is "to ensure the needs of the domestic market in an economically sound supply of energy";
- The draft Energy Strategy of the Russian Federation for the period until 2035 as amended on October 21, 2019, the goals of which are also "improving the legal framework and planning system in the electric power industry", "reducing electric energy losses until 2035 to the level of 7-7 ,5%";
- The national rating of the state of the investment climate in the constituent entities of the Russian Federation, the key indicator of which is "The quality of the provision of public services - indicators of the effectiveness of the provision of various public services for business, including connection to electric networks";
- A comprehensive plan for the modernization and expansion of the trunk infrastructure for the period until 2024 of September 30, 2018, the target indicators of which include "reducing the excess installed capacity of power plants of the UES of Russia".

IOP Conf. Series: Materials Science and Engineering **860** (2020) 012026 doi:10.1088/1757-899X/860/1/012026

In [1], it was experimentally proved that the actual load for apartment buildings differs from the estimated one by 1.5-2.5 times. Below is a diagram, Figure 3, the dependence of the average specific power consumption of 32 apartment buildings on the month and year of operation.

The diagram, Figure 3, confirms that power consumption changes not only during the year (the maximum falls on the winter and the minimum summer months for the city of Kazan), but also on the service life. In the first two years of operation, power consumption is very different, so it is necessary to carry out measurements, starting from the third year of operation, for at least 3 years [7-10].

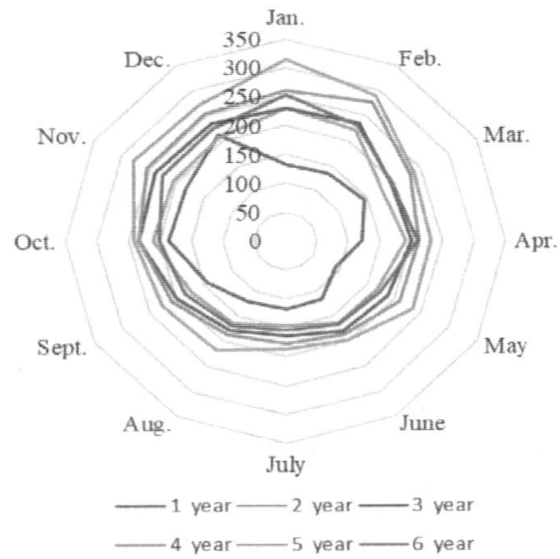

Figure 3. Diagram of the dependence of the electricity consumption of apartment buildings on the month and year of operation

Figure 4 shows the annual average schedule for specific electricity consumption by kindergartens (Psp, kW*h/pupil). The study was conducted in 51 kindergartens in the city of Kazan.

In Figure 5 shows the annual average schedule for the specific energy consumption of schools (Psp, kW*h/pupil). The study was conducted in 67 schools in the city of Kazan.

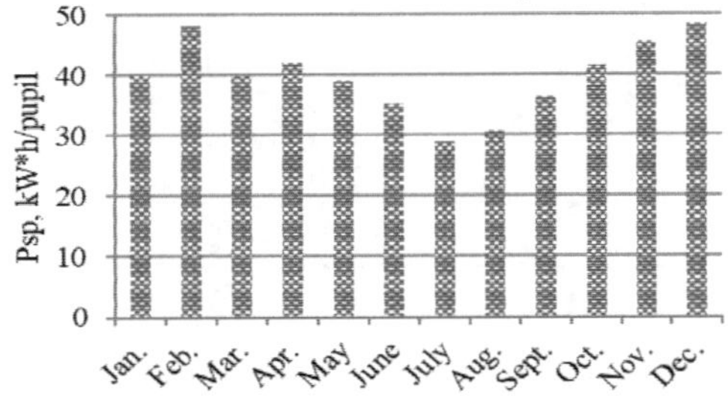

Figure 4. Kindergarten's annual specific power consumption schedule

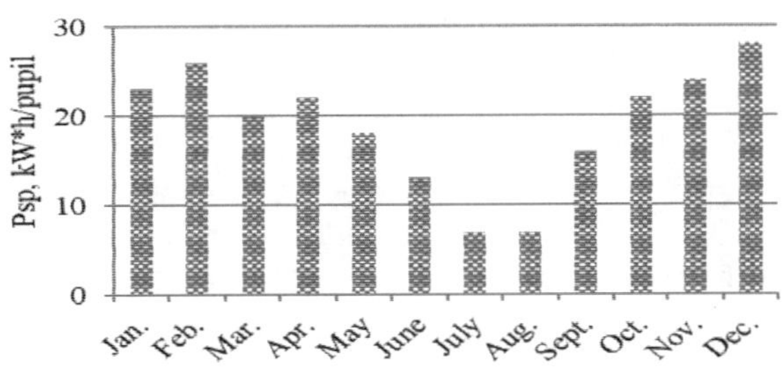

Figure 5. School's annual specific energy consumption schedule

From the graphs Figures 4 and 5, it can be seen that the months of greatest power consumption are December and February, and the minimum is July and August, which is quite expected.

Figure 6 shows the annual average cottage specific electricity consumption schedule (Psp, kW*h/cottage). The study was conducted 375 cottage of one of the villages in the city of Kazan.

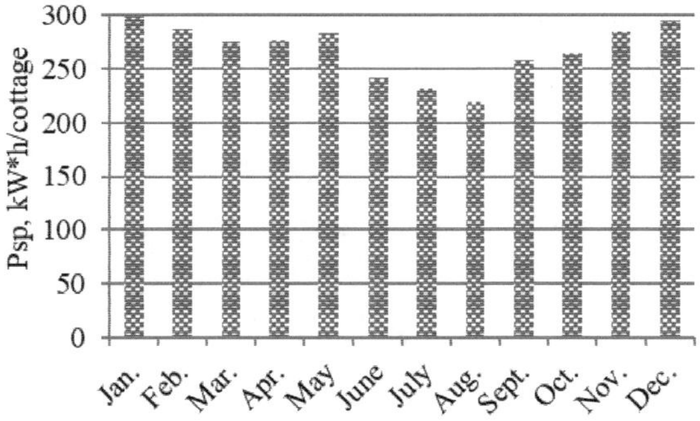

Figure 6. Cottage's annual specific power consumption schedule.

The graph, Figure 6, shows that power consumption varies throughout the year, with a maximum during the winter and a minimum during the summer.

Let us compare the value of the specific load with the declared capacity for kindergartens, which is obtained by calculation and measured in fact, Figure 7.

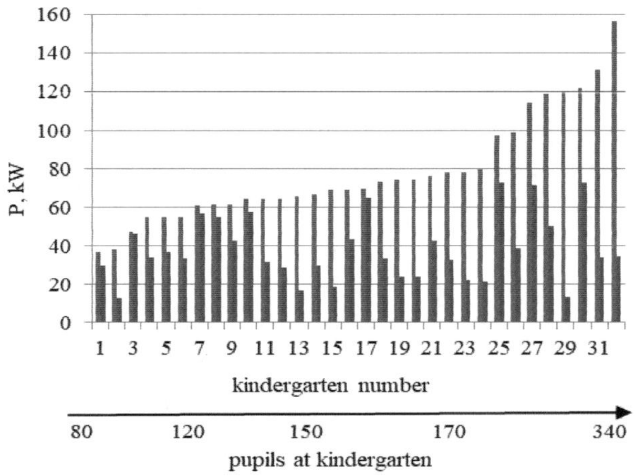

Figure 7. Difference between the actual power and the power calculated by SP

The electrical load of kindergartens calculated using the SP is significantly different from the actual half-hour maximum load measured directly at the facility.

The half-hour maximum loads during the year for kindergartens (Figure 8) and schools (Figure 9). Graph Figure 8 shows the dependence of the half-hour maximum values of unit loads on the number of pupils in kindergarten. The values obtained are comparable with 0.49 [kW/pupil] from SP.

The specific values obtained are significantly lower than those presented in SP. This emphasizes the need to revise the specific design SP loads. In addition, the dependence of the specific load depending on the number of pupils in kindergarten is of interest (Figure 8). According to the graph, there is a pronounced tendency to decrease in the unit load with an increase in pupils in kindergarten, but no clear boundary between the values has been identified, which makes it possible to calculate with one value.

IOP Conf. Series: Materials Science and Engineering **860** (2020) 012026 doi:10.1088/1757-899X/860/1/012026

In Figure 9 there is a graph showing the dependence of half-hour maximum values of unit loads on the number of pupils in school.

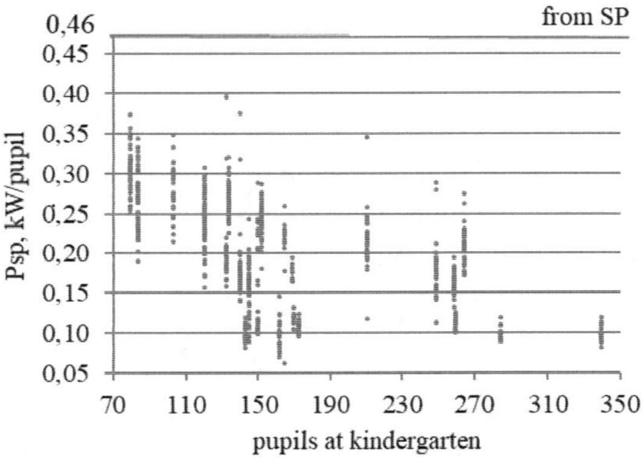

Figure 8. Dependence of the half-hour maximum values of unit loads on the number of pupils in kindergartens

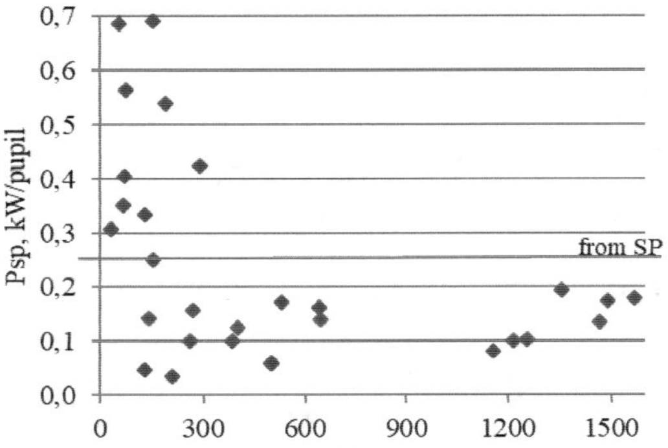

Figure 9. Dependence of half-hour maximum values of unit loads on the number of pupils in the school

The graph in Figure 9 shows the border in the school district of up to 300 pupils and more than 300. One can distinguish two areas on the graph with the corresponding two values of the unit load - 0.7 kW/pupils and 0.19 kW/pupils. The obtained values are comparable with 0.25 [kW/pupils] for SP. In this section, we can conclude that it is advisable to divide secondary schools into two groups - up to 300 pupils and over 300 pupils.

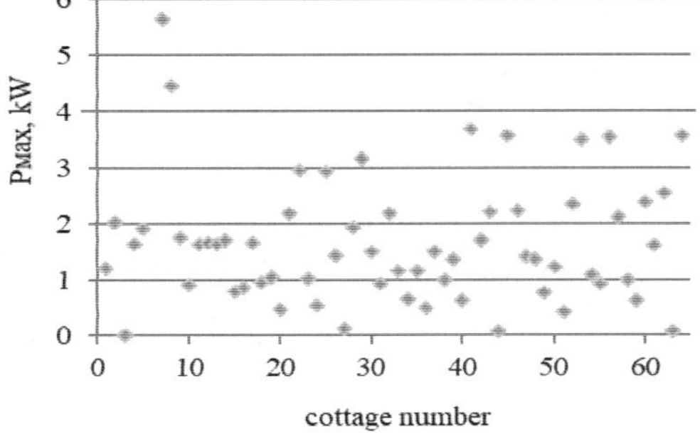

Figure 10. Maximum active power of the cottages

The justification of the need to solve the situation when 15 kW is requested for connecting a cottage, but actually used much less, is shown in the graph, Figure 10, from which it can be seen that the actual half-hour load during maximum hours in winter for 95% of cottages does not exceed 4 kW.

3. Conclusion

Based on the results of scientific research, conclusions are drawn:

- It is necessary to update the calculated specific electrical loads of residential and public buildings;
- Since the regions differ in climatic characteristics, it is necessary to revise the specific electrical load values for residential and public buildings in each region separately;
- The choice of the number and power of power transformers, taking into account the actual loads, will ensure their cost-effective operation;
- For the first time in the Russian Federation, work was done to update the specific rated electric load by the Association “Raselectromontazh” with changes in the specific rated electric loads of multi-apartment buildings in the Cabinet of Ministers of the Republic of Tatarstan No. 805 dated 09.09.2019. “On Amendments to the Resolution of the Cabinet of Ministers of the Republic of Tatarstan dated December 27, 2013 No. 1071 “On Approval of the Republican Standards for Urban Planning Design of the Republic of Tatarstan”.

References

[1] Soluyanov Y I, Fedotov A I and Ahmetshin A R 2019 Calculation of electrical loads of residential and public buildings based on actual data *IOP Conference Series: Materials Science and Engineering* **643** 012051 DOI 10.1088/1757-899X/643/1/012051

[2] Suslov K, Piskunova V, Gerasimov D, Ukolova E, Akhmetshin A, Lombardi P and Komarnicki P 2019 Development of the methodological basis of the simulation modelling of the multi-energy systems *E3S Web of Conferences* **124** 01049 DOI 10.1051/e3sconf/201912401049

[3] Mendeleev D I, Marin G E and Akhmetshin A R 2019 The Implementation and Use of Gas Turbines with Absorption Refrigerating Machine in the Technological Schemes of Thermal Power Plants *2019 International Multi-Conference on Industrial Engineering and Modern Technologies* Publisher IEEE INSPEC 19229407 DOI 10.1109/FarEastCon.2019.8934431

[4] Bakhteev K, Fedotov A and Misbakhov R 2019 The improving efficiency of electric receivers on the industrial enterprises in case of short-term power outages *Proceedings of the 2019 20th International Scientific Conference on Electric Power Engineering* Ostrava Czech Republic 347-352 DOI 10.1109/EPE.2019.8778181

[5] Gracheva E I and Alimova A N 2019 Calculation Methods and Comparative Analysis of Losses of Active and Electric Energy in Low Voltage Devices *2019 International Ural Conference on Electrical Power Engineering* INSPEC 19080057 DOI 10.1109/ URALCON.2019.8877627

[6] Gracheva E I and Alimova A N 2019 Calculating Probability of Faultless work of Shop Nets with the Help of Coefficients of Ratio *2019 International Russian Automation Conference* INSPEC 19090885 DOI 10.1109/RUSAUTOCON.2019.8867727

[7] Beccali M, Cellura M, Lo Brano V and Marvuglia A 2004 Forecasting daily urban electric load profiles using artificial neural networks *Journal on Energy Conversion and Management* **45 (18)** 2879 -2900

[8] Opoleva G N 2017 *Power supply of industrial enterprises and cities* (Moscow Publishing House) *FORUM INFRA-M 2017* 416

[9] Nadtoka I I, Pavlov AV and Novikov S I 2013 Problems of calculating the electrical loads of household consumers in microdistricts of megalopolises *University News. Electromechanics* **1** 136-139

[10] Nadtoka I I and Pavlov A V 2015 Improving the accuracy of calculating the electrical loads of apartment buildings with electric stoves *Izvestiya Vuzov North Caucasian region Technical science* **2** 45-48

IOP Conf. Series: Materials Science and Engineering **860** (2020) 012027 doi:10.1088/1757-899X/860/1/012027

Implementation of wireless data transmission method on wells equipped with sucker-rod pumping units

G I Bikbulatova, A S Galeev, M Z Valitov, S L Sabanov, Y A Boltneva and O A Shipilova

Almetyevsk State Oil Institute, 2, Lenina Str., Almetyevsk, Republic of Tatarstan, 423450, Russia

E-mail: bikbulatova.ngo@mail.ru

Abstract. When operating wells equipped with SRP units, the parameters of the products are monitored mainly via a wired communication channel in close proximity to the bottom by means of sensors. In the course of the work, it was found that, despite the large number of proposed scientific and engineering solutions, the objective of creating a reliable wireless electromagnetic channel connecting to the well bottom along with a configuration for its implementation is urgent, especially as far as the practical implementation is concerned. A distinguishing feature of the engineering solution proposed is data transmission via a galvanic communication channel in the course of well operation, involving the electric current excitation in the metal string in a well using the ground generator with one terminal connected to the ground part of the metal string and the other one to the pickup electrode on the well surface; switching dielectric loading, splitting the metal string in the downhole into the top and bottom parts, and data receipt from the well bottom depending on the ripple caused by the switching of the dielectric loading, where the data transfer phase synchronized with the upward lift of the rod string providing the possibility to use the feedback in order to track the cycle duration with a constant voltage on the dielectric splitter.

1. Introduction

The most commonly used method of artificial oil lift in brownfields that are characterized by a low production rate of most wells and the presence of viscous oils is the sucker-rod lift method. Two thirds of the total producing well stock is operated using sucker-rod pumping units (SRPU).

In the course of the SRPU operation, its technical condition and basic parameters are subject to inevitable changes. Corrosion, abrasive and mechanical wear, as well as various deposits have a negative impact on the performance of pumping units, while leaks that increase as the pump components wear out result in lower pump delivery rate. The speed of changes depends on the properties of the multiphase fluid extracted, operating modes, design peculiarities and quality of equipment. The simultaneous impact of all these factors makes it much more difficult to identify causes and patterns when analysing the causes of unit component failures [1].

The reservoir fluid parameters also change throughout the well's life cycle, since they in turn depend on various changing factors, such as injection of contour water flood of reservoir pressure maintenance (RPM) systems, water content, as well as gas, salts, paraffin content, etc. [2].

2. Problem statement

Some parameters of the products are evaluated in the immediate vicinity of the reservoir using sensors connected to the receiving station via a cable. Depending on what information is required, various sensors can be placed at the pump suction, such as density, temperature, or pressure sensors. The latter is required for measuring the static level of the fluid column during the sucker-rod pump operation, since it is the main indicator which determines the required borehole fluid pumping rate. It is implemented either by the cyclic pumping method, or by adjusting the number of strokes of the SRPU plunger.

The transmission of measurement data from the bottom to the mouth via a wired communication channel is complicated by a narrowing cross-section of the tubing as well as cable breakdowns caused by the movement of rods and paraffin deposits [2]. Patent studies carried out within the framework of implementing the monitoring of the SRPU technical condition and the reservoir fluid parameters at the bottom of the well to adjust the production process have shown that the development trends tend towards the use of cheaper and more reliable data transmission channels.

3. Engineering solution

In the course of performing the applied R&D works, a monitoring system was developed for SRPU with a submersible module and the downhole data transmission via a galvanic communication channel. The system which includes a ground station, a galvanic communication channel and a submersible module, analyses the technical condition of the drive and the sucker-rod pump, monitors the main well parameters, and transmits information to the relevant departments of the enterprise.

Each of the stages is controlled by software installed on controllers of the system's downhole and ground modules. Software, which is based on a mathematical model of the SRPU drive with a transforming mechanism in the form of a lever of the 1st kind, is designed to calculate the correlation between the phase of the rod string movement and time and establish the relationship between wattmetrogrammes and dynamogrammes. The system's SM-SRPU Monitoring software is designed for collecting, storing, processing and interpreting sensor measurement data, evaluating their levels against the limit set points, and is installed on a personal computer. Built-in algorithms for detecting defects in the technical condition provide the possibility to issue and distribute among the subscribers of the relevant services diagnoses of the condition, forecasts of trouble-free operation and recommendations concerning the timing of repair and preventive actions.

A general layout of the SRPU monitoring system with a submersible module and a galvanic communication channel for transmitting downhole data is shown in Fig. 1.

The submersible module of the monitoring station transmits downhole measurements via a galvanic communication channel (GC) represented by a tubing and rod string. The galvanic isolation unit isolates the communication line from the grounded ground equipment of the sucker-rod pumping unit. Further, data from downhole and ground sensors is transmitted to the ground monitoring and control module (ground station) via a wired line or radio communication [4].

The submersible module is located in the immediate vicinity of the downhole pump suction (see Figure 1) and may be equipped with various sensors, in this case, pressure and temperature sensors and the device transmitting the measured information via a galvanic communication channel.

The electrical resistance of the circuit consisting of a rod string, a downhole pump, and tubing is estimated assuming that the circuit consists of homogeneous, electrically isolated rods and tubing with a uniform cross section. The electrical resistance is calculated as follows:

$$R = R_{SH} + R_{NKT} + R_H = \rho \cdot \frac{L}{S_{SH}} + \rho \cdot \frac{L}{S_{NKT}} + R_H,$$

where p is the resistivity of the conductor substance,
L- conductor length,
S_{SH} - rods cross section area,
S_{NKT} - tubing cross section area,

R_H - electrical resistance that occurs in the downhole pump.

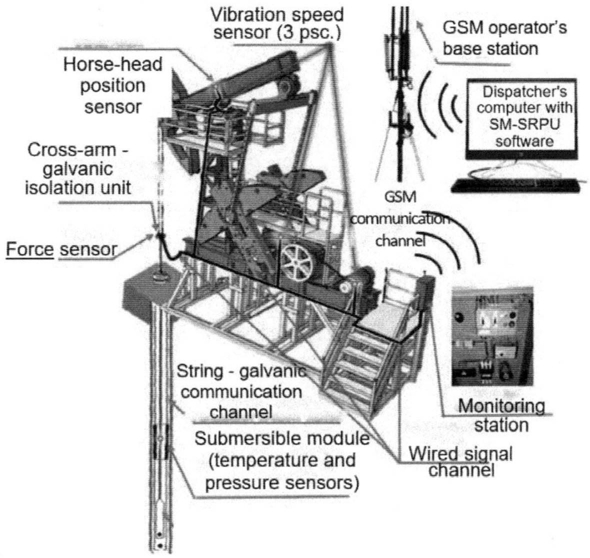

Figure 1. SRPU monitoring system with galvanic communication channel and submersible module.

Given that the distance between the plunger and the cylinder is small and the contact area is large, the electrical resistance that occurs in the pump can be taken as zero. For a well with a pump running depth of 1,500 m, specific resistance ρ = 0.14 Ohms * mm2/m, and rod and tubing cross-sectional areas equal to 380 mm2 and 243 mm2, respectively, the calculated circuit electrical resistance was 1.41 Ohms.

The electric resistance of an open circuit was calculated on the assumption that the tubing is filled with homogeneous medium, and the rod string in the tubing is positioned concentrically. For calculations, the formula for finding the resistance of the cylindrical grounding electrode was used. According to [3], the cylindrical grounding resistance is calculated as:

$$R_3 = 0{,}367 \cdot \frac{\rho}{L} \lg(\frac{L}{d_m})$$

where ρ is the medium resistivity;
L - length of the cylindrical grounding (rods),
d_m - diameter of the cylindrical grounding (rods).
In experimental tests conducted on a well with medium resistivity of 106 Ohms·m, a pump running depth of 1,770 m, and a rod string diameter of 22 mm, the calculated resistance of the medium between the rods and the tubing was 200 Ohms. The general layout of the submersible module is shown in Figure 2.

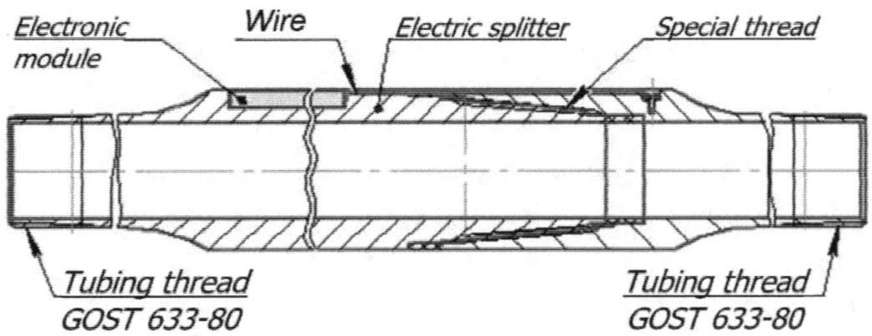

Figure 2. Submersible module

The submersible module (SM) consists of an electric splitter with dielectric loading, an electronic module with batteries, sensors for measuring the downhole fluid parameters, an electronic board with units processing and transmitting data to the ground control station via the GC. The electronic module performs parameter measurement sessions, records the results of measurements in the internal memory and transmits them to the earth's surface at a set frequency via the GC.

To enable the data transmission from the pump to the well mouth, the submersible module is equipped with an electric splitter for the upper and lower parts of the strings, and an electrical key transmitting a signal via the GC.

The electric splitter of the submersible module contains:

- housing elements - a nipple and a coupling made in the form of adapter subs with connecting threads that fit the tubing;
- dielectric insulation layer between housing elements;
- electronic module, which measures parameters and transmits the data via a galvanic communication channel.

The ground-based data receiver comprises a power source with one terminal connected to the ground part of the rod string and the other terminal - to the tubing, in which a metal rod string is installed concentrically with dielectric scratchalisers.

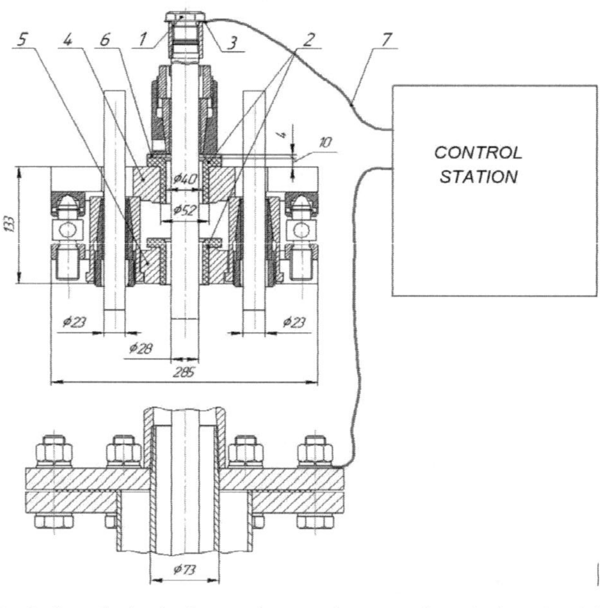

Figure 3. Galvanic isolation unit: 1- plug; 2 - insulation bushing; 3 - slip ring; 4 - upper cross-arm; 5 - lower cross-arm; 6 - washer; 7 - connecting wire

To ensure the electrical isolation of the polished rod and the rod string from the rest of the ground equipment of the sucker-rod pumping unit, a galvanic isolation unit with insulation bushings has been developed, its design shown in Figure 4. Bushings 2 made of insulating material are installed in the central holes of the upper and lower cross-arms 4 and 5. The threaded end of the polished rod coupling contains a plug 1 with a slip ring 3 in order to establish an electrical contact of the wire connection line 7 with the polished rod (rod string). The second wire connection line, which provides contact with the tubing string by means of the slip ring 3, is fixed to the well mouth.

Equipped with an electric splitter, which contains a dielectric loading, the tubing string is an electrically open conductor. The submersible module, together with the tubing string, the rod column and the galvanic isolation unit, transmits borehole data via the galvanic communication channel with the modulation of the useful signal by switching the electric splitter according to a specified algorithm.

4. Results and Discussion

The submersible module and other elements of the monitoring system were successfully tested in the laboratory and on the test bench.

Pilot tests of the monitoring station with a galvanic communication channel were performed in accordance with the approved SM-SRPU acceptance test program and procedures that involved the system installation activities and the start of software, including SM-SRPU Monitoring.

During the tests, data was transmitted and registered in the client's software with the indication of the e-mail address, the parameter and the type of mailing. The SRPU technical condition data viewing by date and time was tested along with saving incoming reports and viewing the data received.

When testing the submersible module parameters, large interference was detected in the control station's static power supply unit, obscuring the signal from the submersible module in the course of a communication session when measurements data was transmitted via the galvanic communication channel to the monitoring and control unit.

The acceptance tests of the monitoring station with a galvanic communication channel were interrupted and the submersible module was lifted. When analysing the reasons for the lack of a stable signal and the connection between the submersible module and the monitoring and control unit, it was found that the production string contacted the rod and tubing strings.

To prevent contact between the rods and the production string, it was decided to equip the rod strings with centralizers, and the tubing with rubber insulation elements (Figure 4.)

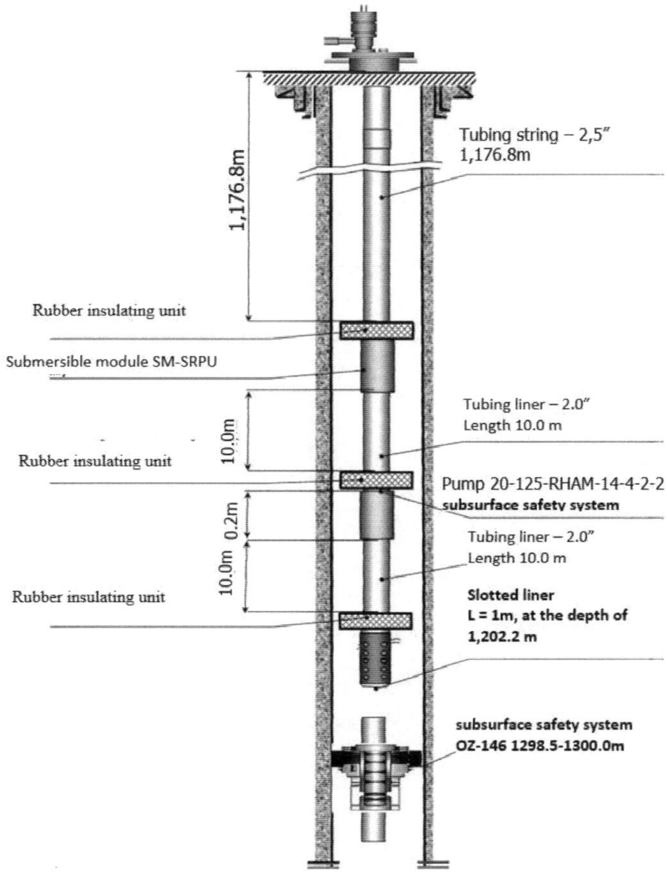

Figure 4. Flow diagram of the downhole pumping equipment configuration

5. Summary and conclusions

The submersible module and other elements of the monitoring system with a galvanic communication channel were successfully tested in the laboratory and on the test bench. Some changes were made based on the first results of pilot tests. So far, tests have continued on PJSC TATNEFT oil field with high-viscosity oil.

The research was conducted within the framework of the Federal Special-Purpose Program for Research and Development in Priority Areas of Development of the Russian Scientific and Technological Complex for 2014-2020 under the Grant Agreement No. 14.610.21.0019 of 23.10.17 on the topic of Creating a Set of Process Solutions for Enhancing Oil Recovery on Reservoirs Containing High-Viscosity Oil, the unique research identifier RFMEFI61017X0019.

References

[1] Polygin V V and Lekomtsev A V 2011 The dynamics of wear of sucker rod pumps during well operation *The journal "Oil industry"* **9** 27-312
[2] Urazakov K R, Bogomolny E I, Seytpagambetov Zh S and Gazarov A G 2003 *Pump production of highly viscous oil from deviated and flooded wells* (M.: Nedra-Business Center LLC)
[3] Galeev A S, Arslanov R I, Ermilov P P and Kuzmin I A 2012 Monitoring the technical condition of the SHSNU during periodic operation *Electronic scientific journal oil and gas business* **1** 24-29
[4] Stepanov V S 2010 A non-generator method for transmitting downhole telemetry data *TUSUR Reports* **2** 22

IOP Conf. Series: Materials Science and Engineering **860** (2020) 012028 doi:10.1088/1757-899X/860/1/012028

Monitoring of technological processes and assessment of technical condition of pumping units

A S Galeev, G I Bikbulatova, Y A Boltneva and S L Sabanov

Almetyevsk State Oil Institute, 2, Lenina Str., Almetyevsk, Republic of Tatarstan, 423450, Russia

E-mail: ssgaleev@mail.ru

Abstract. Remote monitoring of pumping equipment operation in real mode is being widely introduced at oil and gas industry enterprises. The controllers applied in this process enable preventing progressive destruction of equipment and emergency failures and implementing automatic control of machines' operation. The actual data collected by these controllers is transferred to the company's servers and makes it possible to effectively set and address various tasks on equipment maintenance. However, there is still insufficient number of applications to address the problems of pumping equipment operation. The presented paper is aimed at resource efficient operation of equipment by obtaining the information on both actual and predicted technical state of the equipment in real-time mode. The main performance characteristics of a pump deteriorate in the process of its operation, followed by an increase in specific power consumption for injection (with the reservoir properties unchanged). Economic losses due to pump's wear and tear are determined by the price of electricity and the cost of overhauling. Implementation of the designed monitoring system will make it possible to automatically decide whether further operation of the pump is unprofitable and if it needs to be overhauled. Based on the analysis of the trend of monthly unit costs, the pumping unit will be timely removed for major overhaul when the trend reaches its minimum, which shall provide an average annual economic benefit up to 4,000,000 rubles per pumping unit.

1. Introduction

In recent years, the problem of the efficient operation of equipment has been particularly challenging for the oil industry.

The technologies being introduced to optimise production processes can not be used without control and prediction systems requiring modern criteria for evaluation of the technical condition that can be easily adapted to automated monitoring systems [1,2].

2. Problem statement

It is common knowledge that the operating costs borne by the upstream companies to maintain their pump stock are quite high. The existing preventive maintenance (PM) system does not guarantee efficient operation of pumping units (PU), since its scope includes time- and volume-based maintenance and repair works, without consideration for the actual condition of equipment, real operating loads and specific process characteristics that affect the performance of pumps. The PM system has a preventive function and is aimed at preventing the failures. At the same time, the PM design standards do not take into account the wear rates of the equipment's operating components that depend on a variety of interdependent factors.

Pumping equipment monitoring is growing more relevant as the energy prices increase, so enterprises in the region turn towards servicing equipment based on its technical condition along with the application of predictive diagnostics. Objectives in this area become a priority for professionals seeking to reduce energy consumption and prevent downtime of the company's process-related assets resulting from equipment breakdowns. In this regard, systems designed for technical diagnostics of pumping equipment in real time, based on controlled parameters, are widely used in all areas of activity that involve the use of pumps. The main controlled parameters of pumping units are the following:

1. Pressure at the pump suction and discharge lines;
2. Flow rate of the pumped fluid;
3. Pump unit efficiency;
4. Per-unit energy consumption for pumping;
5. Vibration;
6. Temperature of bearings and electric motor winding;
7. Electric motor currents and voltages;
8. Pump and motor operating time.

Controllers installed on the units are designed to prevent "avalanche" (progressive) destruction of equipment and emergency failures, as well as perform automatic control of the units. The continuous data collected by the controllers is sent to the company's servers and may be used to set and address various tasks associated with equipment maintenance. However, the services addressing the problems of pumping equipment operations are quite limited at the moment; therefore, the development of services is of practical importance.

The most important task for mechanical engineering teams is the timely pump shutdown for overhaul.

3. Justification of the time for pump shutdown for overhaul

In the course of pump operation, the pump mechanical performance deteriorates, which is accompanied by an increase in the per-unit electricity consumption during injection (with the reservoir characteristics unchanged). Pumps are usually rejected based on such technical parameters as reduced flow, pressure, or efficiency, while the economic indicators are ignored. On the other hand, it is clear that given expensive overhaul and cheap electricity, it makes sense to operate pumps even with very significant wear, i.e. at low efficiency. In a reverse situation, when the electricity cost is high and the overhaul is cheap, a better option would be to conduct overhaul more often, avoiding significant wear of the pumps. Thus, the question of whether it is time to send a pump for overhaul should be addressed upon consideration of both technical and economic factors.

The mechanical characteristics of sectional centrifugal and screw pumps can only be restored during overhaul. The pump maintenance cycle (time between overhauls) may include many types of maintenance, but none of them affect the mechanical characteristics of the sectional pump, except for overhaul. During pump operation, the pump mechanical performance deteriorates and, consequently, the per-unit electricity consumption for fluid pumping increases. At a certain point, the steady increase in power costs lead to the situation when further operation of the pump unit is inefficient without restoring the operating parameters.

The works [3-5] discuss the efficient operation of high-pressure pumping units, and in particular, propose and justify the choice of the optimal time between overhauls. The overhaul duration should be planned the way which would minimise the average total cost of excessive electric power consumption caused by the wear of the pump's fluid end (diffusers, impellers, seals), as well as the cost of the works related to overhaul (dismantling, installation, transportation, overhaul) for the planned period. Wear costs (costs associated with the wear of the fluid end of the pump ( ) are determined by the price of electricity , pump delta pressure , the efficiency , the volume of fluid pumped and, if calculated per one hour of operating time, are calculated as follows:

$$W_{wear}^{cvs}(\Pi) = \frac{CT_{overhaul.} + \sum_{t=1}^{\Pi} Ц_{ep} \cdot \left(\frac{\Delta P^{t}}{\eta_{agr}^{t}} - \frac{\Delta P^{1}}{\eta_{agr}^{1}} \right) \cdot \Delta V_t}{P}$$

Assuming that the pump is a perfect match for the reservoir and the reservoir head remains unchanged in the course of operation. The per-hour costs caused by pump wear depend on the duration of the time between overhauls (TBO) and have a minimum value. By building this dependence, we can obtain both the optimal pump overhaul period and the minimum possible costs associated with pump wear.

The proposed method can be still applied; however, since each pump has its own wear rate, it gives only an average TBO. In addition, at the time of planning, there is no data on the overhaul and electric power prices at the time of pump shutdown for overhaul. In the current situation, on the back of the introduction of pumping equipment monitoring systems, it becomes possible to automatically track the operation of each pump unit and automatically detect the inefficiency of further pump operation, i.e. the need for overhaul.

To make the ultimate decision, the controlled parameters must include the following:

1. pump operating time;
2. active power consumption;
3. fluid flow and head.

Considering that the unit efficiency is determined by the pump hydraulic horsepower and the electric power consumption:

$$\eta_t = \frac{\left(P_t^{flow} - P_t^{\text{intake}} \right) \cdot Q_t}{W_t}$$

where P_t^{flow} is the pressure in the discharge line;

P_t^{intake} - pressure in the suction line;

Q_t - fluid flow rate;

W_t - active power consumption.

and per-hour consumption ,
where is 1 hour,
we obtain:

$$W_{wear}^{hour}(\Pi) = \frac{CT_{overhaul} + \sum_{t=1}^{\Pi} Ц_{ep} \cdot \left(\frac{W_t}{Q_t} - \frac{W_1}{Q_1} \right) \cdot Q_t \cdot \Delta t}{\Pi}$$

Setting a certain time step T (day, week, decade, month), at the end of each step, we calculate the wear costs, taking the current pump operating time as TBO. Then we track the parameter trend, and as it starts showing a steady growth (several steps in a row), we issue a recommendation to shut down the pump for overhaul. Prices for overhaul and electricity are taken as current prices at the time of calculation.

Thus, regular automated monitoring of the pump unit operating parameters enables us to automatically detect the inefficiency of further pump operation and the need for overhaul.

The developed algorithm for the pumping equipment shutdown for overhaul assumes the following possible optimum criteria:

1. Minimum costs associated with the wear of the pump's fluid end per 1 hour of operation:
2. Minimum costs associated with the wear of the pump's fluid end per 1 cubic meter of pumped fluid.

The time of the unit shutdown for overhaul is identified as follows:

1. The costs of excessive power consumption associated with pump wear are calculated for each month. To do this, the difference between the electricity actually consumed and the electricity that a new pump would have consumed for the same work is calculated based on the monthly summary of electricity costs, flow rate, and pump head data. These overruns are added up from the start of the pump operation.

1. For each month, the financial losses associated with pump wear and tear (excessive electricity consumption and the costs of overhaul) are calculated as the total of the pump overhaul cost and the cost of excessive electricity consumption. The prices of overhaul and electricity valid at the time of decision-making are taken for calculation.

Depending on the selected criterion, the resulting losses are divided by the total pumping volume or

$$C^{wear}_{volume} = \frac{C_{overhaul} + Ц_{el} \cdot \sum_{n=1}^{N} \left(A_П - \frac{(P_n^{flow} - P_n^{\text{int}\,ake}) \cdot \Delta V_n}{\eta} \right)}{\sum_{n=1}^{N} \cdot \Delta V_n}$$

operating time

$$C^{wear}_{hour} = \frac{C_{overhaul} + Ц_{el} \cdot \sum_{n=1}^{N} \left(A_П - \frac{(P_n^{flow} - P_n^{\text{int}\,ake}) \cdot \Delta V_n}{\eta} \right)}{\sum_{n=1}^{N} t_n}$$

where C^{wear}_{hour} is the cost per unit of time determined by the wear of the pump's working parts;

$C_{overhaul}$ – the cost of overhaul (including the costs of pump transportation, installation and dismantling);

An - electricity consumption within the step;

Vn - volume of fluid pumped within the step;

Цэл - price of electricity;

tn - pump operating time within the step;

and per-unit operating costs associated with the pump wear are obtained.

The monthly trend of per-unit costs is built, and once the trend attains its minimum, a decision is made to shut down the unit for overhaul.

4. Algorithm implementation

The developed algorithm has been implemented in the pump unit automated monitoring system. Its main features:

1. Visualisation of the current state of pumping units
2. Monitoring of the pumping unit technical condition
3. Evaluation of compliance of the pumping equipment installed with operating conditions at the sites;
4. Selection of pumping equipment matching the site parameters;
5. Predicting the needs for maintenance and repair;
6. Predicting the need for diagnostics;
7. Justification of the need to shut down pumping equipment for overhaul;
8. Analysis of energy consumption by pumping units.

Figure 1 shows a photo of the automated monitoring system implementation.

Figure 1. Flow diagram of the downhole pumping equipment configuration. 1- SCP unit 180-1422, 2 -vibration sensor, 3 - monitoring station, 4 - display of the main unit operation parameters in the pump box, 5 - computer in the control room (corporate network).

5. Discussion

The implementation of system will offer a new way to develop a preventive pump unit maintenance schedule using the method of determining the optimal pump unit overhaul period [6,7], assuming that the pump efficiency will be restored after attaining a certain critical value. Table I provides the calculation of the yearly cost advantage due to the implementation of this method on a single SCP 180x1422 pump unit for different efficiency loss intervals.

Table 1. Calculation of the yearly cost advantage due to the implementation of the method per pump unit

No.	Initial efficiency, %	Final efficiency, %		Time between overhauls, days		Yearly cost advantage, thousand rubles
		without implementation	after implementation	without implementation	after implementation	
1	69	60	65.4	365	167	1,680
2	60	50	56.1	365	143	2,900
3	50	40	46.7	365	119	4,910
4	44	35	42.1	365	107	6,930

6. Summary and conclusions

Modern level of diagnostic tools development allows continuous monitoring of pump unit parameters in order to control their technical condition. At the same time, regulation is required in respect of the monitoring and analysis of the vibration level, temperature, leaks, pressure parameters, efficiency, and power consumption. According to the pre-calculated feasibility study (Table I), the implementation of monitoring system based on the method for determining the optimal pump unit overhaul period will produce an average yearly cost advantage to the amount of 4,000,000 rubles per pump unit.

The pump unit automated monitoring system is designed to ensure continuous monitoring, collection, storage and visualisation of process and technical data of operating units with the possibility to display current operational parameters and prevent deviations from the parameters set.

References

[1] Galeev A S, Suleymanov R N and Bikbulatova G I 2005 To the problem of increasing the efficiency of pumping units *Technologies of the Fuel and Energy Complex* **2** 92-97

[2] Galeev A S, Bikbulatova G I, Boltneva Yu A, Kuzmin I A, Ermilov P P and Ryazantsev A O 2016 The experience of using the autonomous system of control and registration of technological parameters "ASKiR" *Oilfield business* **9** 43-47

[3] Suleymanov R N, Galeev A S and Bikbulatova G I 2004 *The efficiency of pumping units* (Ufa: UNTU).

[4] Suleymanov R N, Filimonov O V, Galeeva F A and Ryazantsev A O 2002 Vibroacoustic diagnostics of pumping units (Ufa: publishing house of UGNTU)

[5] Suleymanov R N 2004 A technique for interpreting measurements of the hydraulic efficiency of pumps based on a comparison procedure with passport data *TEK Technologies"* **1** 61-66

[6] Sultanov B Z and Suleymanov R N 2004 Interpretation of hydraulic efficiency measurements pumping units *Oil Management* **4** 95-98

[7] Galeev A S, Sultanov B Z, Suleymanov R N and Kaminsky S K 2003 On the issue of choosing the optimal time for preventive overhaul of pumps *Fuel and Energy Technologies* **5** 14-17

IOP Conf. Series: Materials Science and Engineering **860** (2020) 012029 doi:10.1088/1757-899X/860/1/012029

Determination of lower sucker-rod breakage in wells equipped with sucker-rod pumps

A S Galeev, G I Bikbulatova, Y A Boltneva and S L Sabanov

Almetyevsk State Oil Institute, 2, Lenina Street, Almetyevsk, Republic of Tatarstan, 423450, Russia

E-mail: mailto:ssgaleev@mail.ru

Abstract. Development of mature oilfields of the Ural-Volga region is performed primarily by means of downhole sucker-rod pumping units (UShSN). The most characteristic failures of these installations are breakdowns and twist-offs and pump valve failures [1-5]. The ways of the elimination of these accidents vary significantly: rod breakage or twist-off implies pulling out of downhole equipment, and elimination of "sticking" of the pump's shut-off valve assembly requires rehabilitation of downhole equipment through flushing with water, hot petroleum or a solvent. Failure to identify a malfunction leads to incorrect planning of wellbore equipment rehabilitation activities and, eventually, to economic losses.The cause of downhole equipment malfunctions is usually determined through analysis of dynamometer cards (DC). However, in the majority of cases, the dynamometer cards do not allow to distinguish the rods' tail end twist-off from the rod pump valve failure. In the paper presented the method is considered for rapid detection of rod breakage and twist-offs (RBT) in the wellbore, which involves creation of an electrical circuit "rod string - tubing string" and monitoring of its integrity. To determine the failure type, the synchronization module measures the system's resistance on the dielectric coupling via an electromagnetic conduit. In case of RBT, the electrical resistance of "tubing - pump - rod string" circuit will be much higher (over 2 Ohm) than in case of absence of this failure (0...2 Ohm).

1. Introduction

Nowadays, sucker-rod pumps operate a significant part of oil-producing wells. These pumps can be divided into two parts: surface motor and downhole assembly.

Control on the technical condition of the surface part does not cause objective difficulties, at the same time, accuracy control of downhole equipment is usually realized indirectly. Such a situation occurs because of the impossibility of visual control, the inaccessibility of downhole equipment and difficult working conditions, for example, the cyclic load of the rod string. In addition, a significant proportion of emergency failures occur because of the rod breakage or back-off (RBB) [1-5].

2. Research objective

Monitoring the technical condition of the downhole equipment is usually carried out by analyzing dynagraphs (DG).

Dynamometry allows determining the majority types of emergency failures, comparing the received dynagraph with the reference one, and determining deviations in the SPR operation.

However, despite the known advantages of this method, there are also certain disadvantages, since the technical condition is evaluated using indirect parameters. One of the significant drawbacks is the practical difficulty of determining the lower RBB, which graphically coincides with the disrepair of the SPR valves.

The error in fault recognition results in incorrect planning of work to repair the downhole equipment. Thereby, there are extra costs associated with downhole equipment "recovery" (washing with water, hot oil, solvent).

3. Proposed solution

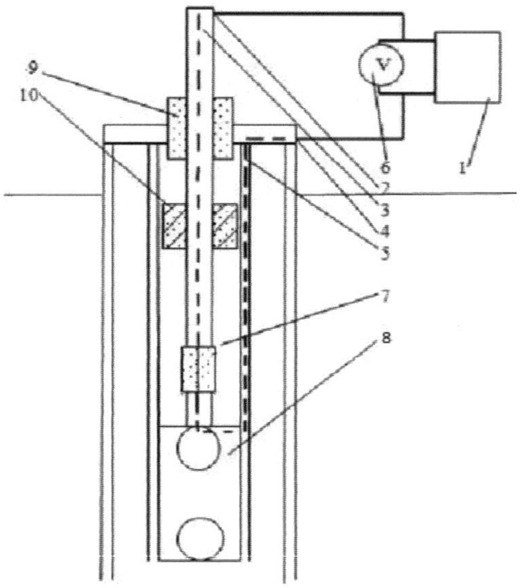

Figure 1. The lower RBB detection device. 1 - the source of the stabilized current, 2 - contact to surface part of string, 3 – metallic rod string, 4 - receiving electrode, 5 – tubing string, 6 - voltage change registration unit, 7 - separator of known electric resistance, 8 - well pump, 9 - dielectric material sealing, 10 – centralizing scraper from dielectric material; dash line shows the created electrical circuit.

For detecting the lower RBB, we suggest the following [6, 7] (Figure 1): at the wellhead, by using a stabilized current source the constant stabilized current I_{stab} is delivered to the wellbore. The stabilized current source connected by one contact (with clasp) to the surface part of the rod string, and with other (with clasp) to the receiving electrode at the wellhead, for which a tubing string is used, that creates a potential difference U_{divide} at the ends of the ohmic resistance, which separates the metallic electrical circuit and has a finite known resistance R_{divide}.:

$$U_{divide.} = R_{divide} \bullet I_{divide} \tag{1}$$

Herewith, the voltage at the wellhead $U_{downhole}$ (at the source of stabilized current) is equal to:

$$U_{downhole} = \left(\left(\frac{R_{divide.} \cdot R_{fluid}}{R_{divide.} \cdot R_{fluid}} \right) + Rkko. \right) \cdot I_{stab.} \tag{2}$$

where R_{kko} is the resistance of the rods string, pump, and tubing,

R_{fluid} is the resistance of the reservoir fluid in annular space between the tubing string and the rod string:

$$Rfluid = \frac{\rho_{fluid} \cdot \ln(Din / d_{out})}{2 \cdot \pi \cdot H} \qquad (3)$$

where ρ is the specific resistance of the formation fluid (oil+water), Din. and dout. are the diameters of the inner tubing string and the outer rod string, respectively, H - the length of the rods string from wellhead to the pump.

Moreover, if there is no lower RBB, the resistance of the entire circuit is equal to:

$$U_{downhole.} / I_{stab.} \approx R_{divide} \qquad (4)$$

Since the R_{amount}. is small: $R_{amount} << R_{divide}$., and as the tubing and rod strings are metallic and have a larger cross-section (more than 150 mm sq.), also as the R_{fluid}. is large: P_{fluid}. >> R_{divide}., since the resistivity of water-oil mixtures is close to oil conductivity and greater than 10 -10 (Ohm.m).

4. Experimental result

The purpose of current flow simulation through the "tubing-rod" channel was to justify the possibility of its use in wells equipped with SPR.

The unit consists of 4 coaxial conductor models, an analog of the "tubing - rod" channel which are metal tubes of a certain diameter and length.

Inside each tube (model), there is a coaxially installed metal (copper) wire of a certain diameter, centered with the use of dielectric centralizing plugs (Figure 2). Each tube has a rubber plug with a hole through which a wire passes and connects to give a signal.

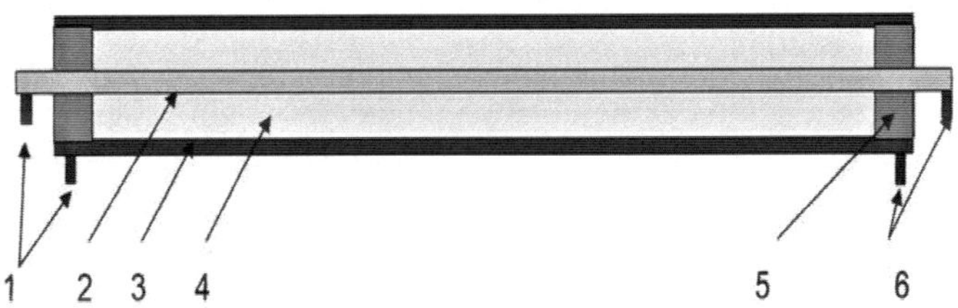

Figure 2. Test unit design. 1-contacts for connecting the signal generator, 2-wire; 3-pipe, 4- pipe filling environment, 5-dielectric plug, 6-power supply connection contacts.

Impulse registration was performed using a two-channel oscilloscope GW Instek gds-71042:

- channel A - signal control at the entrance to the tube,
- channel B – signal control at the exit from the tube.

To evaluate the wave resistance of a coaxial "cable" filled with air, the first series of tests were performed without filling the annular space. For convenience, the impulse frequencies were combined in the 1 . . . 5 GHz and 6 . . . 10 kHz lines.

It is found that at these frequencies the wave resistance does not affect the signal passing and the air is an ideal dielectric.

In the second part of the experiment, the pipes were filled with distilled water.

It is found that the conductivity of the distillate is significantly higher; accordingly, the output signal is ~20% less, regardless of the model number. Also, different frequency signals endure the same blanking (within the measurement error, 5%).

In the third part of the experiment, the pipes were filled with fresh water.

For freshwater, the complex resistance (ohmic + capacitive) is significant and there is a sizable signal blanking (from 3.5 times for model 1 to 11 times for model 4), regardless of the signal frequency. At the same time, we can see that the 6...10 kHz line is blanking more.

The experiments have shown the possibility of using a current flow through the "tubing-rod" channel to control the integrity of the electrical circuit and the rod string, respectively.

Table 1. Experimental results

	The signal amplitude at the exit, B			
Trial model No.	1	2	3	4
Pipe diameter, mm	6.2	10.3	14.2	16.6
Filled with air				
Line 1…5 GHz	8.8	8.8	8.8	8.8
Line 6…10 GHz	8.8	8.8	8.8	8.8
Filled with distilled water				
Line 1…5 GHz	6.8	6.2	6.4	6.4
Line 6…10 GHz	6.4	6.2	6.4	6.8
Filled with fresh water				
Line 1…5 GHz	2.4	1.2	1.0	0.8
Line 6…10 GHz	2.2	1.0	0.85	0.7

5. Results discussion

When the lower RBB occurs, the electrical resistance of the "tubing-pump-separator- rod string" circuit will increase sharply, because at that moment:

$$R_{amount.} \rightarrow \infty \tag{5}$$

which will cause a synchronous increase at the $U_{borehole}$ to maintain the $I_{stab.}$, that means:

$$U_{borehole.}/I_{stab.} \gg R_{divide.} \tag{6}$$

This will serve as a criterion for detecting the lower RBB.

Accidental closures of the rods string during bends (for example when going down), and consequently, the resistance drop to almost zero,

$$R_{divide.} \rightarrow 0 \tag{7}$$

do not affect the reliability of lower RBB determining by this method, since the synchronization block which is included in the wellhead block is configured to register an increase in resistance of the circuit quite more than R_{divide}.

6. Approbation

Practical tests were performed in 3 wells: 68A; 4051; 2823 of the Yelnikovsky oil field with a preliminary cause of failure -breakage of the rod.

Well 68A of the Yelnikovsky field. The operator found that there was no flow in the well. After the flow test (according to the dynagraph the loads were about 3 tons), it was decided to perform an unscheduled washing with hot oil. After flushing the flow at wellhead did not appear, the dynagraph also was showing non-working valves (loads after washing with hot oil were about 2.5 tons). It was quickly decided to invite a workover crew.

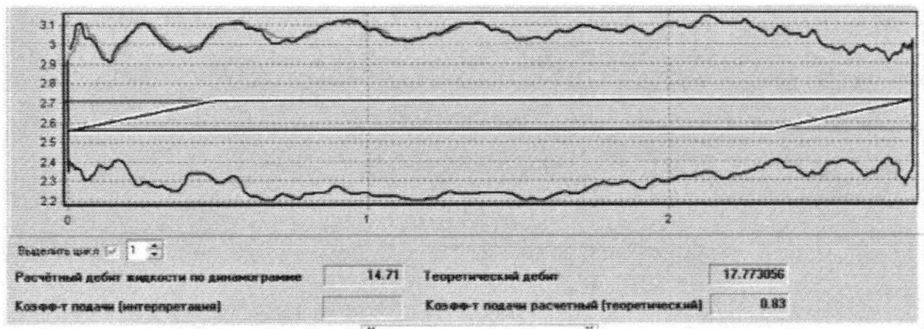

Figure 3. Well 68A dynagraph.

Before lifting the pump rods, the resistance of the rod string-pump-tubing string circuit was measured. For this, one of the contacts was fixed on the rod string, the other contact was attached to the BOP. The insulation resistance was 3.79 m*Ohm which means a possible rod breakage.

During the lifting, a break on the joint between the 95 and 96 pump rods was found. The reason for the break was the high dogleg severity in this section. After lifting of downhole pump equipment, a pump of the previous standard size (H-57) was run in hole, and in the high dogleg severity interval was equipped pumping rods with centralizers. The economic costs associated with hot oil washing, as well as excessive downtime of the well could be avoided if the lower breakage of the rod was detected timely, which is impossible to determine from the dynagraph.

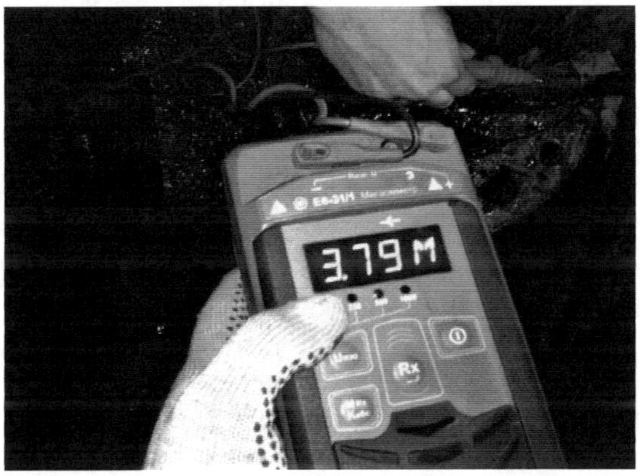

Figure 4. Measuring the isolation of tubing -pump- rod string circuit on the well 68A.

Well 4051 was stopped in February 2019 with no flow. According to the dynagraph (Figure 4) we can see that both valves are not working. After fracking in May 2018, well's operating time was 212 days. After washing the well, the pump did not work, and it was decided to set-up a workover crew.

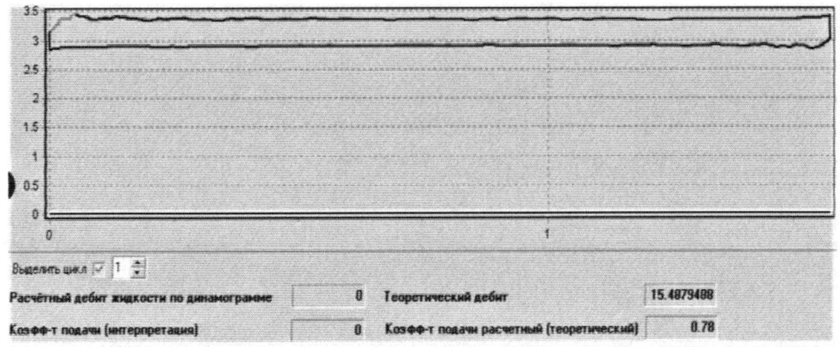

Figure 5. Well 4051 dynagraph

During measuring the isolation of the «tubing-pump-rod string» system, the isolation resistance was 0 Ohm which shows that there is no rod breakage. The lift of downhole equipment proved the absence of break.

Well 2823 was stopped with no flow in September 2018. According to the dynagraph, the pump valves were not working. The well was flushed with 30 m3 hot water. After flushing, the flow at the wellhead did not appear and it was decided to put the workover crew on this well.

During the measuring of the isolation resistance, it was close to 0 Ohm, which shows that there is no rod breakage. After lifting the downhole equipment, a break of the plunger was found at a distance of 50 cm from the discharge valve.

7. Conclusion

Today rod breakage and back-off (RBB) are the main problems of rod pumps. The reasons for this equipment failure may be different: abrasion of the rod string on the tubing string due to wellbore deviation, too large dynamic loads on the rod string for various reasons from deposits of asphalt, resins, and paraffin deposits before sticking of the plunger in the pump cylinder. All breakage failures of the rods are associated with an increase in the specific load per unit cross-section area of the pump rod.

All the "problems" of SPR downhole equipment can be easily recognized through dynamometry. However, it is difficult to determine the lower RBB from the dynagraph which almost does not differ from the non-working valve of the SRP. The error in fault recognition results in incorrect planning of work to repair the downhole equipment. Thereby, there are extra costs associated with downhole equipment "recovery" (washing with water, hot oil, solvent).

To determine the RBB via the electromagnetic channel, the synchronization unit measures the system resistance on the dielectric insert. In other words, if RBB is detected, the electrical resistance of the "tubing-pump-rod string" circuit will be much higher than in the absence of this failure.

The disadvantage of this method is that pumping rod string can contact the tubing string in the intervals high deviation which leads to the closure of the electric circuit "tubing-pump-rod string" before the dielectric separator. This significantly reduces the reliability of the channel, the solution to this problem can be the use of pumping rods with plastic scrapers.

This performing of RBB diagnosis will eliminate costs for "unnecessary" interventions of "well recovery" and lower downtime of well equipment due to the timely set-up of the workover crew if this problem occurs.

8. Acknowledgement

The work was carried out within the Federal Target Program "Research and development in priority areas of the scientific and technological complex of Russia for 2014-2020" under the Grant Agreement No. 14.610.21.0019 from 23.10.17 on the topic "Creating a set of technological solutions for increasing oil recovery from high-viscosity oil reservoirs", the unique identifier of the work is RFMEFI61017X0019.

References

[1] Sultanov B Z and Shagaleev R K 2009 The breakage of the rods of plunger borehole pumping units at OJSC TATNEFT *Oil and gas case* **7** 68-70

[2] Kochekov M A, Sitdikov M and Ishemguzhin I E 2010 Analysis of cliff breaks in NGDU "Chekmagushneft" *Innovative oil and gas equipment: problems and solutions: Sat. tr / All-Russian scientific and technical. conference* (Ufa: UGNTU) 151-154

[3] Halimov F G, Zubairov S G, Zagorsky V K, Salikhov I A and Kuramshin A A 2011 Analysis of the causes and prevention of tops of sucker rods and pipes *Electronic scientific journal Oil and Gas Business* **6** 457-464

[4] Polygin V V and Lekomtsev A V 2011 The dynamics of wear of sucker rod pumps during well operation *The journal "Oil industry"* **9** 22-28

[5] Ibragimov R R 2019 Prevention of turn-ups of a rod string by applying an anti-flap flexible coupling and applying an anaerobic sealant to "bashneft-production llc" *Oil and gas case* **17** **(1)** 50-55.
[6] Galeev A S, Grigoriev V M, Arslanov R I and Ermilov P P 2012 On the possibility of using a wireless galvanic communication channel in wells equipped with sucker rod pumps *International scientific publication. Modern fundamental and applied research* **4** 44-48
[7] Galeev A S, Suleymanov R N, Arslanov R I, Bikbulatova G I and Boltneva Yu A RU, Patent No. 2696954 *A device for transmitting information through an electromagnetic communication channel during well operation*

IOP Conf. Series: Materials Science and Engineering **860** (2020) 012030 doi:10.1088/1757-899X/860/1/012030

Methods for control of crude oil refinery unit

L G Tugashova

Almetyevsk State Oil Institute, 2, Lenina Str., Almetyevsk, Republic of Tatarstan, 423450, Russia

E-mail: Tugashova@yandex.ru

Abstract. We have reviewed the features of the oil rectification process in the low-power crude oil refinery unit. The features of heat removal by intermediate pumparound in comparison with a large-capacity unit have been distinguished. The input and output parameters of the object have been selected. The possibility of representing a multi-loop controlled object as single-channel objects using the EOTF method of effective transfer functions has been reviewed. To improve the quality of unit control, it has been proposed to use the model in practice. For this purpose, a dynamic Hammerstein model for the low-power crude oil refinary unit has been developed to identify the nonlinear part. The proposed model allows studying the influence of various effects (live reflux consumption, diesel oil consumption and superheated steam consumption) on the controlled parameters (points of the temperature profile of the atmospheric tower). The Matlab/Simulink software has been chosen as the tool for implementing the model.

1. Introduction

Control of oil refining units as nonlinear multidimensional objects should be carried out when the requirements for the quality of the produced petroleum products are met and when the technological parameters are maintained at the specified values under the influence of disturbances.

Research by Russian and foreign scholars A. P. Verevkin [1], S. Skogestad, M. Morari [2], A. A. Stopakewich, I. B. Furtat etc., has been devoted to the synthesis of multi-loop systems for controlling the rectification process:

Standard controllers are used in domestic oil rectification unit control systems, and improvements are achieved by using multi-loop automatic control systems. Advanced control systems such as Profit Controller (Honeywell), exaSMOC (Yokogawa), and etc. are used to control large oil refineries based on predictive models. The process control objective while meeting the abovementioned requirements is also relevant for low-capacity oil refining units.

The problem of building robust control systems for the rectification process of a binary mixture in the industrial-type column is solved in the work [2], in order to increase the efficiency of the separation process under uncertainty caused by changes in the composition of the supply stream, the temperature of the streams, etc.

The article [3] proposes a system for controlling quality indicators of primary crude oil processing products with fuzzy process controllers. Models of virtual analyzers have been obtained in a number of works to determine the quality indicators of distillation column products.

To stabilize the product quality of primary oil refining units, two-loop circuits of rectification process regulation are used.

IOP Conf. Series: Materials Science and Engineering **860** (2020) 012030 doi:10.1088/1757-899X/860/1/012030

When controlling refinery facilities (distillation columns, heat exchangers, furnaces, etc.), along with a typical solution that involves one-loop and cascade automatic control systems (ACS), control using mathematical models is used.

A review of foreign articles has shown that dynamic control object models are often represented in the State Space [4], in the form of a system of ordinary differential equations [5], by an artificial neural network.

When controlling the rectification process, an Internal model controllers (Internal model control - IMC) are used. The controller is an inverted model of the control object, which is not always possible to get. The Skogestad IMC (SIMC) method [6] sets the desired type of transition characteristic of a closed-loop control system, described by an first order inertial element The SIMC-method setting is the expected time constant of this element .

The use of a reference model is proposed in a number of works on distillation column control. A method for controlling small refineries using the model in conditions of variable consumption and composition of raw materials is proposed in the article [7] by narrowing down the issue of controlling a petroleum product sampling to controlling the temperature conditions of the installation and stabilizing intermediate parameters.

The possibility of using disturbance compensators for the consumption and composition of raw materials and cross-link compensators for controlling the distillation column has been reviewed in the article [8]. When using compensators, practical implementation is a challenge, and there can be quite a lot of disturbances in the system.

The application of the EOTF (effective open-loop transfer function) method has been reviewed in the work [9] for an object containing two inputs and two outputs. This method is applicable for converting a multi-loop control object to a set of single-channel objects.

The object of the study is the atmospheric tower with auxiliary equipment that operates as part of a low-power oil refining unit (up to 500 thousand tons). The main factors that cause difficulties in controlling the object: multi-connectivity of parameters, functioning under the the influence of disturbances, the presence of nonlinear static characteristics.

To avoid multi-connectivity of parameters, we set the task to determine the main control and disturbance channels and try to apply the *EOTF* method and the disturbance suppression method.

In order to register the nonlinear characteristics of an object, first of all you need to identify their structure and parameters, and then to apply object controlling according to the model. We propose to use a nonlinear Hammerstein model to improve the quality of object control process.

A distinctive feature of a low-capacity oil refining unit is the arrangement of pumparounds and petroleum product sampling. Standard heat removal schemes with intermediate pumparound (as in large-capacity units) are not used, and we take into account this fact when selecting the parameters of the object of study.

2. Materials and methods

For an atmospheric tower with auxiliary equipment operating as part of a low-power oil refining unit (up to 500 thousand tons), the following input and output parameters are selected: top reflux rate *LT*, superheated steam flow rate *VB*, diesel fraction flow rate taken from the atmospheric tower *DT*, tower overhead temperature (naphtha) *T*1, diesel fraction temperature *T*2, column bottom temperature (fuel oil) *T*3, raw material consumption *F*.

Let us define the main control and disturbance channels. Testing dynamic characteristics on a production facility can be a quite difficult expirement. The dynamic model of the oil rectification process at a small oil refinery is presented in the work [10], built on the basis of a system of material balance and heat balance equations. Adequate accuracy of the built dynamic model is shown. Therefore, we consider this circumstance to be the basis for its application as a source for obtaining experimental data of the control object.

The parametric sensitivity of the object has been studied using a dynamic model of the rectification process [10]. By the relative sensitivity coefficient, the sensitivity of the temperature profile has been determined from *T*1, *T*2, *T*3 column height to *LT, VB, DT* controlling actions, and the dynamic control channels has been selected.

As a matter of convenience in the further implementation of the object mathematical model, the controlling actions are combined into the **u** vector, disturbances – into the **d** vector, the intermediate parameters into the **x** vector, and outputs into the **y** vector:

$$\{LT, DT, VB\} \in \mathbf{u}; \{f(LT), f(DT), f(VB)\} \in \mathbf{x}; \{F\} \in \mathbf{d}; \{T_1, T_2, T_3\} \in \mathbf{y}.$$

Thus, it has been determined that an atmospheric tower with auxiliary equipment operating as part of a low-power oil refining unit is a multi-dimensional *MIMO* (*Multiple Input Multiple Output*) object with three control inputs, three controlled outputs, and one disturbance.

The *EOTF* method is applied to a linear object model. The formula that determines the effective transfer function of an open loop is following [9]:

$$W(p)^{EOTF} = \frac{W_{ii}(p)}{DRGA_{ii}}, \tag{1}$$

where *DRGA* is a dynamic relative gain array; $W_{ii}(p)$ is the object's transfer functions for the control channel; *i* is the channel number.

The effective transfer functions of the $W(p)^{EOTF}$ object has been determined using the formula (1) for the case of 3 inputs – 3 outputs for the following control channels: "reflux rate – overhead temperature", "diesel fraction flow rate – diesel fraction temperature", "superheated steam flow rate – bottom temperature".

The controlled disturbance is the consumption of raw materials supplied to the atmospheric oil rectification unit. The "virtual experiment" was used to identify parameters of object models using disturbance channels: "raw material consumption – temperature of naphtha", "raw material consumption - temperature of diesel fuel", "raw material consumption – temperature of fuel-oil residue (bottom)". As a result of step disturbance the curves of the transition characteristics for the above-mentioned channels has been obtained.

Models of disturbance compensators are determined by formulas known from the classical theory of automatic control.

Dependencies for non-linear parts for each control channel has been derived in the form of static characteristics. The model presented as a system of material and heat balance [10] is difficult to apply in unit controlling. In order to implement the dynamic model in the controller and single out the non-linear part, the model of the rectification process [10] has been transformed to a suitable form for implementation. The Hammerstein model which containing nonlinear and linear parts is used for this purpose.

To determine the type of nonlinearity for each control channel, the following "experiment" has been performed on a dynamic model. To measure static characteristics, they raised the control signal, they waited for a stable temperature setting, and then the input step was increased sequentially with a constant increment. Therefore, as a result of the abovementioned steps, a step change in the controlling action in time and a change of the output parameter in time has been obtained.

With adequate accuracy, the static characteristics are approximated by a truncated cubic parabola:

$$x_i = k1 \cdot u_i + k2 \cdot u_i^3. \tag{2}$$

After determining the type of nonlinearity (2), an Input-Output model is recorded for each channel. In the image area, the description is performed using transfer functions (TF) for each channel.

For the case of a *MIMO* system, each output is defined as follows:

$$y_i(t) = \frac{B_{i1}(z^{-1})}{A_{i1}(z^{-1})} \cdot x_1(t) + \frac{B_{i2}(z^{-1})}{A_{i2}(z^{-1})} \cdot x_2(t) + \frac{B_{i3}(z^{-1})}{A_{i3}(z^{-1})} \cdot x_3(t), \tag{3}$$

where A_{i1}, A_{i2}, A_{i3}, B_{i1}, B_{i2}, B_{i3} – polynoms; z^{-1} – first-order lag operator.

Then the parameters of the linear part (3) has been determined, for which the ordinary least squares method (OLS) is applied. The input of the linear model is fed to the output of the nonlinear model for each channel, and the delay should be taken into account as well.

3. Results of experiments

For the case of 3 control inputs – 3 controlled outputs according to the formula (1) using Matlab, the TF$W11(p)^{EOTF}$, $W22(p)^{EOTF}$, $W33(p)^{EOTF}$ have been defined. As a result of identification using OLS method, coefficients of mathematical models of the object have been obtained for the $Wd1$(p), $Wd2$(p), $Wd3$(p) channels of disturbances. For example, figure 1 shows the result of identifying model parameters using the "raw material consumption – diesel fraction temperature" channel in comparison with the "experimental" curve. The experimental values are indicated as separate readings, and the model values are indicated as a solid line.

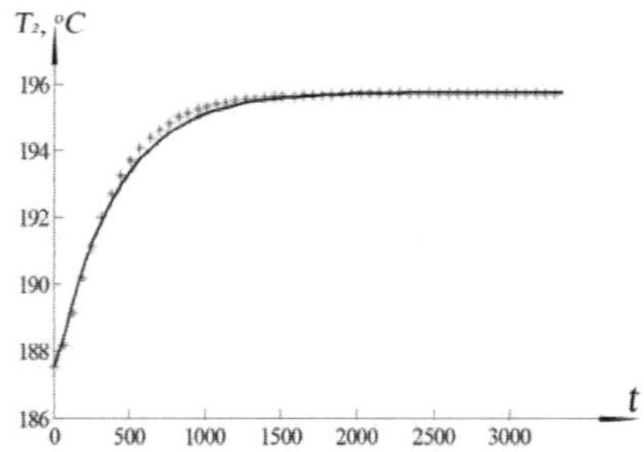

Figure 1. Approximation using the "raw material consumption – diesel fraction temperature" channel.

Figure 2 shows the structure of the ACS that contains the effective transfer function $W(p)^{EOTF}$, naphtha recovery temperature controllers, diesel fraction, $P1$, $P2$, $P3$ fuel oil residue; $WK1$(p), $WK2$(p), $WK3$(p) disturbance compensators; the transfer function of the object for the $Wd1$(p), $Wd2$(p), $Wd3$(p) disturbance channels; $SP1$, $SP2$, $SP3$ controller setting values.

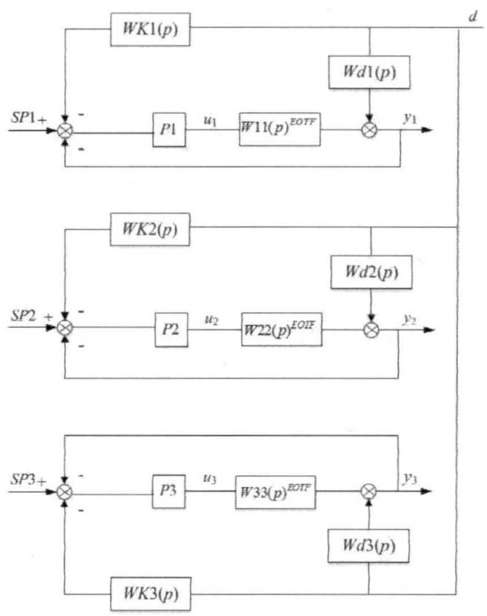

Figure 2. ACS containing effective transfer functions

The $W(p)^{EOTF}$ functions we got are of high order. The Pade approximation was used to exclude lagging elements which also have led to an increase in the order. There is a *pade* function for this in *Matlab*.

Due to the complexity of the obtained expressions, we conclude that it is difficult to apply the *EOTF* method to a multidimensional multiloop system containing 3 control inputs and 3 controlled outputs.

Arrays of controlled parameter and control action values have been formed from the "experimental" values of the dynamic model of the oil refining unit atmospheric tower [10]. The structure and parameters of the Hammerstein model`s linear and nonlinear parts have been determined.

Static characteristics that are visually nonlinear have been built. The *k*1, *k*2 coefficients for the cubic parabola for each control channel have been calculated.

Table 1 shows the coefficients obtained for the cubic parabola (formula 2) for each control channel.

Table 1. Coefficients of the cubic parabola

Control channel	*k*1	*k*2
Reflux rate – overhead temperature	77.11	–3.344
Diesel fraction flow rate – diesel fraction temperature	21.46	2.355
Superheated steam flow rate – bottom temperature	–390.3	31.23

After finding the parameters of the Hammerstein model nonlinear part, the type and coefficients of transfer functions of the linear part of the model have been determined.

The objective of finding coefficients using the OLS method is to minimize the mismatch between the output signal of the model and the experimental one at any time. As a result of identification, the Hammerstein model has been presented in the following way:

$$T1 = W_{11}(p)f(LT) + W_{12}(p)f(DT) + W_{13}(p)f(VB);$$
$$T2 = W_{21}(p)f(LT) + W_{22}(p)f(DT) + W_{23}(p)f(VB);$$
$$T3 = W_{31}(p)f(LT) + W_{32}(p)f(DT) + W_{33}(p)f(VB); \quad (4)$$

where $W_{ij}(p)$ is the transfer function from the *j*-th input to *i*-th output; *f*(*LT*), *f*(*BT*), *f*(*VB*) are nonlinear static characteristics; *T*1, *T*2, *T*3 – temperatures at the draw-off trays

4. Discussion of results and conclusions

An attempt to use the EOTF method for controlling an object containing three inputs and three outputs has resulted in an almost unrealizable form of effective transfer functions.

When conducting experiments using the Matlab software, we have managed to identify the Hammerstein model`s linear and nonlinear parts. The maximum percentage error of the produced Hammerstein model (4) is 4.8% against "experiments".

Thus, on the basis of an accurate dynamic model of the oil rectification process using the proposed algorithm, the Hammerstein model is produced, which allows isolating the nonlinear part of the model and, importantly, use the model to control the oil rectification process over a larger range of input parameters.

References

[1] Verevkin A, Murtazin T and Pavlova Z 2019 Simplified adaptation for the dynamic models objects of the one class *Proceedings of the 7th Scientific Conference on Information Technologies for Intelligent Decision Making Support (ITIDS 2019)* **166** 57-60

[2] Skogestad S, Morari M and Doyle J C 1988 Robust control of ill-conditioned plants: high-purity distillation *IEEE Transactions on Automatic Control* **33 (12)** 1092-1105

[3] Borhan M S and Karim S 2007 Online Multivariable Identification of a MIMO Distillation Column Using Evolving Takagi-Sugeno Fuzzy Model *Proceedings of the 26th Chinese Control Conference* 328-332
[4] Maldonado J L B, Valarezo J M and Peralta D T 2015 Multivariable Control of aBinary Distillation Column *International Journal of Engineering and Applied Sciences (IJEAS)* **2 (12)** 18-23
[5] Al-Shatri A H, Ahmad A, Oladokun O and Al-Shanini A H 2015 Control and Optimization of Aromatic Compounds in Multivariable Distillation Column *Chemical engineering transactions* **45** 468-474
[6] Skogestad S 2001 Probably the best simple PID tuning rules in the world *AIChE Annual Meeting* **77** 1-27
[7] Dmitrievsky B S, Zatonskiy A V and Tugashova L G 2018 Task of controlling oil rectification and the method of its solution *Bulletin of the Tomsk Polytechnic University Geo Assets Engineering* **329 (2)** 136-145 (in Russian)
[8] Giwa A and Karacan S 2012 Decoupling PID Control of a Reactive Packed Distillation Column *International Journ. Eng. Research & Technology* **1 (6)** 1924-1933
[9] Vu T N L and Lee M 2010 Independent design of multi-loop PI/PID controllers for interacting multivariable processes *Journal of process control* **20** 922-933
[10] Zatonskiy A V, Tugashova L G, Alaeva N N and Gorshkova K L 2017 Controlling the Oil Rectification Process in a Primary Oil Refining Unit Using a Dynamic Model *Petroleum Chemistry* **57 (12)** 1121–1131

IOP Conf. Series: Materials Science and Engineering **860** (2020) 012031 doi:10.1088/1757-899X/860/1/012031

Application of unlimited and limited anisotropic core compression results for wellbore stability calculations

M M Aliev, S V Shafieva, N G Karimova and S R Gilyazova

Almetyevsk State Oil Institute, 2, Lenin Str., Almetyevsk, 423250, Russia

E-mail: mmaliev@rambler.ru

Abstract. The study touches upon the issue of wellbore stability in wells drilled in anisotropic rocks. Initial data for solving this task is the results of anisotropic core samples testing for unlimited and limited compression taken from one section of the whole core and drilled in different directions. These data are taken from the works of foreign authors that are collected and presented in [1]. The problem of wellbore stability drilled in multi-layer highly anisotropic rocks is solved by applying a strength criterion that includes stress tensor components and the angle between the normal line to the slip area and the direction of lamination. Given that the values of the elastic modulus and Poisson's ratio for the anisotropic rocks under consideration in the main directions of anisotropy differ insignificantly, the stress distribution around the wellbore is accepted without taking into account the anisotropy of the elastic properties. It is believed that strength anisotropy is the cause of rock destruction. Thus, the components of the stress tensor are determined using the superposition method by summing the solutions of the classical Kirsch and Lame problems.

1. Introduction

Sedimentary rocks have laminar bedding by origin regardless of their mineralogical composition. The variability of the physical and mechanical properties of these rocks is determined by the direction of external loads. This property of rocks is called anisotropy. The anisotropy is mainly related to its texture. When the borehole intersects anisotropic rocks, the well trajectory may deviate from the planned one due to uneven drilling of the bottom hole.

The task of mud density determining that ensures the wellbore stability is complex when the inclination is more than 20°÷30° and requires the use of strength criteria that takes into consideration anisotropy.

The problem of ensuring wellbore stability in the case of anisotropy is solved by considering two cases of rock failure. In the first case, it is assumed that the shear fracture condition is for the main rock with contribution of stresses by Mohr-Coulomb linear criterion for isotropic rock, and in the second case, the shear fracture condition for weakened surfaces is tested using the same criterion, but with consideration of strength parameters that according to the weakening [2,3]. This approach eliminates the possibility of rock shear fracture on a certain surface, where the shear process covers both the main rock and the weakened layer.

For the first time, this approach for determining the destructive load was proposed by V. V. Sokolovsky [4], who applied the Coulomb-Mohr criterion for multilayer rocks. He called the condition

IOP Conf. Series: Materials Science and Engineering **860** (2020) 012031 doi:10.1088/1757-899X/860/1/012031

when a shear fracture occurs in the weak direction as "special condition of the limit equilibrium", and the fault in an unweakened direction as a "normal condition".

In foreign and Russian literary sources, the influence of anisotropy on the stress-strain state is considered according to the research of S. G. Lekhnitskii [5]. This includes the works of B. Amadea [6], N. R. Rabinovich [7] in [2].

2. Research methods

In our work, the multilayer rock is assumed in such a way that the fracture condition is generally provided in directions that do not coincide with the layering direction. Only in the special case, the shear areas coincide with the direction of the least resistance.

To solve the task of boundary state of stress and wellbore stability we use two strength criteria. The first criterion obtained is based on the Mohr nonlinear regularity and takes place in the form [8]

$$\sigma_1 - \sigma_3 = 2\sqrt{T_1^2 + T_2^2}, \tag{1}$$

where

$$\mathrm{T}_1 = \sqrt{(\sigma + \mathrm{T}_2)k + \mathrm{c}};$$

$$T_2 = (z_1 - \sqrt{z_1^2 + (\sigma k + c)ctg^2(\psi + \delta)})^2.$$

k and c are the strength characteristics of the rock, depending on the rock strength along, across the layering and at 45° angle to the layering direction; ψ is the angle between the normal line to the slip area and the layering direction; δ is the angle between the main stress and the x-axis; $z_1 = 0{,}5k \cdot ctg^2(\psi + \delta)$.

The second criterion is based on the Coulomb-Mohr linear dependency and has the following compact form [9]

$$\sigma_1 - \sigma_2 = \left[F_1^2(\psi) + F_2^2(\psi)\right]^{\frac{1}{2}}, \tag{2}$$

where

$$F_1(\psi) = 2\left\{s\left[\sin(2\psi - \rho) - 0{,}5\rho'\mathrm{sec}\rho\sin 2\psi\right] + 0{,}5s'\cos(2\psi - \rho)\right\}x;$$

$$F_2(\psi) = 2\left\{s\left[\cos(2\psi - \rho) - 0{,}5\rho'\mathrm{sec}\rho\sin 2\psi\right] + 0{,}5s'\sin(2\psi - \rho)\right\}x;$$

$$x = x(\psi) = \frac{\cos\rho}{(1 - 0{,}5\rho')}; \quad s = c(\psi) + k(\psi)\sigma;$$

$$\sigma = 0{,}5(\sigma_x + \sigma_y), \quad s' = c'(\psi) + k'(\psi)\sigma.$$

here $c(\psi)$ and $k(\psi)$ are the grip and angle of internal friction; σ_1, σ_2, σ_3 are main stresses.

To determine the strength parameters included in the linear criterion variables of conjunction areas and angle of internal friction, a method of their determination was developed, it is based on results of core sample testing for unlimited compression (no lateral pressure) and limited compression (core samples are pressed by the initial slight lateral pressure).

We assume that during the test for uniaxial compression, the rock shear occurs in the direction of stratification according to the Coulomb-Mohr in the following form:

$$\tau_n = \sigma_n k_{90} + c_{90}, \tag{3}$$

where k_{90} and c_{90} are strength characteristics along with weakened layers determined from two conditions

$$\tau_{n_1} = \frac{\sigma_1 - \sigma_3}{2} \sin 2\beta;$$

$$\sigma_{n_1} = \frac{\sigma_1 + \sigma_3}{2} + \frac{\sigma_1 - \sigma_3}{2} \cos 2\beta,$$

where β is the angle between $\sigma_1$and the direction of the layers.

When $\sigma_3 = 0$, we give the values of σ_1 and β for two cases where it is possible and define the characteristics k_{90}, c_{90}.

In [10-12], the results of experimental studies for mechanical characteristics determination for various anisotropic rocks are presented. The same experimental studies are given in [13] for the Carbona Phyllite rock in uniaxial and triaxial stress states. In this case, the following minimum initial data are accepted for the breaking stress $\sigma_1(\beta)$:

$\sigma_1(0°) = 11533\,psi = 79.5$ MPa; $\sigma_1(30°) = 7045\,psi = 48.6$ MPa;

$\sigma_1(60°) = 5427\,psi = 37.4$ MPa; $\sigma_1(65°) = 5584\,psi = 38.5$ MPa;

$\sigma_1(90°) = 9446\,psi = 65.1$ MPa.

When $\beta = 60°$ $\sigma_1 = 37.4$ MPa and $\beta = 65°$ $\sigma_1 = 38.5$ MPa we calculate $c_{90} = 10.7$ MPa, $k_{90} =$ 0,586.

By taking the values $k_{90} = k_{cp} = 0.586$ from criterion (2) it is possible to calculate $c_0 =$ 26.2 MПa.

After, when $k_{90} = 0.586$ and $\sigma_1 = 79.5$ MPa when $\beta = 0$ we calculate $k_0 = 0.6$.

Using the same method, we determine the initial data of the rock as the conjunction and the angle of internal friction along and across the stratification.

In the same sequence, we calculate the initial parameters for the Austen Slate [14]:

$$c_{90} = 21.5 \text{ MPa}, c_0 = 113.8 \text{ MPa}, \rho_0 = 0.3575, \rho_{90} = 0.34.$$

It should be noted that for calculation of these characteristics, the following data was taken: $\sigma_3 =$ 35 MPa, $\sigma_1 = 328.6$ MPa (when $= 0°$), $\sigma_1 = 346.5$ MPa ($\beta = 90°$), $\sigma_1 = 195$ MPa ($\beta = 40°$), $\sigma_1 = 149.3$ MPa ($\beta = 60°$).

3. Analysis and results

To verify the reliability of the proposed approach, we will use the results of the mentioned above testing two types of anisotropic rocks for unlimited (Carbona Phyllite) and limited (Austen Slate) compression.

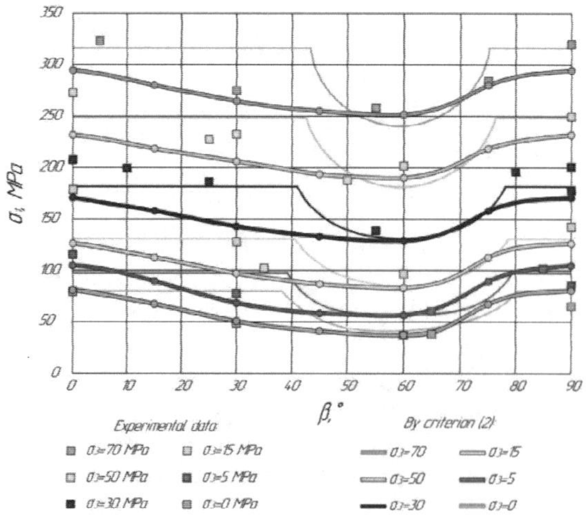

Figure 1. Dependency graphs of the ultimate stress on β for Carbona Phyllite

Figure 1 shows dependency graphs of σ_1 on the angle β for different values of lateral pressure, based on the criterion (2) for Carbona Phyllite. Also, dashed lines show the same dependencies, according to the J. C. Jaeger criterion [15].

In another example, we observe anisotropic Austen Slate rock whose strength characteristics under compound stress were studied in [14]. The results of core samples testing at low (up to 35 MPa) and high (up to 276 MPa) lateral pressures are presented. No results for the unlimited pressure test. Based on characteristics according to criterion (1), it is possible to determine the extreme breaking loads for Austen Slate (Figure 2, solid lines).

Using the criterion (2), theoretically, we can get breaking stress at unlimited compression for this rock (figure 2, bottom line). Taking these characteristics as initial parameters, it is possible to determine destructive stresses depending on the inclination angle of the layers at different lateral pressures from criterion (1) (Figure 2).

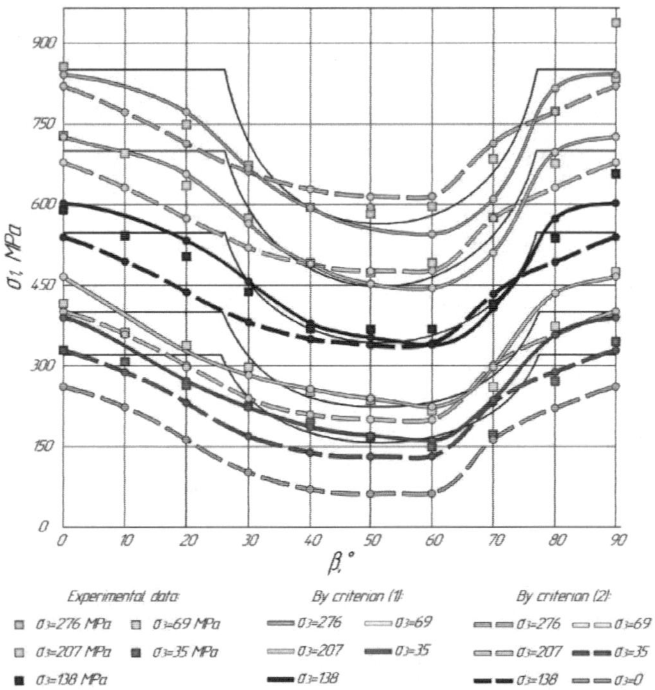

Figure 2. Dependency graphs of the ultimate stress on β for Austen Slate

The dash-line is constructed according to the J. C. Jaeger criterion [15].

The stress distribution around the wellbore is complying with laws that determine the rock behavior in the elastic or elastoplastic stage. Another factor that affects stress distribution is the anisotropy of elastoplastic properties, the different modularity of rocks elastic characteristics (for example, the difference in elastic modulus during tension and compression), and others.

In case of well inclination in a multi-layer with an arbitrary layer orientation, the stress tensor components can be determined according to S. G. Lehnitsky's decision [5], taking the rock model as linearly elastic. This model allows the influence of elastic characteristics on the stress distribution around the wellbore.

As the research results show [2] stresses in the circumferential direction are slightly depend on the anisotropy of the rock elastic properties, while in some cases the anisotropy can significantly affect the stress state around the wellbore.

However, considering that change the energy distribution occurs during the transition from elastic to a plastic state and such a transition cannot be tracked, in the first approximation, we don't have to consider the influence of elastic constants on the stress distribution. The rate of elasticity dynamic modulus for different anisotropic rocks in the main anisotropy directions usually slightly different from each other within 30%. The study of N. R. Rabinovich [7] shows that even if $E/E_1 = 2$ (E is determined along and E1 is defined across the stratification), anisotropy does not affect to change in the polar angle for the circumferential stress.

Based on this, assuming that the elastic characteristics, in this case, do not affect the stress distribution around the well, stress tensor components can be determined from the total solutions of Kirsch and Lame's classical tasks. Then, we get for effective stresses [16]

$$\left.\begin{aligned} \sigma_r' &= p_w - \beta_B p_r, \\ \sigma_\theta' &= \sigma_x + \sigma_y - 2(\sigma_y - \sigma_x)\cos 2\theta - p_w - \beta_B p_r, \\ \sigma_z' &= \sigma_z - \left(2(\sigma_x - \sigma_y)\cos 2\theta + 4\tau_{xy}\sin 2\theta\right) - \beta_B p_r, \\ \tau_{\theta z} &= 2(-\tau_{xz}\sin\theta + \tau_{yz}\cos\theta), \end{aligned}\right\} \quad (4)$$

p_r – formation pressure; p_W – well pressure; $\sigma'_r, \sigma'_\theta, \sigma'_z, \tau_{\theta z}$ – stress in cylindrical coordinates; β_B – Biot's coefficient; θ – polar angle measured from the x-axis coinciding with the direction σ_H; $\sigma_x, \sigma_y, \sigma_z, \tau_{xy}, \tau_{xz}, \tau_{yz}$ – stress tensor components defined as functions from σ_V (main full vertical normal stress) σ_H (main full horizontal normal stress) and σ_h (minimal horizontal normal stress), and the wellbore inclination angle i and from the azimuth angle α associated with the c σ_H.

Consider two options of the well direction axis: along the σ_H and the σ_h.

The main effective stresses acting on wellbore are calculated by this formula

$$\sigma_{1,3}' = \frac{\sigma_z' + \sigma_\theta'}{2} \pm \frac{1}{2}\sqrt{(\sigma_z' - \sigma_\theta')^2 + 4\tau_{z\theta}}, \quad (5)$$

the direction of σ_1' and σ_3' is calculated from

$$tg2\alpha_0 = \frac{2\tau_{z\theta}}{\sigma'_z - \sigma'_\theta}, \quad (6)$$

α_0 is the angle between σ'_z and σ_1.

Inclination angle towards rock layers σ_1' is found by the formula

$$\delta = i + \alpha_0.$$

The sequence of mud density determining that provides inclined wellbore stability:

1. For the given well inclination angle i (figure 3), also known values of σ_V, σ_H and σ_h for any well angle inclination from σ_H, by given angle α stresses are calculated (4).
2. For different polar angles θ the main stresses and their directions are determined by (5) and (6).
3. Mud density is calculated by substituting the obtained stresses in criteria (1) and (2).

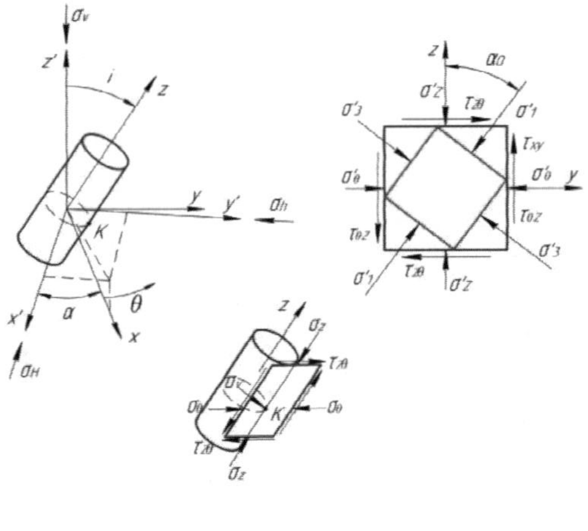

Figure 3. Wellbore position

In a particular case, we assume that a well with a given inclination angle was drilled in a multi-layer rock and the layers are located horizontally. It is necessary to determine mud density.

Example of mud density calculation that provides wellbore stability with the following initial data: $\sigma_V = 52\ MPa$, $\sigma_H = 42\ MPa$, $\sigma_h = 35\ MPa$, $\sigma_c(0) = 79.53\ MPa$, $\sigma_p(0) = 7.95\ MPa$, $\sigma_c(90) = 65.14\ MPa$, $\sigma_p(90) = 6.51\ MPa$, $\sigma_c(45) = 42.75\ MPa$, $\sigma_p(45) = 4.28\ MPa$. The depth is $h = 2000\ m$. Formation pressure $p_r = 20\ MPa$, well inclination angle $i = 50°$. The ratio of the rock strength at separation (rupture) to the rock strength at unlimited compression $\frac{\sigma_p}{\sigma_c}$ is assumed to be 0.1.

Substituting this data into the nonlinear criterion (1), we determine the mud density during the drilling in direction σ_H is $\rho_m = 1448\ kg/m^3$. During the drilling in direction σ_h we get $\rho_m = 1206\ kg/m^3$.

At $\sigma_c(0) = 113.8\ MPa$, $\sigma_c(90) = 21.5\ MPa$, $\rho_0 = 0.3575$, $\rho_{90} = 0.34$ for $\alpha = 0°$ (drilling in σ_H direction) is calculated from non-linear criterion (1) $\rho_m = 1454\ kg/m^3$.

At $\alpha = 90°$ (drilling in σ_h direction) $\rho_m = 1327\ kg/m^3$.

Thus, with such a tectonic regime, drilling in the direction of the minimum horizontal stress is more appropriate.

4. Conclusions

1. The method for calculating the mechanical parameters included in the strength criteria based on the results of unlimited and limited compression is given.
2. The extreme ultimate stresses for two types of anisotropic rocks in complex stress are determined theoretically.
3. An example of wellbore stability calculation based on the strength criteria developed by the authors in multi-layer highly anisotropic rocks is given.

References

[1] Ambrose J 2014 *Failure of Anisotropic Shales under Triaxial Stress Conditions,* PhD thesis, Imperial College
[2] Ashimkhin S G, Kashnikov Y A, Shustov D V and Kukhtinskiy A E 2018 Influence of elastic and strength properties anisotropy on the directional wellbore stability *Neftyanoe khozyaistvo* **2** 54-57
[3] Karev V I and Kovalenok Y F 2011 Oil and gas wells geomechanics *Lobachevsky university reporter* **4-2** 448-450
[4] Sokolovsky V V 1942 Flowing medium statics (Moscow: USSR Acamedy if sciences)
[5] Lekhnitsky S G 1977 Elasticity theory anisotropic body (Moscow: Science)
[6] Amadei B 1996 Importance of anisotropy when estimating and measuring *in situ* stresses in rock *Int. J. Rock Mech. Min. Sci.* **33 (3)** 293-325
[7] Rabinovich N R 1989 Engineering tasks of continuum mechanics in drilling (Moscow: Nedra)
[8] Aliev M M, Lutfullin A A, Shafieva S V and Gilyazova S R 2019 Modified nonlinear Mohr criteria of multilayer rocks *Society of Petroleum Engineers SPE-196905-MS*
[9] Aliev M M and Geniev G A 2001 Bearing capacity calculation of anisotropic based structures *News of higher educational institutions: construction* **6** 18
[10] Amine S, Latifa O and Lahcen B 2018 Adjusted anisotropic strength model for metasiltstones and prediction of ucs from indirect tensile tests *International Journal of Civil Engineering and Technology* **9** 598-611
[11] Singh M, Samadhiya Kumar A, Kumar V 2015 A nonlinear criterion for triaxial strength of inherently anisotropic rocks *Rocks Mechanics and Rocks Engineering* **48** 1387-1405
[12] Kawa M 2015 Reliability Analysis of Bearing Capacity of Square Footing on Soil with Strength Anisotropy Due to Layered Microstructure *Studia Geotechnica et Mechanica* Vol **37 (4)** pp. 19-28

IOP Conf. Series: Materials Science and Engineering **860** (2020) 012031 doi:10.1088/1757-899X/860/1/012031

[13] Ramamurthy T, Rao G V and Singh J 1993 Engineering behavior of phyllites *Eng. Geol* Vol. **33** (Issue **3**) pp. 209-225
[14] McLamore R T and Gray K E 1967 A strength criterion for anisotropic rocks based upon experimental observations *SPE Annual Meeting of the American Institute of Mining, Metallurgical, and Petroleum Engineers. 19-23 February Los Angeles* Paper SPE 1721
[15] Jaeger J C 1960 Shear failure of anisotropic rock *Geol. Mag* **97** pp. 65-72
[16] Fjaer E, Holt R M 1992 Petroleum Related Rock Mechanics (Amsterdam: Elsevier).

IOP Conf. Series: Materials Science and Engineering **860** (2020) 012032 doi:10.1088/1757-899X/860/1/012032

Research of long-term stability of the near-wellbore zone

M M Aliev, M M Baiburova and Z F Ismagilova

Almetyevsk State Oil Institute, 2, Lenin Str., Almetyevsk, 423250, Russia

E-mail: bayburovam@gmail.com

Abstract. The authors used the criterion of short-term rock strength, transformed for the case of long-term strength in order to assess the long-term stability of the borehole wall and determine the permissible pressure in the wellbore. The long-term strength criterion is written according to the method of P.P. Balandina. As a result of the study, it was proved that the determination of the long-term strength of a rock is reduced to a decrease in the values of the short-term strength limits. A curve of the long-term strength of an inclined well is constructed according to a generalized dependence. The expression is obtained that allows determining the time of long-term strength.

1. Introduction

Prognostication of well wall stability is a prerequisite for achieving a reduction in drilling complications. The rocks that compose the walls of the wells are in a difficult state. During the process in the well, the pressure is constantly changing, the number of hydraulic pulses is increasing [1,2]. The cyclic loading resulting from these operations leads to loss of rock stability due to fatigue failure. Practice has established that even if the pressure of the liquid column in the wellbore is lower than the tensile strength of the rocks surrounding the wellbore, after some time the walls of the wells collapse.

The condition for the transition of a loaded solid to a state of destruction when its bearing capacity is exhausted is called the strength criterion. In a complex stress state, the equation of limit equilibrium should contain invariants characterizing its stress state. In the general case, the strength criterion is determined by some ultimate surface.

Today, many different particular criteria for strength under difficult stress conditions have been proposed. [3,4,5]. In engineering calculations, strength hypotheses are used, based on the assertion that the destruction process depends on a change in the shape of the body. When the potential energy of body shaping reaches its ultimate state, its destruction or transition to plastic deformation occurs.

2. Materials and methods

To construct the criterion of long-term strength, the criterion of short-term strength K.V. Zakharova is used in the form

$$a_1\sigma_x^2 + a_2\sigma_y^2 + a_3\sigma_z^2 + a_4\sigma_x\sigma_y + a_5\sigma_x\sigma_z + a_6\sigma_y\sigma_z + a_7\sigma_x +$$
$$+ a_8\sigma_y + a_9\sigma_z + a_{10}\tau_{xy}^2 + a_{11}\tau_{xz}^2 + a_{12}\tau_{yz}^2 + a_{13} = 0. \quad (1)$$

Consider a rock for which the criterion K.V. Zakharova turns into the criterion of P.P. Balandina [6] and is written as

IOP Conf. Series: Materials Science and Engineering **860** (2020) 012032 doi:10.1088/1757-899X/860/1/012032

$$\sigma_1^2+\sigma_2^2+\sigma_3^2-\sigma_1\sigma_2-\sigma_1\sigma_3-\sigma_2\sigma_3+(\sigma_c-\sigma_p)\times$$
$$\times(\sigma_1+\sigma_2+\sigma_3)=\sigma_c\sigma_p. \tag{2}$$

We will carry out the long-term strength criterion according to the technique proposed in [6]. From this technique, it follows that in the case of constantly acting stresses, one can obtain such a record of the strength criterion for which the expression of the short-term strength criterion remains on the left side, and some damped time function remains on the right side.

Using the methodology described above, is writen down the criterion of long-term strength for materials that obey criterion (2). The damping function of time will enter both the left and the right side.

$$\sigma_1^2+\sigma_2^2+\sigma_3^2-\sigma_1\sigma_2-\sigma_1\sigma_3-\sigma_2\sigma_3+[\sigma_c(t)-\sigma_p(t)]\times$$
$$\times(\sigma_1+\sigma_2+\sigma_3)=\sigma_c(t)\sigma_p(t). \tag{3}$$

Thus, the determination of the long-term strength of the rock is reduced to a decrease in the values of short-term strength.

To determine the strength characteristics, it is necessary to conduct a series of experiments that should be planned so that as a result it was possible to record the dependence. It is proposed to approximate the long-term strength curve in the form

$$\sigma_c(t)=\sigma_{c\infty}+(\sigma_c-\sigma_{c\infty})e^{-a_0 t}, \tag{4}$$

where a_0 – approximation parameter,

σ_c - compressive strength with short-term compressive load,

$\sigma_{c\infty}$ - compressive strength if $t\to\infty$.

Next, getting $\kappa=\sigma_{c\infty}/\sigma_c$ equation (4) can be reduced to

$$\sigma_c(t)=\sigma_c(\kappa+(1-\kappa)e^{-a_0 t}). \tag{5}$$

Parameter $\kappa=\sigma_{c\infty}/\sigma_c<1$ can vary over a wide range. The reason for the change in this parameter will be an increase in rock damage over time, as well as a weakening of structural intermolecular bonds. The considered parameter κ is usually determined by conducting a long-term test of rock samples under various deformation conditions.

The parameter a_0 included in (5) determines the shape of the long-term strength curve

$$y=\sigma_c(\kappa+(1-\kappa)e^{-a_0 t})=f(t).$$

We construct a curve of long-term strength according to the obtained dependence. To do this, we assume that $\kappa=0{,}5$. By changing the parameter a_0 we build lines $y=f(t)$.

Analyzing the obtained graphs, we observe the influence of the approximation parameter on the limiting value of the compression stress.

To derive the equation for finding the time of long-term strength, we use the strength criterion given by N. Rabinovich [7].

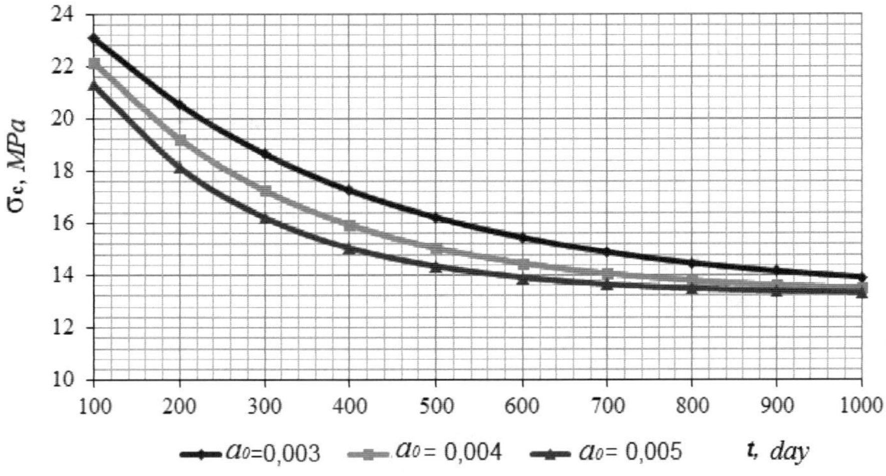

Figure 1. Graph of compression stress σ_c versus time t for various parameters a_0.

Consider the long-term strength of an inclined well wall filled with fluid. In the polar coordinate system, the stress intensity

$$\bar{\sigma}_u = \frac{\sqrt{2}}{2}\left[(\bar{\sigma}_r - \bar{\sigma}_\theta)^2 + (\bar{\sigma}_r - \bar{\sigma}_z)^2 + (\bar{\sigma}_\theta - \bar{\sigma}_z)^2 + 6\left(\bar{\tau}_{rz}^2 + \bar{\tau}_{r\theta}^2 + \bar{\tau}_{z\theta}^2\right)\right]^{\frac{1}{2}}. \quad (6)$$

takes maximum value at angle $\theta = \pi/2$.

The voltage components according to [7] are defined as

$$\bar{\sigma}_r = -q, \quad \sigma_\theta = q - 2a - 4b, \quad \bar{\sigma}_z = -c - 4\nu b,$$

$$\bar{\tau}_{r\theta} = \bar{\tau}_{rz} = 0, \quad \bar{\tau}_{\theta z} = d, \quad q = \frac{p_c - p_n}{p_2 - p_n},$$

where p_c – retaining fluid pressure,

p_2 – overlying rock pressure,

p_n – reservoir (pore) pressure.

q - reduced well pressure.

Hence we find that

$$(\bar{\sigma}_r - \bar{\sigma}_\theta)^2 = [2(q-a) - 4b]^2 = 4(q-a)^2 - 16(q-a)b + 16b^2;$$

$$(\bar{\sigma}_r - \bar{\sigma}_z)^2 = [(q-a) + K]^2 = (q-a)^2 + 2K(q-a) + K^2;$$

$$(\bar{\sigma}_\theta - \bar{\sigma}_z)^2 = [(q-a) - K - 4b]^2 = (q-a) + 8Kb + K^2 +$$
$$+ 16b^2 - 8b(q-a) - 2K(q-a);$$

$$6\tau_{\theta z}^2 = 6d^2;$$

where $K = a - c - 4\nu b$;

$$a = \frac{1 + 2\nu - (1 - 2\nu)\cos 2\alpha_0}{4(1-\nu)} \cos\alpha_0;$$

$$b = \frac{1 - 2\nu}{4(1-\nu)} \sin\alpha_0 \sin 2\alpha_0;$$

$$c = \frac{1 + (1 - 2\nu)\cos 2\alpha_0}{2(1-\nu)} \cos\alpha_0;$$

$$d = \frac{1 - 2\nu}{2(1-\nu)} \sin 2\alpha_0,$$

where ν - rock Poisson's ratio.

Substituting the obtained values in (6), we write

$$\overline{\sigma}_u = \frac{\sqrt{2}}{2}[6(q-a)^2 + 8(q-a)b + 2K^2 + 32b^2 +$$

$$+ 8Kb + 6d^2]^{\frac{1}{2}} = [3(q-a)^2 + (c - a + 4\nu b)^2 + 3d^2 -$$

$$- 4(q-a)b + 16b^2 + 4Kb]^{\frac{1}{2}}. \quad (7)$$

With these components, the average pressure is

$$\overline{\sigma} = \frac{1}{3}(\sigma_r + \sigma_z \sigma_\theta) = \frac{1}{3}(-q - c - 4\nu b + q - 2a - 4b) =$$

$$= -\frac{1}{3}(c + 2a + 4\nu b + 4b). \quad (8)$$

We write criterion (3) in the form:

$$\sigma_u^2 + 3(\sigma_c(t))\overline{\sigma} = \sigma_c(t)\sigma_p(t), \quad (9)$$

where

$$\sigma_c = \alpha\sigma_p; \quad \sigma_c(t) = \alpha\sigma_p(t); \quad \sigma_p(t) = \frac{\sigma_c(t)}{\alpha};$$

The stress intensity has the form

$$\sigma_u = \left[3(q-\delta)^2 + (1-\delta)^2\right]^{1/2}. \quad (10)$$

Criterion (9) can be written as follows

$$\sigma_u^2 - 3\overline{\sigma}\sigma_c(t)\left(\frac{1-\alpha}{\alpha}\right) = \frac{\sigma_c^2(t)}{\alpha}. \quad (11)$$

From here we define

$$\sigma_c(t) = -1{,}5\bar{\sigma}(1-\alpha) + \sqrt{2{,}25\bar{\sigma}^2(1-\alpha)^2 + \sigma_u^2}. \tag{12}$$

Denote $\sigma_c(t) = A$, then (12) we write in the form

$$A = -1{,}5\bar{\sigma}(1-\alpha) + \sqrt{2{,}25\bar{\sigma}^2(1-\alpha)^2 + \sigma_u^2}.$$

Moreover, according to (5), the time of long-term resistance is determined as

$$t = -e^{\theta} / a_0, \qquad \text{where} \qquad \theta = \frac{1}{1-\kappa}\left(\frac{A}{\sigma_c} - \kappa\right).$$

Given that the value of the compressive strength is denoted by A, we write equation (4)

$$A = \sigma_c(\kappa + (1-\kappa)e^{-a_0 t}).$$

From here

$$e^{-a_0 t} = \frac{1}{1-\kappa}\left(\frac{A}{\sigma_c} - \kappa\right).$$

Logarithm, we find the time of long-term strength

$$t = -\frac{1}{a_0} \cdot \ln\left[\left(\frac{A}{\sigma_c} - \kappa\right)/(1-\kappa)\right].$$

This formula makes it possible to find the desired value.

3. Results and discussion

Thus, it is proved that the determination of long-term strength reduces to a decrease in the values of short-term strength. An expression is obtained that allows determining the time of long-term rock strength surrounding the well wall. The curves constructed on the basis of the obtained data allow estimating the dependence of the compression stress on time.

4. Conclusion

The study of the stress-strain state of the near-wellbore zone of the well, the results expand the possibilities of applying the criteria for fatigue failure of the walls of the wells from cyclic hydrodynamic loads, allows reasonable choice of the methods of drilling wells, methods, mode of tripping operations, increase the stability of the walls of the wells.

References

[1] Aliev M M, Baiburova M M, Ibragimov I I, Ismagilova Z F and Lutfullin A A 2017 Investigation of relationships of geomechanic properties of layered rocks depending on hydrostatical pressure. *Society of Petroleum Engineers – SPE Russian Petroleum Technology Conference* 16-18 October Moscow

[2] Aliev M M, Ismagilova Z F and Baiburova M M 2017 Determination of elastic characteristics of anisotropic rocks under volumetric compression *Scientific notes of Almetyevsk State Oil Institute* **16** 6-10

[3] Zaharov K V 1961 Strength criterion of laminated plastics *Plast. Massy* **8** 61-67

[4] Goldenblat I I, Kopnov V A 1965 Strenght criterion of anisotropic materials *Izv. AN SSSR Mekhanika* **6** 77-83

[5] Nasseri M H B et al. 2003 Anisotropic strength and deformational behavior of Himalayan schists *International Journal of Rock Mechanics&Mining Sciences* **40** 3-23

[6] Balandin P P 1937 On the question of strength hypotheses *Bulletin of engineers and technicians* **1** 19-24
[7] Rabinovich N R 1989 *Engineering problems of continuum mechanics of drilling* (Moscow: Nedra)

IOP Conf. Series: Materials Science and Engineering **860** (2020) 012033 doi:10.1088/1757-899X/860/1/012033

Warm insulation of exhaust pipe when taking gas from the "dome - separator" installation from large depths

A S Chiglintseva[1], A A Nasyrov[2] and I A Chiglintsev[2]

[1] Ufa State Petroleum Technological University, 1, Cosmonauts ave., Ufa, 450062, Russia
[2] Birsk branch of Bashkir State University, 10, International ave., Birsk, 452450, Russia

E-mail: nasaza@mail.ru

Abstract. The article describes the impact of the use of polyurethane insulation on the tubes sclerosis by hydrate when gas is taken from the "dome-separator", installation that designed for emergency hydrocarbon collection at the bottom of the ocean. Such an installation collects hydrocarbons directly above the breakthrough point at the bottom of the ocean, and their transportation is carried out through pipes to the collecting vessel. Currently, in the conditions of the development of the Arctic shelf in order to increase the level of oil production, such breakthroughs can cause enormous damage to the ecosystem, in particular, in the Arctic Ocean, since living organisms that exists in the harsh conditions of these latitudes are more sensitive to changes in the environment. But during the passage of gas through the exhaust pipe in the presence of moisture and corresponding thermodynamic conditions, a hydrate formation process occurs. This phenomenon can cause the pipeline to overlap and impede the normal operation of the selection of hydrocarbons from great depths. To prevent these consequences, work suggests possible options for warming the pipeline. It is shown that sclerosis of a pipe extending to great depths can be prevented even with small values of polyurethane foam insulation.

1. Introduction

The "dome-separator" installation considered in this work can be used to prevent the consequences of oil and gas accidents at the bottom of the oceans [1, 2, 3]. Similar accidents related to well breakthroughs and hydrocarbon releases to the open ocean occurred in the Gulf of Mexico in 2010 and in the North Sea in 2012. The Dome-Separator collects hydrocarbons directly above the breakthrough site at the bottom of the ocean, and they are transported by pipes to the picking vessel.

This work is a continuation and further development of the work [2, 4].

2. Basic equations. Impulse and Energy equations

The formation of gas hydrates occurs when it corresponding thermodynamic conditions and there is water and light hydrocarbons [5, 6, 7].

We introduce the indices l and υ, which will correspond to the liquid phase and the vapor, w and g to water and the gas flow as a whole, and the index s to the parameters in the equilibrium state.

Let us suppose that in the process of gas movement through the outlet pipeline, the temperature of the liquid and gas phases are equal over the entire cross section of the channel; phase transitions occurs in an equilibrium state and the gas flow is quasi-steady.

We considered mass gas flow m_g is constant:

$$m_g = m_{g0} = Const, \tag{1}$$

$$m_g = \rho_g \vartheta_g S\,, \quad S = \pi a^2\,.$$

Here ρ_g and ϑ_g – the gas density and velocity; S и a – the area and radius of the "live section" of the pipeline. a and S are functions of the coordinate z and time t, because during gas extraction, hydrate will form on inner surface of shell of the discharge tube. The z coordinate will be counted from the input section into the pipeline. The pipeline is strictly vertical. To describe the thickness of hydrate near the shell there is $\delta_h = \delta_h(z,t)$. The radius of the pipeline is a_0. δ_{pp} is thickness of the polyurethane foam insulation. There is insulation between the steel shell (δ_{st}) and a layer of polyurethane (δ_p) (Figure 1).

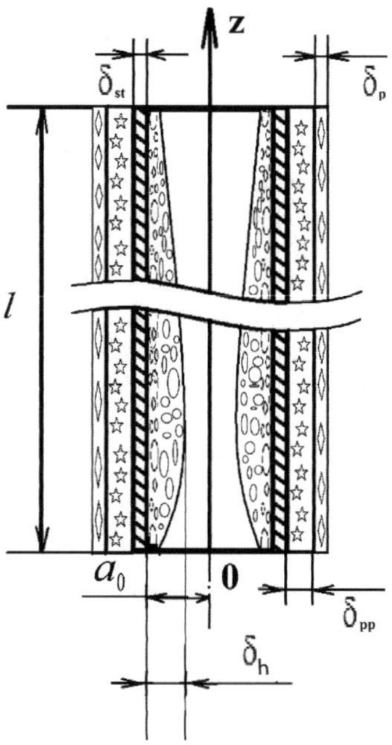

Figure 1. Scheme of pipeline

There is the water in gas in two forms: liquid (k_l) and gas (k_υ) phase. The mass concentration of water k_w is defined as:

$$k_w = k_l + (1 - k_l) k_\upsilon\,. \tag{2}$$

The gas flow in the stationary approximation is described by the impulse equation

$$m_g \frac{d\vartheta_g}{dz} = -S\frac{dp}{dz} - \rho_g g S - f\,. \tag{3}$$

Here p is the gas pressure.

The friction force generated during the selection of gas between its flow and the shell of the pipeline is written as [8, 9, 10]:

$$f = 2\pi a \tau, \qquad \tau = \lambda \frac{\rho_g \vartheta_g^2}{8},$$

where τ is hydraulic friction force, referred to the unit area of the inner surface of the pipeline and λ is coefficient of hydraulic resistance. If $\mathrm{Re} < 10^5$ the hydraulic resistance coefficient is determined by the Blasius equation [10]:

$$\lambda = \frac{0{,}3164}{\mathrm{Re}^{0{,}25}}.$$

There is heat generation because of steam condensation and the energy equation is written as:

$$m_g c_g \frac{dT_g}{dz} = \frac{m_g}{\rho_g}\frac{dp}{dz} + m_g l_w \frac{dk_l}{dz} - Q_{gen}, \tag{4}$$

$$Q_{gen} = 2\pi a q_{gen},$$

where l_w is specific heat of vaporization; T_g is gas temperature in the stream; q_{gen} is general heat transfer per unit area from the stream to the shell; c_g – specific heat of gas. The expressions for heat transfer from stream to the external environment:

$$q_g = \beta_g\left(T_g - T_h^{/}\right),\ q_h = \beta_h\left(T_h^{/} - T_h^{//}\right),\ q_{st} = \beta_{st}\left(T_{st}^{/} - T_{st}^{//}\right),$$

$$q_{pp} = \beta_{pp}\left(T_{pp}^{/} - T_{pp}^{//}\right),\ q_p = \beta_p\left(T_p^{/} - T_p^{//}\right),\ q_w = \beta_w\left(T_{p1}^{/} - T_w\right)$$

$$q_{gen} = q_g = q_h = q_{st} = q_{pp} = q_p = q_w, \tag{5}$$

$$q_{gen} = \beta_{gen}\left(T_g - T_w\right),$$

$$\frac{1}{\beta_{gen}} = \frac{1}{\beta_g} + \frac{1}{\beta_h} + \frac{1}{\beta_{st}} + \frac{1}{\beta_{pp}} + \frac{1}{\beta_p} + \frac{1}{\beta_w},$$

$$\beta_g = \frac{\lambda_g \mathrm{Nu}_g}{2R},\ \beta_h = \frac{\lambda_h}{\delta_h},\ \beta_{st} = \frac{\lambda_{st}}{\delta_{st}},\ \beta_{pp} = \frac{\lambda_{pp}}{\delta_{pp}},\ \beta_p = \frac{\lambda_p}{\delta_p},\ \beta_w = \frac{\lambda_w \mathrm{Nu}_w}{l}.$$

where T_i^j are temperatures; index $i = g, h, st, pp, p, w$ indicate respectively gas, hydrate, steel, polyurethane foam, polyurethane, surrounding water; index j with one hatch is for temperature on the inner surface of the layer and with two hatch is for temperature on the external surface of the layer; q_i are heat transfer rates per unit area; λ_i are thermal conductivity coefficients, l – length (height) of the pipeline.

Heat transfer from the gas stream to the pipe shell occurs in three modes [11, 12]. The first mode is implemented in the area where there is no hydrate formation. Then $q_h = 0$ и $\frac{1}{\beta_h} = 0$.

The second mode is characterized by heat sink:

$$q_g = \beta_g\left(T_g - T_s(p)\right), \tag{6}$$

where $T_s(p)$ is equilibrium hydrate formation temperature.

The third mode is describes by the diffusion:

$$q_g = \beta_g\left(T_g - T_h^{/}\right),$$

$$q_g + l_h j_h = q_w, \tag{7}$$

$$T_h^l = \frac{\beta_g T_g + \beta_w T_w + l_h j_h}{\beta_w + \beta_g},$$

where j_h is mass hydrate formation rate per unit area; l_h is specific heat of hydrate formation.

To describe the heat transfer, we determine the Nusselt number for gas (Nu_g) and water (Nu_w) flows [8, 10, 12, 13]:

$$\mathrm{Nu}_g = 0.021\,\mathrm{Re}^{0.8}\,\mathrm{Pr}^{0.43},\ \mathrm{Nu}_w = C\,\mathrm{Pr}^n\,\mathrm{Re}_w^m \tag{8}$$

where $\mathrm{Re}_w = \frac{w_w D_{equl}}{\nu_w}$; Pr is Prandtl number; w_w – flow rate for water; D_{equl} is equivalent diameter; ν_w is kinematic viscosity of sea water; n =0.35 is empirical coefficient; C and m *are* flow velocity and cylinder shape coefficients. To simplify, we assume that the velocity of the sea water flow current along the entire length of the outlet pipe is constant and varies from 0.25 m/s to 1 m/s. And that is why the coefficients will be C= 0.0026, m=0.8 [13].

The written equations (1), (3), (4), taking into account the Mendeleev - Clapeyron equation, the heat removal intensity Q_{gen} and hydraulic friction f, will form the equations system:

The obtained equations can be used for numerical calculations to describe changes in the temperature, pressure and the growth of hydrate deposits on the inner shell side of the pipe over time.

3. Results

Let us consider a deep-water well with parameters for the outlet pipe: a=5 cm; l=1500 m. We accept the following thermo-physical parameters: T_w=277 K; m_g=1 kg/s; R_g=520 J/(kg·K); μ_g=1028·10^{-8} kg/(m·s); c_g=2365 J/(kg·K); λ_g=0.03 W/(m·K); λ_p=0.25 W/(m·K); λ_{pp}=0.024 W/(m·K); λ_{st}=86 W/(m·K); l_w=1.7·10^6 J/kg; R_υ=461 J/(kg·K); p_{w*}=9.34·10^9; T_{w*}=4228; l_h=5·10^5 J/kg; k_υ=3·10^{-3}; ρ_h=910 kg/m^3; T_{h0}=283 K; p_{h0}=6.95 MPa; T_*=10 K. The calculation was carried out at the following thicknesses: δ_{st}=1 cm; δ_p=0.5 cm; δ_{pp}=0.5; 1 cm. The initial temperature and gas pressure at the inlet have been taken T_g^0=310 K; P_g^0=15 MPa.

Figures 2 and 3 shows the calculation results for an uninsulated and insulated pipeline, corresponding to the circuit shown in Figure 1 at various thicknesses of polyurethane foam. For three cases, three lines of different colors are shown: black for a steel pipe without insulation, red – δ_{pp}=0.5 cm and green – δ_{pp}=1 cm. The gas evacuation time from under the dome-separator is 35 hours, the flow rate of sea water w_w=0.25 m/s.

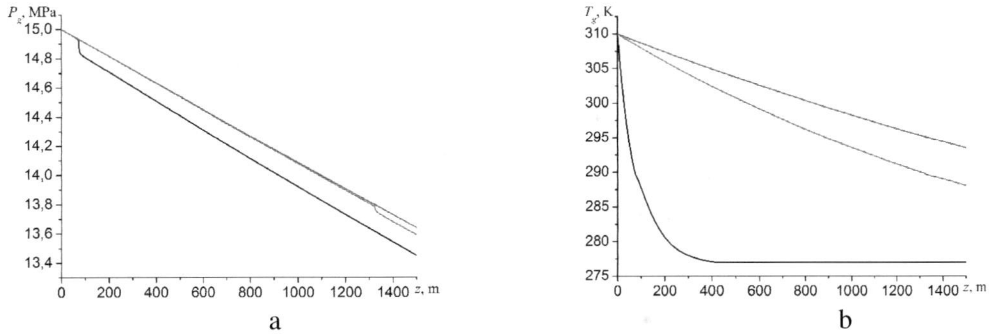

Figure 2. Change of pressure (a) and temperature (b) of the gas along the outlet pipe at different thicknesses of insulation: δ_{pp} =0.5 cm – red line; δ_{pp} =1 cm – green line; without insulation – black line

It can be seen from the results that in the case of using a heater for a steel pipe in the form of a layer of polyurethane foam, the gas cools more slowly, unlike an insulated pipe, where its temperature drops quickly enough to ambient temperature.

The gas pressure along the pipeline pumping gas to the surface from under the dome-separator experiences a sharp drop in some areas, which can be explained by the adiabatic expansion when gas is passing through the narrowing region in the channel formed due to gas hydrate deposits on the walls. When using a heater with a thickness of 1 cm, such a sharp decrease is not observed, in contrast to the case of an insulated pipe, where it falls almost at its turn.

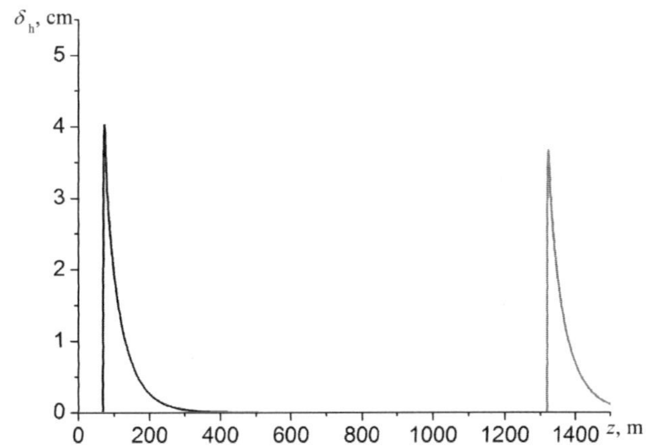

Figure 3. Change in the thickness of gas hydrate along the discharge pipe at different thicknesses of insulation: δ_{pp} =0.5 cm – red line; δ_{pp} =1 cm – green line; without insulation – black line.

Figure 3 shows the calculation results for the thickness of the gas hydrate layer formed on the channel walls. It can be seen that when the gas is pumped out after 35 hours, the pipe with a radius of 5 cm is almost blocked by gas hydrate formations. In this case, in the absence of insulation, a blockage forms in the pipe almost at the very bottom of the ocean, in the case of insulation, it is displaced to the surface and when the thickness of the polyurethane foam is 1 cm, sclerosis can be completely prevented along the entire length of the pipeline.

4. Conclusions

We consider the process of gas movement through an exhaust heat-insulated steel pipe, which contains moisture in the form of steam from large depths of the ocean from the dome-separator. This process is occurs into conditions for the formation of hydrates on the inner shell of the pipeline.

The calculation results for the gas pressure, temperature and growth of the hydrate layer on the channel shell are shown. It was found that during thermal insulation of a steel pipe with polyurethane foam insulation, even with small values of its thickness of 1 cm, channel obstruction by hydrates formed during gas evacuation from great depths can be prevented.

It can be concluded that when pumping gas from great depths from the dome-separator, thermal insulation of a steel pipeline with polyurethane foam insulation is an effective way to prevent the formation of gas hydrates and blockage of the pipeline.

5. Acknowledgments

The research was funded by a grant of the President of the Russian Federation for state support of young Russian scientists - doctors of sciences (Competition - MD-2020), according to the research project No. MD-2179.2020.1.

References

[1] Zhukov A V, Zvonarev M I and Zhukova Yu A 2013 Method for gas production from deep-water gas hydrate deposits *International Journal of Applied and Fundamental Research* **1 (10)** 16 - 20

[2] Chiglintsev I A and Nasyrov A A 2016 Modeling of the Process of Filling a Dome Separator with the Decomposition of a Gas Hydrate Formed During the Mounting of the Installation *Journal of Engineering Physics and Thermophysics* **89 (4)** 854–863 https://doi.org/10.1007/s10891-016-1446-0

[3] Gimaltdinov I K and Kildibaeva S R 2015 About the theory of initial stage of oil accumulation in a dome-separator *Thermophysics and Aeromechanics* **22 (3)** 387-392 https://doi.org/10.1134/S0869864315030130

[4] Urazov R R, Chiglinstev I A and Nasyrov A A 2017 Formation Of Sclerotic Hydrate Deposits In A Pipe For Extraction Of A Gas From A Dome Separator *Journal of Engineering Physics and Thermophysics* **90 (5)** 1162 -1169 https://doi.org/10.1007/s10891-017-1670-2

[5] Bilyushov V M 1984 A mathematical model of the formation of hydrates during the flow of wet gas in pipes *Engineering Physics Journal* **1** 57-64

[6] Bondarev E A, Gabysheva L N and Kanibolotsky M A 1982 Modeling of hydrate formation during gas movement in pipes *Izvestiya AN USSR. Mechanics of fluid and gas* **5** 105-112

[7] Istomin V A 1992 *Phase equilibria and physicochemical properties of gas hydrates: analysis of new experimental data (* Moscow: VNIIEgazprom press)

[8] Bekirov T M and Shatalov A T 1986 *Collection and preparation for the transport of natural gases* (Moscow; Nedra press)

[9] Bilyushov V M 1984 Mathematical model of hydrate formation during the flow of wet gas in pipes *Engineering Physics Journal* **1** 57-64

[10] Guzhov A I, Titov V G, Medvedev V F and Vasiliev V A 1978 *Collection, transport and storage of natural hydrocarbon gases* (Moscow: Nedra press)

[11] Shagapov V Sh, Chiglintseva A S and Syrtlanov V R On the possibility of leaching gas from a gas hydrate array through the circulation of warm water *Applied mechanics and technical physics* **50 (4)** 100-111

[12] Kutateladze S S 1979 *Fundamentals of the theory of heat transfer* (Moscow: Atomizdat press)

[13] Kutateladze S S and Borishansky V M 1958 *Heat Transfer Reference* (Moscow: Gosenergoizdat press)

IOP Conf. Series: Materials Science and Engineering **860** (2020) 012034 doi:10.1088/1757-899X/860/1/012034

Separation of analytical signal for atomic-emission microwave plasma measurements

V V Drokov, V Vl Drokov, A D Kazmirov, A V Sinitskaya, N A Ivanov and A Y Hodunaev

Irkutsk National Research Technical University, 83 Lermontov Str., Irkutsk, 664074, Russia

E-mail: alex.hodunaev@gmail.com

Abstract. The analytical signal for atomic-emission microwave plasma (AEMP) spectral analysis method is studied. It is shown that the signal consists of two parts: background signal, proportional to content of the analytical element in a sample shaped by submicron particles and metalorganic compounds; pulse signal, proportional to content of analytical element in coarse discrete particles. The "dynamic discrimination" algorithm is suggested that allows measuring the element content in a liquid or powder ore sample independently of the content shape using AEMP method. It is shown that atomic absorption (AA) and AEMP measurements of Fe, Cu agree within error for samples both low and high in content of element in submicron particles.

1. Introduction

Atomic-emission microwave plasma (AEMP) [1] spectral analysis method, also called scintillation method, is based on continuous introduction of the sample into the spectrum excitation source. The sample usually consists of powder of discrete particles [2]. The atomized sample matter produces light flashes (scintillations), which are registered as pulses by the equipment. The equipment counts pulses that have amplitudes exceeding specified threshold. Thus, equipment noise is not registered. This approach to signal extraction was developed by researchers during almost 40 years for precious metals analysis in powder ores [3-5]. The main task was to develop a method to express measurement of parameters of precious metal content. In particular, mass fraction, fineness and particle size of gold was measured.

The task was solved partially. The matter is that metallic impurity can take shape of uniformly dissolved (metalorganic) particles with size so small, that produced pulses do not exceed equipment noise level. The work [6] shows that this impurity consists of particles below 2 um in diameter (submicron particles). Inability to account for submicron impurity results that method lies in category of semi-quantative analysis.

Liquid samples are analysed using atomizer (sprayer) for sample introduction [7]. Oil samples and filter wash samples of aircraft engine systems are analysed for tribodiagnostical purposes [8]. In these samples, mass fraction of submicron particles can exceed 50 %. Therefore, the precision of method can be increased significantly by accounting for submicron impurity in a sample.

For AEMP measurements so called "scintillation principle" is defined. The principle is usually formulated as such – the probability of simultaneous presence of multiple discrete particles in analytical field of spectrum excitation source must be minimal, i.e. particles must reach the plasma one by one. In other words, adherence to the principle enables the ability to register the signal from each particle individually.

In real samples, particles are distributed lognormally and most common case is that the signals from multiple submicron particles overlap over each other and "scintillation principle" is violated. The edge case is registering uniformly distributed impurity, i.e. when metal is present in sample in shape of true solution (metalorganic compounds). In these conditions each solid particle for powder ore samples or each drop of liquid sample can be viewed as separate "portion'' that contains analytical element. As these "portions" are introduced to spectrum excitation source, the background signal value increases and coarse particles will produce significant pulses on the increased background.

This paper aims to develop mathematical description for separate extraction of analytical signal for both submicron elemental impurity and impurity in coarse discrete particles; develop corresponding software and perform experimental verification of the developed method for signal extraction.

2. Analytical signal for uniformly distributed impurity

Let the sample introduction device generate N "portions" in a time unit. A number of similar pulses with different amplitude but same shape forms the analytical signal in this case. The time of a pulse beginning is an evenly distributed random variable. The expression for the signal can be written as

$$S(t) = \sum_{i=0}^{\infty} a_i v(t - t_i) \tag{1}$$

where a_i – pulse amplitude (depends on the size of the "portion"), v – pulse shape (same for every "portion"), t_i – the sequence of random time moments of pulse beginning (Poisson point process with mean rate N). Since the pulses overlap continuously, this process can be approximately described as a Gaussian process with characteristics

$$M\left[S_u(t)\right] = NM\left[a_i\right]\int_{-\infty}^{\infty} v(t)dt \tag{2}$$

$$D\left[S_u(t)\right] = NM\left[a_i\right]^2 \int_{-\infty}^{\infty} v^2(t)dt - M\left[S_u(t)\right]^2 \tag{3}$$

where $M[S_u(t)]$ – the mean signal value, $D[S_u(t)]$ – variance.

As such, the mean signal value does not depend on time in this case and is proportional to the amount of generated "portions'' and quantity of metal within. The variance also does not depend on time and increases with mean amplitude ($M[a_i]$) which is proportional to "portion" size.

Since Gaussian random process is ergodic, using (2) we have

$$\overline{S}_u = \frac{1}{T_e}\int_{0}^{T_e} S_u(t)dt = M\left[S_u(t)\right] \infty N \cdot \overline{m} \infty C_u \tag{4}$$

where $\overline{S_u}$ – mean signal value through exposition,

$\overline{m}$ – mean particle mass in a "portion",

C_u – mass fraction of uniformly distributed impurity.

Therefore, mean value of analytical signal through exposition is proportional to content of uniformly distributed impurity and the variance of analytical signal decreases with "portion" size.

3. Analytical signal for discrete impurity

Now we will consider the case where the probability of simultaneous appearance of two or more particles in spectrum excitation source is approaching zero. As such, there will be periods of time when either one pulse is present or there are none (the signal is zero). The area of each pulse will contain information about evaporated mass. The expression the for signal is

$$S_I(t)=\sum_{k=0}^{n\to\infty} I_k(t) \tag{5}$$

where n – amount of particles in sample, $I_k(t)$ – k-th particle pulse.

Most common case in real situation is some intermediate scenario, when signals from submicron particles (present in large quantity in the sample) will overlap and cause the increase of a background signal, indistinguishable from increase, caused by dissolved, uniformly distributed impurity. The coarse discrete particles will produce significant pulses on this background.

The general case for expression (1) can be formulated as follows:

$$S_I(t)=\sum_{i=1}^{\infty} a_i v(t-t_i)+\sum_{k=0}^{n\to\infty} I_k(t)+S_N(t) \tag{6}$$

The first addend describes the signal constituent, shaped by uniformly distributed and submicron impurity. The second – the constituent, caused by discrete impurity, and the third – the noise constituent. The latter is caused by equipment noise, which raises the background signal has its own mean and variance.

The task now is to discriminate significant pulses from noise background. This can be accomplished using different approaches, most trivial being the “thresholding'' approach. The second constituent in expression (5) can be written as

$$S_I(t)=\sum_{k=0}^{n_1\to\infty} I_k(t)+\sum_{j=0}^{n_2} I_k(t) \tag{7}$$

where first addend is the totality of small overlapping pulses, and second is totality of separate non-overlapping pulses. Finally, the analytical signal expression can be written as sum of gaussian and pulse random processes:

$$S(t)=S_G(t)+S_I(t) \tag{8}$$

Gaussian random process is shaped by three constituents: noise background signal (plasma and equipment noise), signal from uniformly distributed impurity and signal from submicron particles

$$S_G(t)=S_N(t)+\sum_{i=1}^{n\to\infty} a_i v(t-t_i)+\sum_{k=0}^{n\to\infty} I_k(t) \tag{9}$$

Mean exposition value of this signal is given by

$$\overline{S}_G=\frac{1}{T_e}\int_0^{T_e} S_G(t)dt=a_N+b\cdot C_U+c\cdot C_M \tag{10}$$

where b, c – propotion coefficients, C_U – mass fraction of uniformly distributed impurity, C_M – mass fraction in submicron particles.

The second addend in expression (7) decribes the signal from random pulse process, discriminating which is the main task of AEMP measurements:

$$S_D(t)=\sum_{k=0}^{n_2} I_k(t) \tag{11}$$

The mean value of this signal is proportional to mass fraction of element in discrete particles C_D:

$$\overline{S}_D = \frac{1}{T_e}\int_0^{T_e} S_D(t)dt \propto C_D \qquad (12)$$

We will note some extra factors that affect the characteristics of analytical signal:

1. Unstable output of the liquid sprayer (mean quantity or size of the drops can differ over time);
2. Parameter fluctuations of spectrum excitation source.

Due to there factors background signal value can slowly change over time. Formally this is equivalent to a pulse with small amplitude and high duration. This part must be excluded from pulse signal.

4. "Dynamic discrimination" algorithm

As it is shown above, the analytical signal in general case is constitued from two random signals:

1. Continuous signal with small noise amplitudes, slow change of the mean over time, which is proportional to the mass fraction of dissolved element and element in submicron particles;
2. Pulse signal – separate non-overlapping pulses with big amplitude, proportional to mass fraction of element in coarse particles.

The figure 1 shows pulse signals (signal intensity over time) overlapped with continuous, slowly changing signal (a) and the picture for signal parts after their extraction, that must be achieved (b, c).

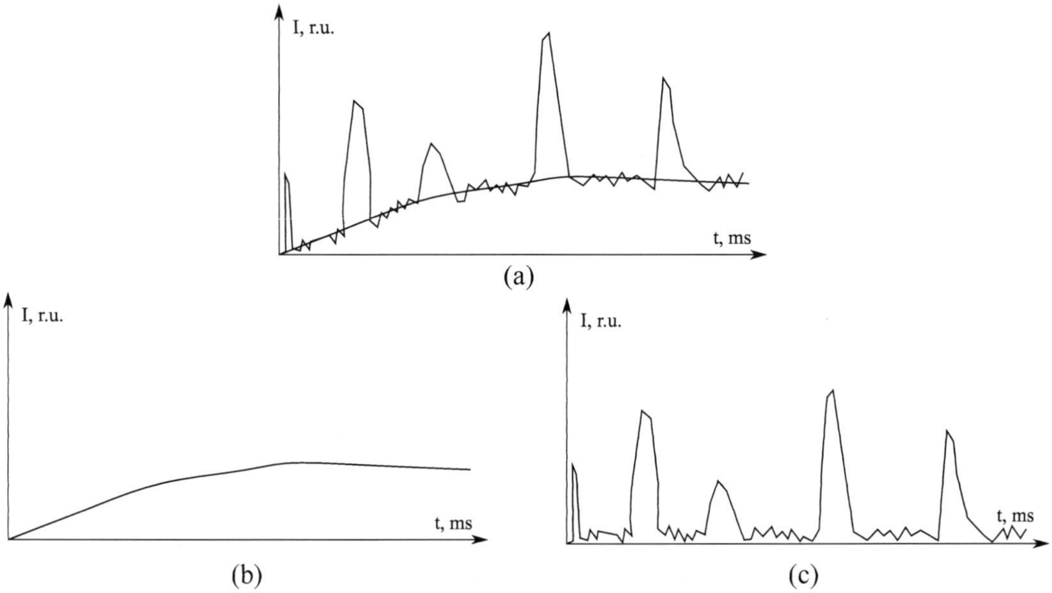

Figure 1. Overlapping background and pulse signals.

The simplest way for this is a software one. The algorhitm for "dynamic discrimination" (Figure 2) works as follows. The overall signal is smoothed out as moving average of 5 samples. The registration begins in "background" mode. Every signal sample is treated as a background one – that is, proportional to mass fraction of submicron impurity. The plasma constituent of the background signal is written at the start of the exposition and is corrected for later. When the signal rate of change exceeds the specified value L_D, the "particle" registration mode begins and signal value at the time is remembered as "background" threshold. The value L_D is called "discrimination level". In "particle" mode, the signal above the "background" threshold is treated as proportional to the mass fraction of coarse particles.

The pulses are additionally discriminated by duration. The pulses durations are dependent on the temperature distribution of spectrum excitation source and the time which impurity spends in analytical (high temperature) zone. The durations vary from hundreds of us [3] to 1's of ms [9]. The spectrum excitation source for AEMP measurements [6] allows full evaporation of particles up to 60 um in size in 1-10 ms. This makes the discrimination by duration an easy task – 1-10 ms thresholds are used for signals in "particle" mode.

The "discrimination level" is set in such way that sample that does not contain coarse particles and does not produce a single pulse. As this sample, the "Conostan" type sample is used, for which elements are introduced to the oil in the form of metalorganic complex.

Using the "dynamic discrimination" approach, registered "particle" mode signal is shown on the fig. 1, c. The "background" mode signal – on fig. 1, b. I.e., required signal separation was achieved.

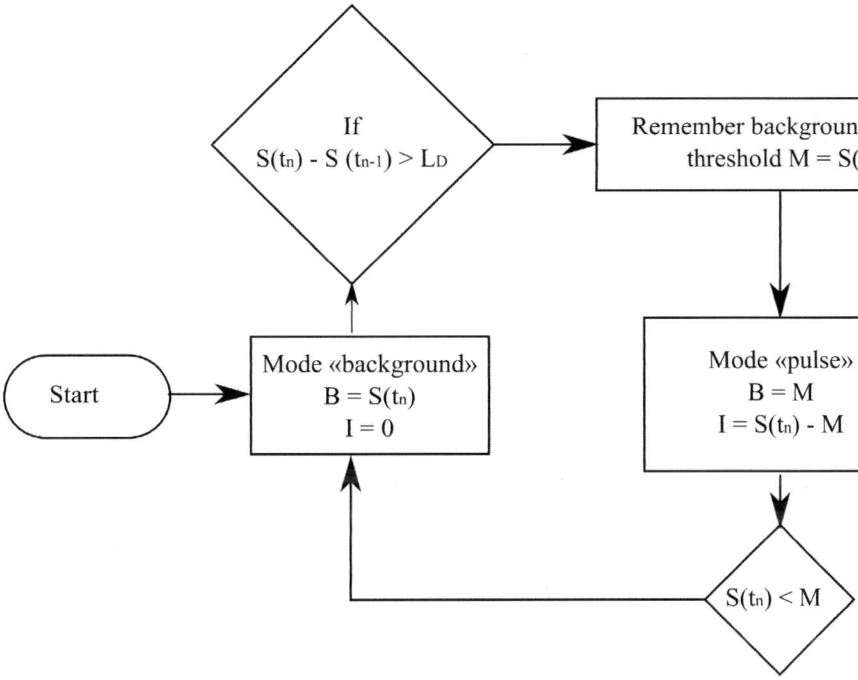

Figure 2. Flowchart for "dynamic discrimination" algorithm,
$S(t_n)$ – signal in time moments t_n;
B – background signal value (output);
I – pulse signal value (output);
M – background signal threshold;
L_D – «discrimination level».

The signal separation results in two analytical parameters – mean background (expression 10) and pulse (expression 12) signal values throughout exposition. The mass fractions in these expressions can be determined after proportion coefficients are found via equipment attunement, i.e. measuring standard samples with known content.

5. Experimental verification for the dynamic discrimination approach

The developed approach was verified through comparison of AEMP and acid dissolution with atomic-absorption finish (AA) measurements. The results are shown in table 1.

C_U is mass fraction of uniformly distributed element and element in submicrion particles. C_D is impurity in discrete particles. In first case (№ 1), the oil sample from D30KP engine (70 um filter cell finesse) was measured. In second case (№ 2), the oil filter wash sample was measured. For each case, 3 concurrent measurements were carried out.

The table 1 shows, that results for AA and AEMP measurements agree within error for both cases. This is evidence that the developed approach to signal extraction allows taking the measurements of samples independently of the content shape, being submicron or coarse particles or metalorganic content.

Table 1. Results of measurements of Fe, Cu in liquid samples using AEMP and AA methods.

№	Element	AEMP method	AA method
		$C_U + C_D = C_{total}$, ppm	Ctotal, ppm
1	Fe	2.15+0.80=2.95 ± 1.04	3.05 ± 0.31
	Cu	0.04+0.01=0.05 ± 0.04	0.06 ± 0.02
2	Fe	2.25+3.00=5.25 ± 1.79	6.07 ± 0.61
	Cu	1.37+1.12=2.49 ± 0.89	2.25 ± 0.21

6. Conclusion

The mathematical study of analytical signal for AEMP measurements showed that the signal consists of two parts. The first part was shaped by submicron (less than 2 um) particles and metalorganic content, described by random process and resulted in uniform signal. The second part was pulse signal from discrete coarse metallic particles. The suggested "dynamic discrimination" algorithm allowed extracting each part from signal individually, which was verified by comparative AEMP and AA measurements.

7. Acknowledgements

This article was prepared using the results of work carried out in the framework of the federal target program "Research and development on priority directions of scientific-technological complex of Russia for 2014 - 2020 years". Topic: "Development and creation of software and hardware microwave plasma system for monitoring, control and safe operation of oil system of engines for ground and air purposes" The agreement with the Ministry of Education for a grant No. 14.577.21.0289, 075-15-2019-1369. Unique identifier PNIER RFMEFI57718X0289.

References

[1] Drokov V, Drokov V, Ivanov N, Myrishenko V, Skudaev Y, and Hodunaev A 2019 Development of microwave plasma method for measurement of wear particle parameters in lubricant oil samples from aircraft gas turbine engines *IOP Conf. Series: Materials Science and Engineering* **560** **(1)** 012059

[2] Raihbaum Y 1980 *Physical fundamentals of spectral analysis* [In russian] (Science)

[3] Krestyaninov A, Raihbaum Y, Koretskaya A, and Markova N 1969 Scintillation spectral analysis method in ores *Applied spectroscopy journal* [In russian] **10** 17–21

[4] Prokopchuk S 1994 *Scintillation spectral analysis method in geology* [In russian] (Institute of Geochemistry SB RAS)

[5] Shabanova E, Busko A, and Vasil'eva I 2012 Arc scintillation atomic-emission analysis of powder samples using MAES with high temporal resolution *Industrial Laboratory. Diagnostics of Materials* [In russian] **78** 24–33

[6] Alhimov A, Drokov V, Kazmirov A and Morozov V 1996 Transfer function of microwave plasmatron as a spectrum excitation source *Analytical chemistry* [In russian] **51** 39–44

[7] Drokov V, Drokov V, Kaloshin A, Skudaev Y, Myrishenko V and Hodunaev A 2019 Ultrasonic sprayer of liquid samples for atomic-emission microwave plasma analyzer *IOP Conf. Series: Materials Science and Engineering* **560** **(1)** 012060

[8] Drokov V, Drokov V, Ivanov N, Myrishenko V, Skudaev Y and Hodunaev A 2019 Technical

state evaluation of oil-lubricated parts and units of aircraft gas turbine engines using microwave plasma method *IOP Conf. Series: Materials Science and Engineering* **560 (1)** 012058

[9] Djuba A, Labusov V, and Babin S 2015 Optimization of base exposition for solid body emission detector of scintillation atomic-emission spectral analysis *Analytics and control* [In russian] **19** 6–12

IOP Conf. Series: Materials Science and Engineering **860** (2020) 012035 doi:10.1088/1757-899X/860/1/012035

Elimination of aperiodic components of measuring elements of relay protection

N S Buryanina[1], Y F Korolyuk[1], M L Koryakina[1], E V Lesnykh[2] and K V Suslov[3]

[1] Chukotka branch of North-Eastern Federal University, 3, Studencheskaya str., Anadyr, 689000, Russia
[2] Siberian Transport University, 191, D. Kovalchuk str., Novosibirsk, 630049, Russia
[3] Irkutsk National Research Technical University, 83, Lermontov str., Irkutsk, 664704 Russia

E-mail: bns2005_56@mail.ru

Abstract. The authors propose the algorithms to calculate the actual values of currents and their derivatives. The algorithms allow a sharp reduction in the aperiodic components of currents in the calculated actual values and their derivatives. The algorithms are developed to calculate the instantaneous values of positive- and negative-sequence currents when the information obtained during a time interval equal to two milliseconds or less is used. An algorithm is devised to calculate the sinusoidal components of the currents in the presence of aperiodic components in the samples. To eliminate the disoperation of digital relay protection and improve the accuracy of fault location, it is necessary to eliminate the influence of aperiodic components in the currents drawn to the measuring devices of relay protection. The algorithm completely eliminates the aperiodic component at any time constants and is implemented by four samples of currents. The four samples of currents separated by three sampling intervals are sufficient for calculating all the parameters of currents, including the positive and negative-sequence currents, in particular in the case when the magnetic cores of current transformers reach saturation. The proposed methods for setting the industrial frequency parameters to avoid the aperiodic components in short-circuit currents are based on the algorithms that process the data from three and four samples of currents.

1. Introduction

The adoption of microprocessors in the control of energy systems enables these functions to be performed during transient processes as well. On the one hand, digital control during transient processes expands the possibilities of automated devices, on the other hand, it requires settings to enable relay protection to cope with interferences arising during short circuits caused by the non-utility frequency conditions. This mainly involves the algorithms of protections performed by electromechanical relays [1 - 5]. The difference between the digital relay protection and analogue one is that the analogue protections receive the data on operating conditions continuously, whereas digital relay protections receive the data discretely at equal time intervals called sampling intervals.

Particular attention is paid to the protection that employs the data on changes in currents [6-8]. Let us consider the application of the principles of analogue protections to digital ones. It takes the overcurrent protection at least the time equal to the period of utility frequency to respond to the integral value of the current. The overcurrent protection performed by microprocessors that respond to the integral values of currents does not reduce the time of relay operation, i.e. there are no important advantages of microprocessors over analogue devices in this case.

The significant advantages of digital protection over analogue protection manifest themselves in the differential protection of generators, transformers, and buses. The response time of the applied analogue protection on the basis of relay with fast satiable transformers and with voltage-restrained overcurrent relay [9] is 40 - 50 ms.

The developers of digital relays [11–14] abandoned the principles laid down in the above relay. It was proposed to use information in the areas of sufficiently accurate transformation, which made it possible to develop algorithms with a fault fixing time of 2–3 ms. These algorithms however faced phenomenon of saturation of the magnetic core of current transformers [15].

The main studies in the field of digital protection are summarized in [15-18], where the authors consider the differences between analogue and digital protections, the advantages of the latter in hardware reliability, the expansion and improvement in the quality of protective functions, totally new protection control capabilities, etc. However, the issue of tuning the protections to cope with the interferences caused by the non-utility frequency parameters is not fully resolved. There is also still no solution to the problem of locating damage on power lines in the presence of interference from aperiodic and high-frequency components [19–22].

2. Research objective

In order to eliminate false operation of digital relay and to more accurately locate short circuits, it is required to eliminate the influence of aperiodic components in the currents supplied to the measuring devices of the relay. Below, we propose the algorithms that reduce aperiodic components dozens of times in the case of three samples of currents during short circuit, and completely eliminate them when the data from four samples are used [24,25].

3. Materials and methods

In the case of three samples from the instantaneous values of current separated by equal sampling intervals, we obtain an actual value of the current at the time of the average sample, and its derivative:

$$i_{ev}(t) = \frac{2 \cdot i(t) - i[t + \Delta t] - i[t - \Delta t]}{4 \cdot \sin^2\left(\frac{\omega \cdot \Delta t}{2}\right)}$$
$$i'_{ev}(t) = \frac{i[t + \Delta t] - i[t - \Delta t]}{2 \cdot \sin(\omega \cdot \Delta t)} \quad (1)$$

where t is the current time; Δt is the sampling interval separating the sample; ω is the angular frequency; i is the instantaneous value of the primary current to be measured; i_{ev} – calculated secondary current.

It is easy to see that $i_{ev}(t + \Delta t)$ exactly repeats the function$i(t) = \sin(\omega \cdot t + \Delta t)$, if one uses the equation $\sin(\omega \cdot t \pm \omega \cdot \Delta t) = \sin(\omega \cdot t) \cdot \cos(\omega \cdot \Delta t) \pm \cos(\omega \cdot t) \cdot \sin(\omega \cdot \Delta t)$.
i.e. equations (1) accurately calculate sinusoidal components and filter aperiodic components. Figure 1 demonstrates the aperiodic components of currents and their derivatives calculated with (1) for different time constants of the protected part of the electrical network, if the primary current at time t = 0 is specified by the aperiodic component equal to one.

The reduction in aperiodic components in the actual values of currents is due to the subtraction of two close instantaneous values of the current, spaced in different directions from the time t. Moreover, the greater is the time constant T of the electrical network part from the equivalent EMF to the short-circuit point, the smaller is the residual part of the aperiodic component after solving equation (1). In the algorithms considered in [17] however the aperiodic component is determined as a half-sum of two close values, i.e. it is practically not taken into account.

4. Research

At time t = 0, the secondary current at a time constant of the protected network area of 0.005 s is equal to 35% of the primary current, and the derivative is 53%. With time constants up to 0.01 s, the saturation of the magnetic cores, as a rule, does not occur. Therefore, one can begin sampling after a period of utility frequency. The residual aperiodic component in the secondary currents will not exceed 1%, which is sufficient for the proper operation of the relay protection. To locate the fault on the power line, the current sampling can also be started in a period from the time of the fault occurrence. For large time constants (more than 0.05 s.), the residual aperiodic component in the secondary currents does not exceed 0.5%.

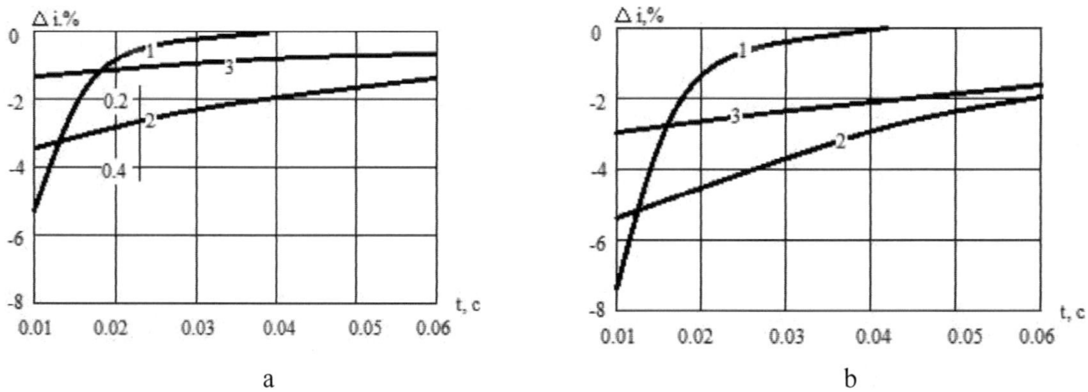

Figure 1. Secondary currents and their derivatives with aperiodic primary with different time constants of the protected area network (1 – time constant T = 0,005 sec., 2 – T = 0.05 sec., 3 – T = 0.1 sec. For the curve 1 in Figure 1a – vertical scale in percent, for curves 2 and 3 scale in tenths of percent).

In order to reduce the error caused by the presence of aperiodic components when determining the parameters of the fundamental frequency, it is proposed to once again determine the actual values of the parameters with equations (1) but use the calculated parameters as the primary ones. Figure 2 shows the obtained derivatives of the currents after the second calculation of the derivatives of the currents. After a period of network frequency, the residual aperiodic components do not exceed 0.4% for any time constants of the protected network section. At the time constants greater than 0.01 s, the error due to aperiodic components does not exceed hundredths of a percent. The minimum number of measured primary parameters with repeated use of equations (1) is five, separated by four sampling intervals. With a sampling frequency of 1600 Hz, the four sampling intervals will be 2.5 ms in total. During this time, as practice shows, the magnetic cores of current transformers do not reach saturation. The error in the calculation due to aperiodic components in the section of ideal transformation is 0.03%. The fault location algorithms on the 400-km-long line will locate the short circuit with the accuracy of 120 m. This allows recommending equations (1) to filter the calculated currents for all the tasks necessary to control the emergency conditions of power systems.

Consider how to determine the positive- and negative-sequence currents and voltages through the instantaneous values of the parameters. At time t, the phasor of the current is obtained as follows:

$$\dot{I}(t) = i_{\text{в}}(t) + i'_{\text{в}}(t) \cdot j \tag{2}$$

In literature (in Russia and other countries) it is proposed to determine the positive- and negative-sequence currents with a special phase A through the phasors:

$$\dot{I_{1B}}(t) = \frac{\dot{I_{AB}}(t) + \dot{I_{BB}}(t) \cdot e^{j\frac{2\pi}{3}} + \dot{I_{CB}}(t) \cdot e^{-j\frac{2\pi}{3}}}{3}$$
$$\dot{I_{2B}}(t) = \frac{\dot{I_{AB}}(t) + \dot{I_{BB}}(t) \cdot e^{-j\frac{2\pi}{3}} + \dot{I_{CB}}(t) \cdot e^{j\frac{2\pi}{3}}}{3} \quad (3)$$

The amplitude and phase of currents are defined as:

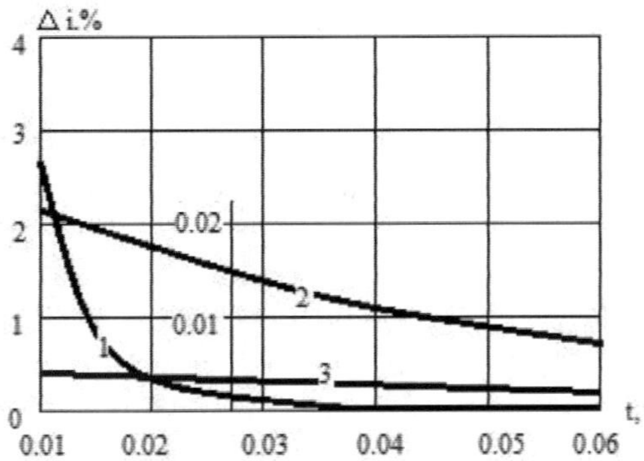

Figure 2. Calculated secondary currents at repeated application of the equations (1).

Since the tangent in the first and third quadrants is positive, and in the second and fourth quadrants it is negative, the angles should be specified according to Table 1.

Table 1. Current phase corrections

$i(t_k) > 0$ and $i'(t_k) > 0$	$\varphi = \varphi_в$
$i(t_k) > 0$ and $i'(t_k) < 0$	$\varphi = -\varphi_в$
$i(t_k) < 0$ and $i'(t_k) > 0$	$\varphi = 180 - \varphi_в$
$i(t_k) < 0$ and $i'(t_k) < 0$	$\varphi = 180 + \varphi_в$

The disadvantage of calculating the positive and negative-sequence currents through phasors is a large error if the short-circuit currents contain considerable aperiodic components.

We propose calculating the instantaneous values of positive- and negative-sequence currents using two formulas:

$$i_{1B} = \frac{i_{AB}(t) + i_{BB}\left(t + \frac{T_c}{3}\right) + i_{Cc}(t - \frac{T_c}{3})}{3}$$
$$i_{2B} = \frac{i_{AB}(t) + i_{CB}\left(t + \frac{T_c}{3}\right) + i_{BB}(t - \frac{T_c}{3})}{3} \quad (4)$$

or via phase currents without zero components:

$$i_{1B} = \frac{i_{AB}(t) + \frac{i_{BB}\left(t + \frac{T_c}{4}\right) - i_{CB}(t - \frac{T_c}{4})}{\sqrt{3}}}{2} \quad (5)$$

$$i_{2\text{B}} = \frac{i_{\text{AB}}(t) - \dfrac{i_{\text{BB}}\left(t + \frac{T_c}{4}\right) - i_{\text{CB}}\left(t + \frac{T_c}{4}\right)}{\sqrt{3}}}{2}$$

where T_c - the period of the frequency voltage.

The equations (4) and (5) suggest the calculation of phases B and C in time intervals that are distant from the time of phase A current measurement by a third of the period before and after the measurement when we use (4), and by a quarter of the period – when we use (5), i.e. the processing time of the currents in the first case is 10.33 ms, in the second - 5 ms. By this time, the magnetic cores of current transformers in some cases may reach saturation, and the calculation results will prove incorrect.

The idealized picture of the saturation of the magnetic core of current transformers is shown in Figure 3. The secondary current over time has two sections in half the period: a section of an ideal transformation and a section with an error where part of the secondary current is missing. If the secondary current is expanded into a Fourier series, the first harmonic has a frequency greater than the nominal one and is smaller in magnitude than the primary current reduced to the secondary one.

Consider the effect produced by the non-utility frequency components in currents on the accuracy of determining an emergency situation during short circuits. The short circuit current in time is determined by the equation:

$$i(t) = \frac{E_{eq}}{\sqrt{R_{eq}^2 + X_{eq}^2}} \cdot \left(\sin(\omega \cdot t + \psi - \varphi) - \sin(\psi - \varphi) \cdot e^{\frac{-t}{T_c}}\right) \tag{6}$$

where E_{eq} – equivalent EMF; R_{eq}, X_{eq} – equivalent active and reactive resistances from the equivalent EMF to the short-circuit point; ω – angular frequency equal to 2πf; f – frequency; T_c – network constant equal to L/R; ψ – voltage angle at which the short circuit occurred, expressed in radians; φ – impedance angle of a short-circuit branch.

Let us calculate the actual values of current at time instants Δt and 2Δt by the formula:

$$i_{ev}(\Delta t) = \frac{2 \cdot i(n \cdot \Delta t) - i\big((n-1) \cdot \Delta t\big) - i\big((n+1) \cdot \Delta t\big)}{4 \cdot \sin^2\left(\omega \cdot \frac{\Delta t}{2}\right)} \tag{7}$$

where n is the sequence number of the sample.

The network constant is defined as:

$$T_c = \frac{-\Delta t}{\ln\left(\frac{i_{ev}(2\Delta t) - i(3\Delta t)}{i_{ev}(\Delta t) - i(2\Delta t)}\right)} \tag{8}$$

The angle φ is determined by a constant network and the angular frequency:

$$\varphi = arctg(T_c \cdot \omega) \tag{9}$$

The aperiodic component is calculated as:

$$i_A(t) = k \cdot \left(-\sin(\psi - \varphi) \cdot e^{\frac{-t}{T_c}}\right) \tag{10}$$

where k is calculated as

$$\frac{E_{eq}}{\sqrt{R_{eq}^2 + X_{eq}^2}} = k = \frac{i(\Delta t)}{\sin(\omega \cdot t + \psi - \varphi) - \sin(\psi - \varphi) \cdot e^{\frac{-\Delta t}{T_c}}} \tag{11}$$

In the event that there is a threat of the current transformer magnetic core saturation, it is proposed to create calculated currents of utility frequency for several periods ahead and ignore the saturation of the magnetic cores of the current transformer.

The algorithm for this task is as follows [23-26]:

1. Make four samples of currents and calculate the actual values of currents using equation (7) at time point's Δt and 2Δt.
2. Determine time constant of the network from equivalent EMF to the fault point using equation (8).
3. Determine the amplitudes and phases of the currents according to (11). The amplitudes of the currents can also be determined by the ratios of the calculated values of the sinusoidal currents to the sines of the angles of the calculated currents.
4. Determine the current values of the sinusoidal components and their derivatives at any time:

The sinusoidal component of the current is equal to:

$$i_{sin}(t) = i(t) - k \cdot i_A(t) \qquad (12)$$

$$i'_{sin}(t) = k \cdot \cos(\omega \cdot t + \psi - \varphi) \qquad (13)$$

where k – the time of calculating the actual value of currents and their derivatives.

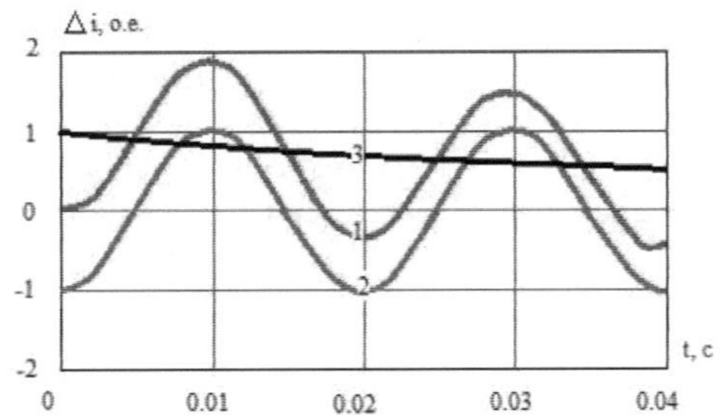

Figure 3. Secondary calculated currents: 1– equal to the sum of the sinusoidal component and aperiodic, 2 – sinusoidal current component, 3-aperiodic component of the current ($T_{network}$ – 0.05 sec.), 4 – the current calculated according to (9).

The equations (13) and (14) can be used to determine the actual values of sinusoidal components of the currents and their derivatives at any point in time, including the points in time when the current transformer magnetic cores reach saturation. In addition, these equations allow determining the parameters of secondary currents in the "future" time. This is also relevant for the development of emergency control actions.

Thus, an algorithm has been proposed to identify the sinusoidal components of the currents for four samples separated by three sampling intervals. These samples should be stored so as not to overload computer with extra calculations, and after the line is cut, perform calculations. The complete elimination of aperiodic components from the short-circuit current equal to the sum of aperiodic and periodic components is proposed for the first time.

Based on equations (8–12), the following algorithm for calculating the fundamental frequency component of current can be also recommended for the protections using the information on currents:

- Make four samples of currents;
- Calculate actual values of currents using equation (8) at time points t = Δt and t = 2Δt;
- Determine the time constant of the electrical network from the short circuit to the generators using equation (9);
- Determine the impedance angle from the equivalent EMF to the short-circuit point;
- Determine aperiodic component of the current using equation (11);

- Determine the component of the current of fundamental frequency (12);
- Determine the parameters of currents using equations (4-8) for the digital current protection. The approximate execution time of the algorithm for the currents of three phases will not exceed the time equal to half the period of the utility frequency.

In order to eliminate the influence of a frequency change, namely, a constant sampling interval, on the error in calculating the actual value of network operating parameters, the sampling interval should be calculated as part of the voltage zero crossing intervals on the resistance and capacitance of the RC circuit [21]. These crossings always correspond to a quarter of the frequency period. If the network frequency is not equal to the nominal one and the sampling rate is equal to 1600 Hz, the sampling interval will be equal to 1/32 of the voltage zero crossing interval of the RC circuit.

5. Conclusions

The algorithm is proposed to calculate the actual values of secondary currents and their derivatives that dramatically reduce the magnitude of aperiodic components. Thus, only periodic components are brought to the relay protection. The algorithm is based on a sample for calculating three values separated by identical sampling intervals, and provides an aperiodic component with a residual aperiodic component at the time of the average measurement. The residual aperiodic components do not exceed units of percent of those present in the measured secondary currents. Moreover, the greater is the time constant of the electrical network section from the short circuit to generators, the lower is the residual in the calculated secondary currents.

The algorithm is proposed to completely eliminate the aperiodic components in secondary currents using only four samples. The application of the algorithm eliminates the influence of the saturation of the magnetic core of current transformers on the operation of the relay protection. The algorithm is proposed for the first time.

The algorithm is proposed to determine the actual values of currents and their derivatives at "future" points in time, including the saturation of the magnetic cores of current transformers. This allows the use of the information received during 0.02 s. from the time of fault occurrence.

Acknowledgments

The study was carried out with the financial support of the Russian Foundation for Basic Research and the Subject of the Russian Federation - the Republic of Sakha (Yakutia) № 18-48-140 010.

References

[1] Bayliss C R and Hardy B J 2012 Chapter 10: Relay Protection Transmission and Distribution Electrical Engineering (Fourth Edition) pp 287-359

[2] IEC 2006 Electromechanical elementary relays – Part 7: *Test and measurements procedures, IEC 61810-7*

[3] Gurevich V 2006 *Electric Relays: Principles and Applications* (first ed.), (CRC Press, Boca Raton)

[4] Akke M and Thorp J 1998 Some improvements in the three-phase differential equation algorithm for fast transmission line protection *IEEE Transl. Power Delivery* **13(1)** 66–72

[5] Gergič B and Hercog D 2019 Design and implementation of a measurement system for high-speed testing of electromechanical relays Measurement **135** 112-121

[6] Saleh K A, Zeineldin H, Al-Hinai A and El-Saadany E F 2015 Optimal coordination of directional overcurrent relays using a new time-current-voltage characteristic *IEEE Transl. Power Delivery* **30 (2)** 537-544

[7] Suslov K, Gerasimov D and Solodusha S 2015 Smart grid: Algorithms for control of active-adaptive network components, *IEEE Eindhoven PowerTech* (Netherlands)

[8] Enríquez A C and Martínez E V 2006 Enhanced time overcurrent coordination *Electric Power Systems Research* **76** 457-465

[9] Drozdov A D 1965 *Electrical curciuts with ferromagnetic cores in relay protection* (Energiya)

[10] Gers J and Holmes E 1998 Protection of Electricity Distribution Networks *The Institution of Electrical Engineers*
[11] Haeg H and Forster M 1965 Elektronischer Sammelschienenschutz *Brown Boveri Mitteilungen* **53** 326-339
[12] Evans E J and Wells G 1970 Use of Sampling to Detect Transient Saturation in Protective Current Transformers IEEE Transl. on Instrumentation and Measurement **19(3)** 144-147
[13] Kuzhekov S L and Nudel'man G S 2009 Maintaining the right function of microproseccor devices of differential protection in case of current transformers saturation *Izvestiya Vuzov Elektromekhanika* **4** 12-19
[14] Gangadharan P K, Sidhu T S and Finlayson G J 2007 Current transformer dimensioning for numerical protection relays *IEEE Transl. on Power Delivery* **22 (1)** 108–115
[15] Zinov'ev D V 2009 *Development of the prosecc information analysis theory in electrical system and its application to relay protection Cheboksary*
[16] Hosemann G, Steigerwald H 1993 Modal saturation detector for digital differential protection *IEEE Transl. on Power Delivery* **8 (3)** 933–940
[17] Macieira G L and Coelho A L 2017 Evaluation of numerical time overcurrent relay performance for current transformer saturation compensation methods *Electric Power Systems Research* **149** 55-64
[18] Shneerson E V 2007 *Digital relay protection* (Energoatomizdat)
[19] Kiliçkiran H C, Şengör İ, Akdemir H, Kekezoğlu B and Paterakis N G 2018 Power system protection with digital overcurrent relays: A review of non-standard characteristics *Electric Power Systems Research* **164** 89-102
[20] Arzhannikov E A, Lukoyanov V Y and Misrikhaniv M Sh 2003 Identification of short circuit place on high voltage lines (Energoatomizdat)
[21] Kulikov A L and Obalin M D 2015 Software development to support of solution in damage elimination on transmission lines *Izvestiya Vuzov "Elektromekhanika"* **2** 70-75
[22] Kulikov A L 2017 Algorhitm of the damage identification of transmission lines with branches *Vestnik NGIEI* **9(76)** 29-38
[23] Buryanina N, Korolyuk Y, Koryakina M, Lesnykh E, Suslov K and Shamarova N 2019 Using a new algorithm of current protection for substations *Proceedings of the 10th International Scientific Symposium on Electrical Power Engineering, ELEKTROENERGETIKA*
[24] Buryanina N S, Korolyuk Y F and Lesnykh E V 2016 RU, Patent No. 2 625 172 *The way to count voltage and current instantaneous values*
[25] Kondrat'ev V V, Govorkov A S, Lavrent'eva M V, Sysoev I A and Karlina A I 2016 Description of the heat exchanger unit construction, created in IRNITU *International Journal of Applied Engineering Research* **11 (19)** 9979-9983
[26] Sysoev I A, Kondrat'ev, V V, Zimina T I and Karlina A I 2018 Simulation of the Energy States of Electrolyzers with Roasted Anodes at Elevated Currents *Metallurgist* **61(11-12)** 943-949

IOP Conf. Series: Materials Science and Engineering **860** (2020) 012036 doi:10.1088/1757-899X/860/1/012036

Wear evaluation methods for friction units of aircraft GTE oil systems: the on-line methods

A Y Hodunaev

Irkutsk National Research Technical University, 83, Lermontov Str., Irkutsk, 664074, Russia

E-mail: alex.hodunaev@gmail.com

Abstract. On-line sensors based on optical, ultrasonic, electostatic and inductive principles are evaluated for use in determination of wear in oil systems of aircraft GTEs. This application is limited due to contraints of the high oil flow rate (40 – 60 L/min) and the flow diameter of 20-25 mm. Under such conditions, most sensors are unable to function efficiently and give incorrect readings. As such, creating on-line wear evaluation systems remains hardware-limited. Decrease of detection limit of non-ferrious particles for inductive type sensors is a possible way to solve the issue.

1. Introduction

Quantifying particles that are suspended in working oil can be useful for determining excessive wear of friction pairs that generated them, prevent further damage and thus reduce maintenance costs. However, in complex systems such as gas-turbine engine (GTE) oil systems there are many friction pairs, each generating normal wear particles, therefore simply measuring the particle content is insufficient to determine whether the wear of a system or a particular friction pair is excessive. As such, the source of each particle must also be localized.

In lab-based approach, this can be done by analysis of the elemental composition of the wear particles. If a material can be identified from analysis, it can be traced to a number of friction pairs that are made of the same material. Thus, most successful lab-based approaches to wear evaluation of complex tribo-systems consist of sensitive spectral equipment.

On-line wear sensors are a subject to rising interest of aircraft operators through last couple decades. Compared to lab-based methods, sensors require less maintenance costs due to no sampling, analysis or delays.

The problem is element-sensing equipment is currently bulky and can't be used as an on-line sensor [1]. However, similar effect can be potentially achieved if multiple particle sensors are installed throughout the system (figure 1).

In the figure, cogs represent friction pairs that come in contact with a particular flow area and cameras represent sensors. Cog 1 generates the particle, which is detected by sensor 1. Cog 2 generates the second particle. Both particles are captured by the filter. If sensor 2 has readings from sensor 1, it would be able to differentiate the number of particles that came from cog 2 and thus excessive wear can be localized to within those areas.

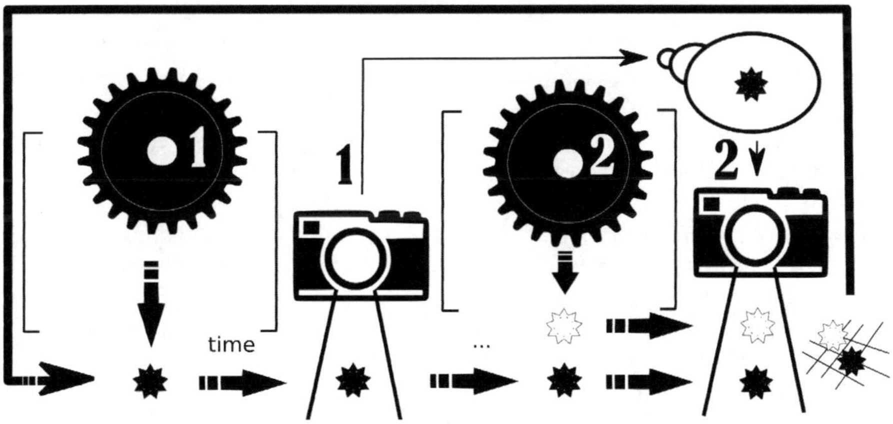

Figure 1. Oil system layout with multiple friction pairs and sensors

2. Sensor types

On-line sensors differentiate mainly by underlying principle by which a particle is detected.

1. In optical [1-3] sensors, oil-particle flow is illuminated by a light source and registered by a photosensitive element or image sensor (figure 2). Particles are detected by either light intensity decrease or by image analysis. As such, size and particle quantity can be evaluated.

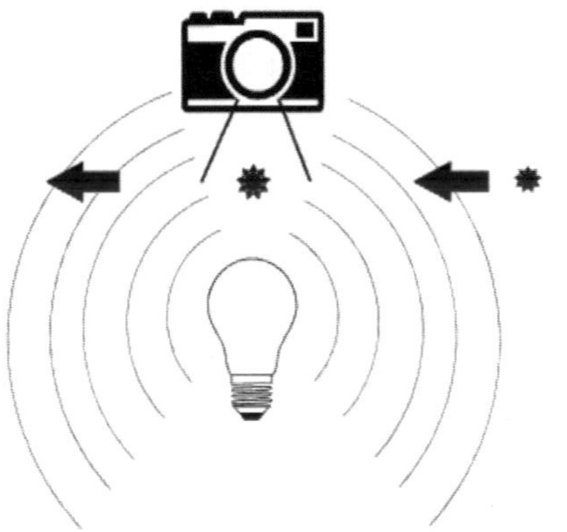

Figure 2. Optical sensor layout

The sensor readings are a subject to oil cleanliness, flow diameter and rate, excessive particle counts and presence of extraneous elements such as air or water bubbles. In small flows (1.2 x 1.6 mm, flow rate up to 1 liter per minute, L/min) particles 5 μm or above can be detected [1]. Oil flows in GTEs, however, can reach diameters of 20-25 mm and flow rates of 40 – 60 L/min. In these conditions, most particles are out of focus and cannot be reliably detected. To counter this, the common technique is to introduce microflow – a separate flow of lesser diameter with controllable oil flow rate. In this case, however, only a slight fraction of total particle count is detected which makes multi-sensor correction, and therefore, wear localization – impossible. This is because wear particle, generated by oil system controlled by one sensor can be ignored by it and detected by another sensor and vice-versa.

These circumstances limit the usefulness of optical sensor for GTE oil system wear control [1].

2. The general layout of ultrasonic sensors [4-6] is similar to the optical sensor layout. The difference is that the acoustic source and the detector are used instead of light-based counterparts.

The size and particle quantity can be measured [5]. In hydraulic or oil systems, particles from 30 μm to 1000 μm can be detected [7]. Oil, air bubbles and metallic particles can be detected separately [8]. Work [4] shows that metallic particles with size of 75 μm or larger can be detected in flow with a rate of 400 μL/min and a diameter of 6.5 mm.

The ultrasonic equipment can be a useful tool for detecting excessive engine vibration. As an online sensor though, the application is limited due to dependency of the sensor reading on oil viscosity, speed, low throughput and mechanical vibration interference [1].

3. Wear particles accumulate charge through friction. This phenomenon is a base for the electrostatic sensor that represents a conductor put in controllable medium (oil).

Initially, sensors of this type were developed for control of the engine gas flow [9]. Then, potential application of electrostatic sensors in oil systems was shown using gear and bearing test machines [10], including roller bearing of the F100 jet engine [11]. One outstanding feature of electrostatic sensors is the ability to detect non-metallic particles, which is important for wear control of plastic and ceramic bearings. The material of the particle itself cannot be determined, though.

The sensor allows one to quantify particles of 20-50 μm in flows of 7 to 9 L/min using test machines [12]. An electrostatic sensor was able to detect roller bearing failure 18 hours before critical failure [11] in overload (200% nominal load) conditions and flow rate of 4 L/min.

In complex tribo-systems, such as the GTE oil system, however, the device did not receive industrial use. This is due to the issue with identifying charge sources and mechanisms in such systems [13].

4. The inductive sensors received most use in aircraft industry [14-16]. The ferrous and non-ferrous particles can be detected by fluctuation in the electromagnetic field of a coil when the particle passes through it.

The GasTOPS company has developed the MetalSCAN sensor which employs three coils, placed sequentially through the oil flow (figure 3) [14].

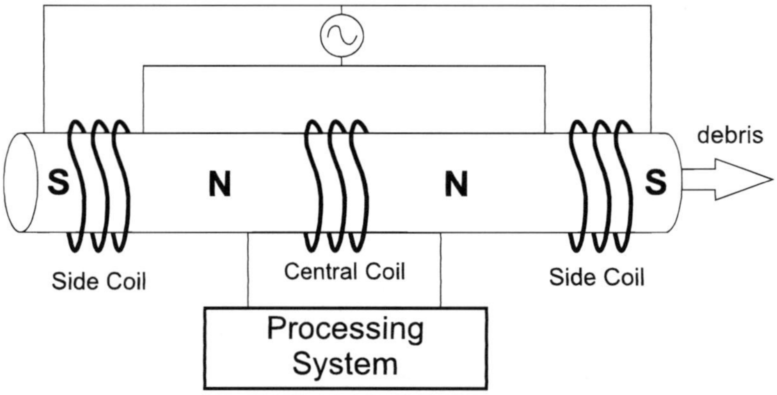

Figure 3. MetalSCAN sensor layout

Side coils are connected to AC source and their electromagnic field is mutually compensated in the centre. When particle passes through, central coil detects field fluctuation of each coil. Signals from ferrous and non-ferrous particles can be differentiated by signal phase [14].

The sensor can determine size, mass of ferromagnetic particle larger than 125 μm, and non-ferrious of 1000 μm in a flow of 0.5 inch (12.7 mm) in diameter and rate of 3.7 L/min (1 Gallon/min). For a flow of 25 mm, minimal diametr of ferromagnetic spherical particle is 275 μm. For a particle of irregular shape, detection limit is 180 μm (of equivalent spherical particle) [16].

The QDM® sensor counter consists of a single coil and a magnet for particle capture, which ensures the ability to extract particles and execute further analysis [17, 18]. The sensor can count particles in sizes ranging from 400 to 600 μm [19].

IOP Conf. Series: Materials Science and Engineering **860** (2020) 012036 doi:10.1088/1757-899X/860/1/012036

Most (>90%) metallic particles in the oil system are in a size range of 10 to 100 μm. But even engines that exhibit normal wear can generate particles 300 μm and above, therefore excessive wear potentially can be detected by QDM and MetalSCAN. High detection limit for ferrous particles makes the sensor blind to wear modes which do not generate particles of big sizes (fatigue). But in Russian aircraft, fatigue wear is rare compared to other wear modes.

Detection limit of non-ferrous particles of 1000 μm makes that feature almost entirely irrelevant, since particles of this size almost exclusively are a product of catastrophic failure.

The error in determining mass of specific particle for this type of sensors can reach 370%; size – up to 170%. This is because a signal from a particle depends on particle shape, trajectory and orientation [20]. Size detection error makes it harder to determine excessive wear by analysis of change in size distribution function.

The detection limit is proportional to flow size and speed. In extremely low flows (3 μL/min, 1 mm), 5 μm ferromagnetic particles can be detected [21]. To increase throughput, multiple sensing channels can be used. As such, using 3x3 sensing array can increase sensor throughput to 460 μL/min [22].

Using permanent magnets instead of coils can decrease detection limit due to higher density of the magnetic field. Device that uses permanent magnets [23] allows one to register ferromagnetic particles 83 μm or larger in flow of 20 L/min with diameter of 12 mm.

For MetalSCAN sensor, changing flow diameter from 7.6 mm to 26.9 mm (3.5 times), detection limit increases from 100 μm to 275 μm (2.75 times) [16]. Assuming the same would be true for permanent magnet sensor, the detection limit for 25 mm flow diameter can be evaluated as 170 to 250 μm i.e., 2-3 times bigger than detection limit for 12 mm flow.

Capabilities of QDM and MetalSCAN were tested in work [19]. Both were simultaneously installed on oil system test machine. Next, ferromagnetic particles of irregular shapes were introduced to the system. The size ranges of the particles were 850-1180 μm, 425-600 μm, 150-500 μm, 10 mg each, which amounted to tens of particles. The tests were repeated 10 times. Flow rate was 60 L/min, diameter – ½ inches (12.7 mm). Both sensors (QDM and MetalSCAN) were detecting 60-70% of particles in their nominal detection limits (250-1000 μm for MetalSCAN, 400-600 μm for QDM). This holds true even for larger particles (600-1000 μm).

This is a major issue, since detecting only a fraction of particles, as was noted above, breaks multi-sensor reading correction and wear localization. Also, accumulated mass will be measured with significant error.

Preliminary tests at UEC "Aviadvigatel" were performed to assess viability of MetalSCAN sensor as wear control device for new generation GTEs [24]. Tests were performed on engine test machines, equipped both with multiple MetalSCAN sensors and standard wear control equipment, such as filter-signalers, flak-signalers and particle detectors. At best, MetalSCAN sensors allowed one to detect impending failure 10 minutes before standard equipment would [25]. As such, currently there is no significant advantages to prefer MetalSCAN over widely available, much older equipment.

3. Conclusion

Despite more than thirty years of development of integrated on-line sensors, the application of sensors for wear evaluation of friction pairs in the aircraft gas turbine engine oil systems remains severely limited. The limitation is attributed to harsh environment that is oil system of GTE, which exhibits big flow diameters (25 mm) and flow-rates (40-60 L/min). Most sensors are unable to function efficiently under these constraints.

For inductive (or other type) sensors one major step to the solution of the problem could be increasing the fraction of detected particles up to 95-99%, making mutual correction of readings of multiple sensors less prone to error.

Another step would be to determine material of the particle, at least roughly. This would allow one to trace source of the particles in a way, used by lab-based methods and can be achieved by decreasing detection limit of non-ferrous 300 μm particles for inductive sensors.

Until at least those problems are solved, lab-based methods remain as the most reliable means for wear evaluation of GTE oil system friction pairs.

4. Acknowledgements
This article was prepared using the results of work carried out in the framework of the federal target program “Research and development on priority directions of scientific-technological complex of Russia for 2014 - 2020 years”. Topic: “Development and creation of software and hardware microwave plasma system for monitoring, control and safe operation of oil system of engines for ground and air purposes”. The agreement with the Ministry of Education for grant No. 14.577.21.0289, 075-15-2019-1369. Unique identifier PNIER RFMEFI57718X0289.

References
[1] Wei H, Wenjian C, Shaoping W and Tomovic M 2018 Mechanical wear debris feature, detection, and diagnosis: A review. *Chinese Journal of Aeronautics* **31** 5 867-882
[2] Wu T, Mao J, Wang J, Wu J and Xie Y 2009 A new on-line visual ferrograph *Tribology transactions* **52 (31)** 623
[3] Wu T, Wang J, Peng Y and Zhang Y 2012 Description of wear debris from on-line ferrograph images by their statistical color *Tribology Transactions* **55 (14)** 606
[4] Du L и Zhe J 2012 An integrated ultrasonic–inductive pulse sensor for wear debris detection *Smart Materials and Structures* **22 (2)** 025003
[5] Xu C, Zhang P, Wang H, Li Y and Lv C 2015 Ultrasonic echo waveshape features extraction based on QPSO-matching pursuit for online wear debris discrimination *Mechanical Systems and Signal Processing* **60** 301-315
[6] Zhe J, Choy F, Murali S, Sarangi M and Wilfong R *2007* Oil debris detection using capacitance and ultrasonic measurements *In ASME/STLE International Joint Tribology Conference* pp 113-115
[7] Edmonds J, Resner MS and Shkarlet K 2000 Detection of precursor wear debris in lubrication systems. *IEEE Aerospace Conference. Proceedings (Cat. No. 00TH8484)* **6** 73-77
[8] Nemarich CP, Whitesel HK and Sarkady A 1998 On-line wear particle monitoring based on ultrasonic detection and discrimination (Tech Report) *DAVID TAYLOR RESEARCH CENTER BETHESDA MD PROPULSION AND AUXILIARY SYSTEMS DEPT*
[9] Powrie H and McNicholas K 1997 Gas path condition monitoring during accelerated mission testing of a demonstrator engine *ASME/SAE/ASEE Joint Propulsion Conf.*
[10] Powrie H, Fisher C, Tasbaz O and Wood R 1999 Performance of an electrostatic oil monitoring system during an FZG gear scuffing test *International Conference on Condition Monitoring* pp 145-155
[11] Harvey T, Wood R and Powrie H 2007 Electrostatic wear monitoring of rolling element bearings *Wear* **263** 1492-1501
[12] Wood R, Browne M and Thew M 1997 Electrostatic charging precursor to scuffing in lubricated contacts
[13] Liu R, Zuo H, Sun J and Wang L 2017 Simulation of Electrostatic Oil Line Sensing and validation using experimental results *Tribology International* **105** 15-26
[14] Miller JL and Kitaljevich D 2000 In-line oil debris monitor for aircraft engine condition assessment *IEEE Aerospace Conference. Proceedings (Cat. No. 00TH8484)* **6** 49-56
[15] Feng S, Yang L, Qiu G, Luo J, Li R and Mao J 2019 An inductive debris sensor based on a high-gradient magnetic field *IEEE Sensors Journal* **19** 8 2879-2886
[16] 2016 MetalSCAN MS4000 oil debris sensor specification sheet, available at: http:// www.gastops.com/wp-content/uploads/2016/03/C008850_000-MS4000-Specification-Sheet.pdf
[17] Centers PW and Price FD 1988 Real time simultaneous in-line wear and lubricant condition monitoring *Wear* **123** 303-312

[18] 2009 *Electronic Oil Debris Monitoring System*, available at: https://www.herberaircraft.com/pdf/library/Glenolden/Eaton_Electronic_Oil_Debris_Monitoring.pdf
[19] Harkemanne E, Berten O and Hendrick P 2018 Analysis and Testing of Debris Monitoring Sensors for Aircraft Lubrication Systems *Multidisciplinary Digital Publishing Institute Proceedings* **2** 461
[20] Zaricky S 1995 – Comparative analysis of sensor characteristics [In Russian] *First internation conference "Energodiagonsics"* **3** 293-306
[21] Du L and Zhe J 2011 A high throughput inductive pulse sensor for online oil debris monitoring *Tribology International* **44** 175-179
[22] Du L, Zhu X, Han Y, Zhao L and Zhe J 2013 Improving sensitivity of an inductive pulse sensor for detection of metallic wear debris in lubricants using parallel LC resonance method. *Measurement Science and Technology* **24** 075106
[23] Hong W, Wang S, Tomovic MM, Liu H and Wang X 2015 A new debris sensor based on dual excitation sources for online debris monitoring *Measurement Science and Technology* **26** 095101
[24] Halliulin V and Muhutdinov F 2016 MetalScan in action [In Russian], available at: http://www.pmz.ru/upload/pages/7915/48_49__MetalSCAN.pdf
[25] UEC Aviadvigatel 2017 MetalSCAN efficiency analysis (Tech Report)

IOP Conf. Series: Materials Science and Engineering **860** (2020) 012037 doi:10.1088/1757-899X/860/1/012037

Wear evaluation methods for friction units of aircraft GTE oil systems: the lab-based methods

A Y Hodunaev

Irkutsk National Research Technical University, 83 Lermontov Str., Irkutsk, 664074, Russia

E-mail: alex.hodunaev@gmail.com

Abstract. The problems before reliable wear evalution using traditional lab based methods, such as ferrography, x-ray and spectrography, are discussed. One of those problems is that only a subset of wear debris features is measured by each method, which does not allow one to determine wear type. Filter sample analysis technology must be developed, because oil sample is devoid of diagnostic information due to fine filtration. The solution to aforementioned problems is proposed in the form of a newly developing microwave plasma method.

1. Introduction

The common consideration is that the wear process can be desribed by rate, type, location and severity. These features can be determined through measurement of concentration, size, morphology and elemental composition of the wear debris [1-6]. Relationship between wear and debris features [1] is shown in table 1, where "x" means related.

Table 1. Relationship between wear and debris features

Debris feature	**Wear feature**			
	Severity	*Rate*	*Type*	*Location*
Concentration	×	×		
Size	×	×	×	
Morphology	×		×	×
Composition				×

The most important task in wear evaluation is determining the wear type. Without clear understanding of the wear type, relationship between debris features "concentration" and "size" and wear "severity" cannot be established. For example, in case of non-periodic pulse generation of debris, the oil sample won't show significant increase of the debris "amount". Debris "size" too won't give much information.

2. Results

Metallic debris that appear in the engine oil system can be described by multiple wear types [8]. It's considered that using ferrograph measurements it's possible to identify wear type using sizes and morphology of the debris [9].

Four most common wear types are normal (rubbing) wear, cutting wear, fatigue wear and sliding wear. The issue is that debris morphology often intersect between different wear types [1, 10]. As such, both rubbing and fatigue wear types may exhibit debris that take shape of flat thin scales [10]. Intersections like this complicate correct identification of wear type and prognosis of its development.

Another example is spherical debris of 2-5 micrometers in size that appear in ferrographic measurements and are considered to be feature, commonly associated with scuffing of roller bearings [11] and occur from welding, cavitation or likewise processes [5]. During prolonged tests of D30-KU154 gas-turbine engine, spherical debris of 2-5 in diameter were found [12]. Yet, there were no issues with bearings. Engine teardown showed no scuffing, and the reassembled engine was operated without bearing issues at least 7972 hours.

Therefore, debris morphology is not reliably tied to wear type and other debris features must be considered.

Regarding particle size, the results are ambigious too. There is information that debris size is related to wear type [5, 12]. Normal wear generates scale-shaped debris less than 15 micrometers in size, and bigger debris are related to excessive wear [5, 10, 12]. On other hand, debris of larger sizes generated on test machines do not relate to excessive wear [13], see figure 1.

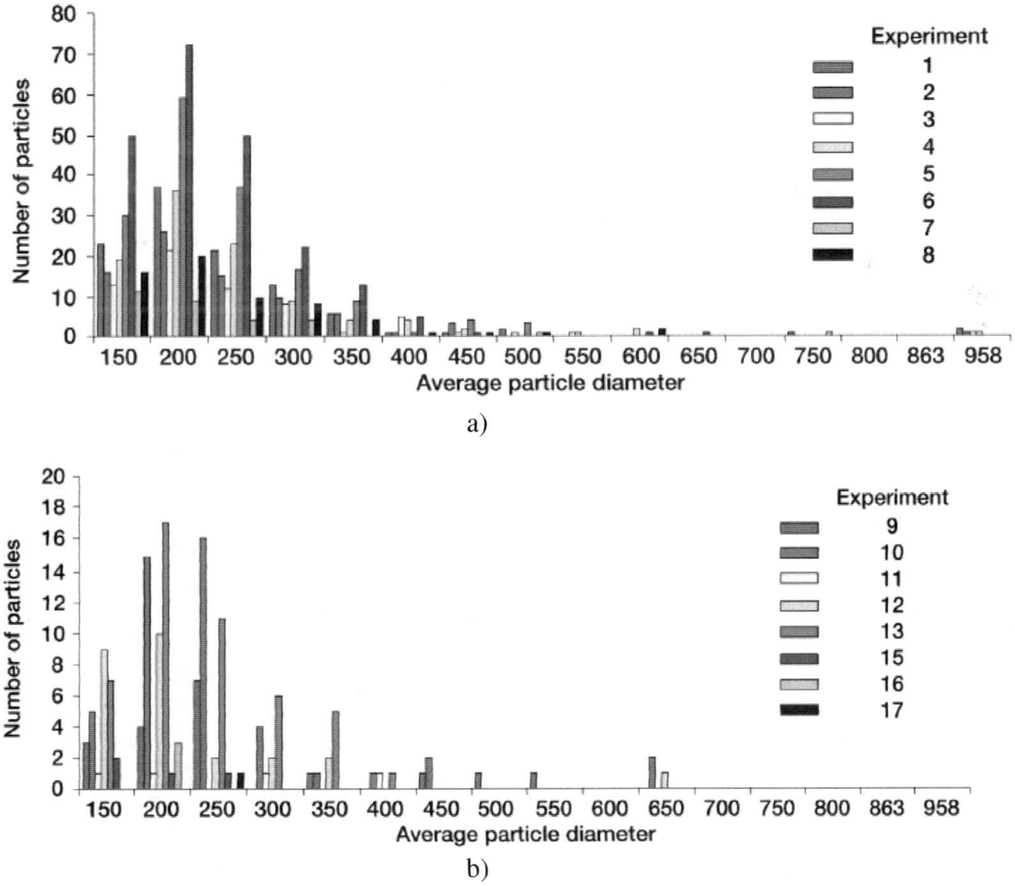

Figure 1. Spur rig experiments a) debris sizes after pitting damage was observed b) debris sizes without observable damage

The figure from work [13] shows that both with and without observable pitting damage, maximum debris size is >450 micrometers. The main difference between a) and b) figures is that amount of the debris increases in every class when the damage becomes visually detectable.

Further complications arise in the system of a real engine. When multiple wear processes coexist, the debris mix together. That can lead to "drowning" of excessive wear process within multiple normal wear processes. As such, work [13] states that "oil debris alone cannot discriminate between bearing and gear fatigue damage when both share a common lubrication system".

Figure 2 shows distribution of the average debris diameter with relation to presence of excessive wear.

The data is accumulated through microwave plasma measurements of the filter sample of different D30-KP gas-turbine engines [14]. The averaging is performed within each sample. The top curve represents distribution for engines exhibiting normal wear, the bottom curve – engines exhibiting excessive wear with confirmed damage.

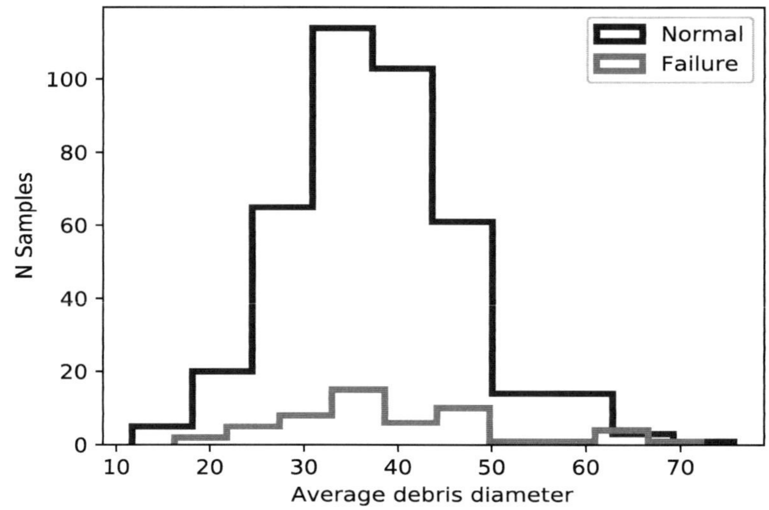

Figure 2. Distribution of average debris diameter within gas turbine engine samples

The data shows that the wear process can develop among generation of debris in a wide size range – from tens to hundreds of micrometers. Therefore, the feature "debris size" should be used carefully through wear evaluation. To distinguish wear sources for debris mixed within one sample, the information on elemental composition of the debris is required.

Distinct sets of debris characteristics from table 1 can be measured using ferrography; x-ray and atomic emission spectral methods. Diagonsis reliability of D30-KP/KU/KU154 gas turbine engines using x-ray spectral measurements of oil samples does not exceed 5-7%, according to data of PJSC "UEC-Saturn" [15]; the damage of PS-90A engine high pressure turbine roller bearing was not ever identified using atomic-emission and ferrographic ground measurements [16]. Similar picture was observed for diagonstic of J52 4 1/2 bearing – atomic-emission measurements weren't able to identify the damage of the bearing even at 10-day sampling period [17, 18].

Therefore, diagnostic miss is highly probable while using only a subset of debris features from table 1. Exception to this rule is when wear process is a frequently repeating one with continous increase in concentration of metallic debris in the oil system. In this case, spectral methods can give fairly complete understanding about wear state of friction pairs in the system. Even more so - predict the remaining lifetime. As such, the damage of the high turbine bearing and middle compressor bearing of the AL-7F-1 engine was identified and monitored up to pre-failure state using atomic-emission method with 92% reliability [19]. More than 280 engines were removed from operation prior to catastrophic failure.

The discussion above was relevant wear evaluation of friction pairs of gas-turbine engine oil system using spectral and ferrographic analysis of the oil sample. Another problem is raised by operation of 15-micrometer fine filters in the engine oil system. The fine filtration increases lifetime of oiled engine units by order of magnitude [20], but leaves the oil sample with no diagnostic information. More than 95% of the wear debris are deposited on the filter [21].

The analysis of the oil sample allowed one to use absolute-value features for diagnostics, such as concenctration or debris count [22]. Yet, for the filter sample, these and other absolute-value features generally depend on filter operation time, engine operation environment, filter washing technique and other non-accountable conditions.

In work [23] the problem was solved as follows. An automatic filter wash system was developed. A washed filter sample was put through MetalSCAN debris counter and deposited on the 20-micrometer fine membrane filter. By-percent elemental composition of the membrane debris was determined using x-ray spectral analysis. This machine (FilterCHECK 300) and the developed technology allowed one to identify the early signs of the wear degradation of 4 ½ bearing in J52 engine with 95% reliability [18, 24, 25]. In words of on of the authors, Gary Humphrey, the developed technology allowed one to "save 58 units of EA-68 Prowler aircraft, which equates to $765 million and 116 air crew" [26].

However, the information about the application of this technology for other units of J52 engine or other types of engine isn't there. Most probably, the developed equipment has narrow specialization.

The same was true for AL-7F-1 bearing wear evaluation techonolgy. When transferred to D30-KP/KU/KU-154, the reliability of the early wear identification dropped from more than 90% to 5-7%. Same could be said for IVU-1M, ultrasonic equipment which allowed to detect the damage D30KP/KU/KU-154 intershaft bearing, undetectable using traditional spectral methods. The damage of other bearings (high pressure turbine, compressor) couldn'be identified using IVU-1M.

Therefore, repeatable wear processes is the simplest way to perform reliably diagnosis. In this case, most probably, a single wear type dominates, which allows one to find effective diagnostic feature and monitor the damage up to a pre-failure state.

There is one more problem that is barely discussed in engine tribodiagnostics literature. Some elements in the oil or filter sample can be present simultaneously as both wear particles and oil additives. For example, molybdenum is a common anti-scuffing additive that is also present in M50 bearing steel; silicium is used as anti-foam additive and can appear into the oil system from environmental dirt; latter is also true for Al_2O_3 and other chemical compounds. An aforementioned problem mixes wear and non-wear sources for some of the debris, confusing the diagnost.

3. Conclusion

The diagnosts possess in their toolbox several laboratory methods that allow measurements of sets of specific features, such as wear debris concentration, size, morphology and elemental composition. Long history of empoyment of these methods shows not enough reliability in wear evaluation. There is several problems that need to be solved for reliable:

1. Wear type identification, which decides further diagnostic actions.
2. Measurements of debris amount and elemental composition among their whole size range.
3. Separate acquisition of information about chemical elements that are present both in shape of wear debris and oil additives.
4. Filter sample diagnostic features must be found for engines that are equipped with fine oil filters that capture most of the wear debris as the diagnostic information.

These problems are currently being solved with newly developed equipment.

One example of such equipment is the new microwave plasma atomic-emission method [14]. The method employs ten to thirty minutes long analysis of the filter/oil sample. The method allows one to determine size and elemental composition of each registered debris that consists of Al, Fe, Cr, Ni, Mg, V, Ti, Mo and the elemental dictionary is being expanded. Partial solution of the aforementioned problems allowed the method to reach 90% diagnosis reliability for D30-KP/KU/KU-154 engines.

Acknowledgements
This article was prepared using the results of work carried out in the framework of the federal target program "Research and development on priority directions of scientific-technological complex of Russia for 2014 - 2020 years". Topic: "Development and creation of software and hardware microwave plasma system for monitoring, control and safe operation of oil system of engines for ground and air purposes". The agreement with the Ministry of Education for grant No. 14.577.21.0289, 075-15-2019-1369. Unique identifier PNIER RFMEFI57718X0289.

References
[1] Wei H, Wenjian C, Shaoping W and Tomovic M 2018 Mechanical wear debris feature, detection, and diagnosis: A review. *Chinese Journal of Aeronautics* **31** 5 867-882
[2] Seifert W and Westcott V 1972 A method for the study of wear particles in lubricating oil *Wear* p 27—42
[3] Yarrow A and Gadd P 1984 — The role of ferrography in the monitoring of helicopter assemblies *Pineridge Press Ltd* 503—524
[4] Dowson D, Dalmaz G, Childs T, Taylor C and Godet M 1992 Wear particles: from the cradle to the grave *Elsevier* **21**
[5] Anderson D P 1982 *Wear particle atlas*
[6] Roylance B and Raadnui S 1994 The morphological attributes of wear particles—their role in identifying wear mechanisms *Wear* **175** 115—121
[7] Eisentraut K J, Newman R W, Saba C S, Kauffman R E and Rhine W E 1984 Spectrometric oil analysis. Detecting engine failures before they occur *Analytical Chemistry* **56** 1086A—1094A
[8] Delly R D 1995 The review of ferrography and its application in technical maintenance [In russian] — First international conference "Energodiagnostics" **3** 120—152
[9] Johns M and Massudi A 1995 The solution of big debris analysis problem — First international conference "Energodiagnostics" **1** 84—102
[10] Stepanov V A 2000 *The development and research of means and methods for complex diagnostics of gas-trubine engine oiled friction units by wear product features* [In Russian] (thesis syn.)
[11] Stepanov V A 2002 *The technical state evaluation of gas-turbine engine transition friction units by wear product features*
[12] Bowen E R and Westcott V C 1976 *Wear particle atlas*
[13] Dempsey P J, Lewicki D G and Decker H J 2004 Investigation of gear and bearing fatigue damage using debris particle distributions (Tech Report) *NATIONAL AERONAUTICS and SPACE ADMINISTRATION CLEVELAND OH GLENN RESEARCH CENTER*
[14] Drokov V G, Drokov V V, Ivanov N A, Myrishenko V V, Skudaev Y D and Hodunaev A Y 2019 Development of microwave plasma method for measurement of wear particle parameters in lubricant oil samples from aircraft gas turbine engines *IOP Conference Series: Materials Science and Engineering* **560** 1 012059
[15] PJSC UEC "Saturn" 2005 The D30-KP/KU/KU-154 engine teardown analysis due to oiled friction unit failures through 2000-2004 (Tech Report №44-596168) [In Russian] *PJSC UEC "Saturn"*
[16] UEC "Aviadvigatel" 1999 Statistics and diagnostic features analysis for high pressure roller bearing (Tech Report № 34676) [In Russian] *UEC "Aviadvigatel"*
[17] Humphrey G, Rojas J, Churchill D, Gorzen D and Cross B 2002 Prototype instruments built for PRAM *JOAP International Condition Monitoring Conference*
[18] Toms AM and Cassidy K 2008 Filter debris analysis for aircraft engine and gearbox health management *Journal of Failure Analysis and Prevention* **8** 183—187

[19] Military Unit №75360 1978 *The development of quantifying techniques for measurement of non-organic deposits of consistent oils and flammable liquids using photo-electric equipment of MFS type* (Tech Report) [In russian]
[20] Frankstein L 2003 The development and integration of prospective devices into the aircraft engine oil system [In russian] *Conversion in machine building* 72—91
[21] Engine PS-90A. Operative check of multi-element spectral analysis for engine diagnostics [In Russian] *Bulletin № 94370* 9
[22] Yudin A and Stepanov V 2002 – The recommendations for norm criterions of transmission techincal state evaluation *Conversion in machine building* 35—41
[23] Humphrey G et al. 2002 Filter Debris Analysis of J52-P-408 Engines, Part I *JOAP International Condition Monitoring Conference*
[24] Toms A, Jordan E and Humphrey G 2005 The success of filter debris analysis for J52 engine condition based maintenance *41st AIAA/ASME/SAE/ASEE Joint Propulsion Conference & Exhibit* 4338
[25] Humphrey GR 2008 Filter debris analysis by energy dispersive x-ray fluorescence applied to J52P408 engines *JCPDS-International Centre for Diffraction Data*
[26] Day L 2008 In the filter, available at: http://www.gastopsusa.com/knowledge_center_documents/2/tlt_filter_debris_analysis_article_2-08.pdf

IOP Conf. Series: Materials Science and Engineering **860** (2020) 012038 doi:10.1088/1757-899X/860/1/012038

Estimation of efficiency of the thermal method in the development of gas-hydrate deposits

M R Davletshina, A S Chiglintseva, and M V Stolpovskii

Ufa State Petroleum Technological University, 1, Kosmonvtov ave., Ufa, 450062, Russia

E-mail: davletshina@yandex.ru

Abstract. In the work on the basis of the proposed mathematical model, the process of decomposition of methane hydrate under thermal action on the reservoir is numerically investigated. Based on the calculations, it was shown that the energy efficiency of the decomposition of methane hydrate, i.e. the ratio of the energy obtained by burning the released methane to the energy spent on heating the formation decreases with increasing temperature of the coolant. The formation temperature distributions are also presented and the dependences of the position of the phase transition boundary on the system parameters are investigated. The problem is solved in a plane-one-dimensional formulation. According to the proposed scheme, a homogeneous porous layer of finite length, containing methane hydrate in the initial state, is considered. In this problem, we believe that the decomposition of hydrate into water and gas occurs at the boundary of phase transitions.

1. Introduction

Gas hydrates are currently giant storages of natural gas, mainly methane, the occurrence of which is tied to offshore deposits, where fairly low temperatures and high pressures are realized. Thus, according to studies [1,2], in 1 m3 of methane hydrate up to 165 m3 of methane can be contained under normal conditions. Experimental studies of the processes of formation and decomposition of gas hydrates were considered in [3]. Theoretical studies of the formation and decomposition of hydrates in porous formations are discussed, in particular, in [4–7]. One of the methods of gas production from the gas hydrate is the thermal method, the essence of which is to supply hot coolant to the gas hydrate formation, as a result of which the gas hydrate decomposes into gas hydrate and water. In the present work, in the plane-one-dimensional approximation, we consider the problem of decomposition of methane hydrate completely saturating a porous medium as a result of exposure to a coolant. Issues of modeling the process of decomposition of gas hydrates are discussed in [9–11].

The energy efficiency of this method is investigated and the dependences of the position of the phase transition boundary on the system parameters are investigated [12].

2. Materials and methods

The conditions for the formation and decomposition of gas hydrates are depicted using phase diagrams in the coordinates "temperature - pressure". For most gas hydrates, the condition of three-phase equilibrium between the hydrate-forming gas, its gas hydrate and water is satisfactorily described by the dependence:

$$T_r = T_{r0} + T_* \ln\left(\frac{p}{p_{r0}}\right); \tag{1}$$

equilibrium hydrate formation temperatures corresponding to pressures p and p_{r0}, T_* are an empirical parameter. In particular, for methane hydrate from the equilibrium curve, $T_{r0} = 280$ K, $p_{r0} = 5.5$ MPa.

Let us consider a porous layer of length L, the pores of which are completely saturated with methane hydrate. The initial pressure will be assumed equal to p_0, the initial temperature of the system will be determined by the condition of phase equilibrium, i.e. $T_0 = T_r(p_0)$. We will assume that at the left boundary of the formation, with the help of a coolant, a constant temperature T_e is maintained with $T_e > T_r(p_0)$. Assuming that the pressure change in the system is insignificant, i.e., $p = p_0 = const$, we assume that in there are two areas in the reservoir: the nearest, adjacent to the left boundary of the formation and saturated with methane and water, and the distant, adjacent to the right boundary of the formation and containing methane hydrate. In addition, a mobile boundary of phase transitions $x = x_{(s)}$ arises, separating the above regions and moving deep into the reservoir.

When describing heat and mass transfer processes accompanied by decomposition of methane hydrate, we will accept the following assumptions: porosity is constant, the skeleton of the porous medium, hydrate and water are incompressible and motionless, the gas is calorically perfect [8]:

$$m,\ \rho_{sk},\ \rho_h,\ \rho_w = const;\ \upsilon_{sk} = \upsilon_h = \upsilon_w = 0;\ p = \rho_g R_g T$$

Here m is the porosity, ρ_j and υ_j ($j = g, sk, h, w$) are the true density and velocity of the j-th phase, respectively; p is the pressure; T is the temperature; R_g is the gas constant. The subscripts sk, h, w and g correspond to the parameters of the skeleton, hydrate, water and gas. Hydrate is considered a two-component system with a mass concentration of gas G.

The system of basic equations for the near field, including the equations of continuity and heat influx, under the assumptions noted above, has the form:

$$\frac{\partial}{\partial t}\left(mS_g\rho_g\right) + \frac{\partial}{\partial x}\left(mS_g\rho_g\upsilon_g\right) = 0, \tag{2}$$

$$\rho c\frac{\partial T}{\partial t} + \rho_g c_g mS_g\upsilon_g\frac{\partial T}{\partial x} = \frac{\partial}{\partial x}\left(\lambda\frac{\partial T}{\partial x}\right). \tag{3}$$

Here S_g, υ_g, and c_g are the pore saturation, velocity, and specific heat of the gas phase, respectively; ρc and λ are the specific volumetric heat capacity and the thermal conductivity of the system. Since the main contribution to the values of ρc and λ is made by the corresponding skeleton parameters of the porous medium, we will consider them to be constant values.

The temperature in the far region saturated with methane hydrate will be assumed constant equal to the initial temperature of methane hydrate.

In addition, we supplement the system of equations (2) and (3) by the relations following from the conditions of mass and heat balance at the phase transition boundary $x = x_{(s)}$:

$$mS_{w(1)}\rho_w\dot{x}_{(s)} = m\rho_h(1-G)\dot{x}_{(s)},\ mS_{g(1)}\rho_{g(1)}\left(\upsilon_{g(1)} - \dot{x}_{(s)}\right) = mG\rho_h\dot{x}_{(s)},$$

$$\lambda\frac{\partial T_{(1)}}{\partial x} = -\rho_h L_h\dot{x}_{(s)}. \tag{4}$$

Here L_h is the specific heat of hydrate decomposition, $\dot{x}_{(s)}$ – the velocity of the phase transition boundary; the subscript $i = 1$ corresponds to the parameters of the near field. From the first equation of system (4) to determine the values of water and gas saturation for the near field, we get:

$$S_{w(1)} = \frac{\rho_h (1-G)}{\rho_w}, \quad S_{g(1)} = 1 - S_{w(1)}.$$

Since changes in the temperature ΔT of the formation associated with the dissociation of the hydrate are small compared to its initial temperature T_0, i.e. $\Delta T \ll T_0$ and in accordance with the assumption of a constant pressure in the system, we find that the true density of the gas phase is a constant and equal to:

$$\rho_g = \frac{p_0}{R_g T_0}.$$

Then, based on the continuity equation (2), the velocity υ_g of the gas phase through the cross section determined by the x coordinate can be obtained from the following relation:

$$\upsilon_g = \upsilon_{g(s)} \tag{5}$$

where $\upsilon_{g(s)}$ is the gas velocity at the boundary of the phase transition.

The initial conditions and conditions on the external boundaries of the reservoir are presented in the form:

$$\begin{gathered} t=0: \quad p=p_0, \ T=T_r(p_0) \ (0 \le x \le L); \\ t>0: \quad p=p_0, \ T=T_e \ (x=0); \\ t>0: \quad p=p_0, \ T=T_0 \ \left(x=x_{(s)}\right). \end{gathered} \tag{6}$$

The presented system (2) – (4) is closed. To solve it, the method of catching the front into the spatial mesh node is used. The essence of this method is that for an unknown and determined in the course of solving the problem time step, the front of the phase transition moves along the grid also by one spatial step.

3. Calculations results

For different times, Figure 1 shows the temperature distribution in the formation with a length of $L = 100$ m at a coolant temperature of $T_e = 286$ K. The initial pressure was assumed to be $p_0 = 6$ MPa, the initial temperature, in accordance with (1), $T_0 = 280.87$ K. The following values are accepted for the remaining parameters characterizing the system: $m = 0.15$, $G = 0.12$, $\rho_h = 900$ kg/m3, $\rho_w = 1000$ kg/m^3, $R_g = 520$ J/(K·kg), $\lambda = 2$ W/(m·K), $c_g = 2600$ J/(K·kg), $L_h = 5 \cdot 10^5$ J/kg. As follows from the figure, over time, the heat received from the well mainly goes to the warming of the frontline area.

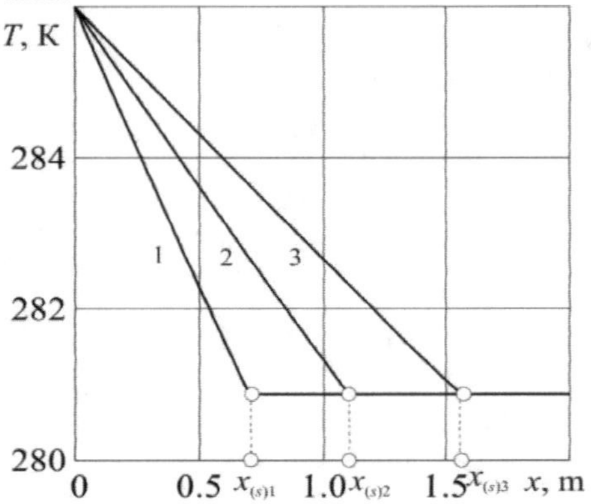

Figure 1. The temperature distribution in the reservoir for a point in time: 1 –$t = 20$ days; 2 – $t = 50$ days; 3 –$t = 100$ days.

Dependences of the coordinate of the phase transition front x = x(s) on time are presented. Line 1 corresponds to the temperature of the coolant $T_e = 285$ K, line 2 – $T_e = 295$ K. As follows from the figure, with an increase in the temperature of the coolant, the velocity of the front of the phase transition increases. This is due to the fact that with an increase in T_e, the intensity of the hydrate dissociation process increases.

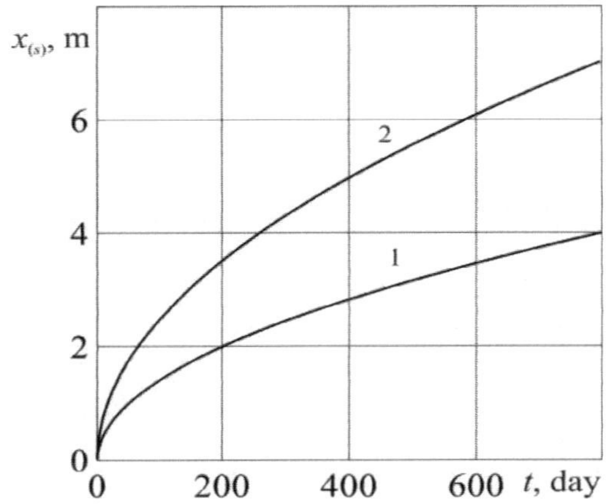

Figure 2. The dependence of the coordinate of the front of the phase transition from time to time: 1 – $T_e = 285$ K; 2 –$T_e = 295$ K

Of particular interest is the determination of the energy efficiency of the thermal method of gas extraction from the gas hydrate composition. We define energy efficiency κ in the form of:

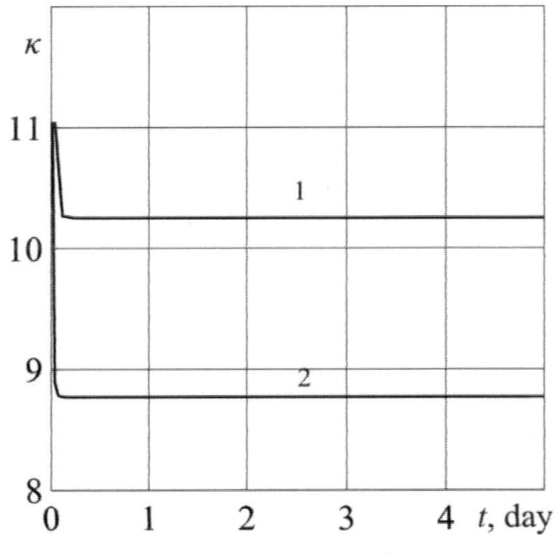

Figure 3. The dependence of the energy efficiency of the thermal method from time to time: 1 –$T_e = 285$ K; 2 –$T_e = 295$ K.

Dependences of energy efficiency on time are presented. Lines 1 and 2 correspond to the temperature of the coolant $T_e = 285$ K and $T_e = 295$ K. From the figure it follows that the energy efficiency of gas extraction at the initial stage of the process decreases, which is associated with the heating of in-situ areas. Moreover, with decreasing temperature of the coolant, energy efficiency increases.

4. Conclusion

The problem of the possibility of decomposition of a gas hydrate formation upon thermal exposure in a plane-one-dimensional formulation with a frontal boundary of phase transitions is solved. For this approximation, solutions are constructed that describe the temperature fields, as well as the dynamics of the hydrate decomposition boundary and the gas mass flow rate. The influence of the temperature difference between the formation and the injected water on the evolution of the hydrate decomposition front and on the energy efficiency of methane hydrate decomposition at different porosity of the formation was studied. The proposed calculation scheme and the results obtained on its basis allow us to analyze the efficiency of hydrate formation development by heating the working zone of the field. A quantitative assessment was made of the effectiveness of the thermal methane extraction method with a change in the heating temperature and pressure drop.

Acknowledgments
The study was supported by grant MD-2179.2020.1

References

[1] Makogon Yu F 2010 The history of the study and development prospects*Geology and minerals of the oceans. Gas hydrates* 25–21
[2] Makogon Yu F and Omelchenko R Yu 2012 Messoyakha – gas hydrate reservoir, role and significance*Geology and Minerals of the World Ocean* 35–19
[3] Yin Z, Moridis G, Chong Z, Tan H and Linga P 2017 Numerical analysis of experiments on thermally induced dissociation of methane hydrates in porous media*Ind. Eng. Chemical Res., Newly Accepted Manuscript*.5–21
[4] Chersky N V and Bondarev E A 1972 On the thermal method of developing gas hydrate deposits*Dokl. USSR Academy of Sciences* 550–552
[5] Verigin N N, Khabibullin I L and Halikov G A 1980 The linear problem of the decomposition of gas hydrates in a porous medium*Izv. USSR Academy of Sciences. MZHG* **1** 174–177
[6] Bondarev E A, Maksimov A M and Tsypkin G G 1989 To mathematical modeling of dissociation of gas hydrates*Dokl. USSR Academy of Sciences* 575–577
[7] Shagapov V Sh, Musakaev N G *of hydrate formation and decomposition in gas production, transportation and storage systems* (Moscow: Nauka)
[8] Nigmatulin R I 1987 *The Dynamics of multiphase media* (Moscow: Nauka)
[9] Verigin N N, Khabibullin I L and Halikov G A 1980 The axisymmetric heat and mass transfer problem in a saturated porous medium *M. Phys.* **38(5)** 581–585
[10] Tsypkin G G 2005 The effect of differences in gas and reservoir production in a free state *Bulletin of the Russian Academy of Sciences. Mechanics of fluid and gas* 132–142
[11] Nigmatulin R I, ShagapovV Sh and Syrtlanov V R 1998 The self-similar problem of the decomposition of gas hydrates under conditions of depression and heating*Applied Mechanics and Technical Physics* **39(3)** 111–118
[12] Chiglintseva A S, Davletshina M R and Stolpovsky M V *XII All-Russian Congress on Fundamental Problems of Theoretical and Applied Mechanics: a collection of works in 4*.**2** 1137–1139

AUTHOR INDEX